This excellent edited volume on the consequences of the Fukushima Daiichi nuclear power plant meltdowns in Japan comes exactly at the right time: when people are being exposed to tremendous misinformation on the subject and when decision makers seek to weigh the lessons learned from the ongoing disaster. Based in the field of science, technology and society, this explicitly comparative book provides insights into how and to what extent a natural-technological disaster has had an impact around the world on energy source development and management. Observers wondering why some countries moved away from nuclear power following the disasters while others kept their atomic energy programs will find this critical reading.

**Daniel P. Aldrich**, author of *Site Fights and Building Resilience*

# The Fukushima Effect

*The Fukushima Effect* offers a range of scholarly perspectives on the international effect of the Fukushima Daiichi nuclear meltdown four years out from the disaster. Grounded in the field of science, technology and society (STS) studies, a leading cast of international scholars from the Asia-Pacific, Europe, and the United States examine the extent and scope of the Fukushima effect. The authors each focus on one country or group of countries, and pay particular attention to national histories, debates, and policy responses on nuclear power development, covering such topics as safety of nuclear energy, radiation risk, nuclear waste management, development of nuclear energy, anti-nuclear protest movements, nuclear power representations, and media representations of the effect. The countries featured include well-established "nuclear nations," emergent nuclear nations, and one non-nuclear nation to offer a range of contrasting perspectives.

This volume will add significantly to the ongoing international debate on the Fukushima disaster and will interest academics, policy-makers, energy pundits, public interest organizations, citizens, and students engaged variously with the Fukushima disaster itself, disaster management, political science, environmental/energy policy and risk, public health, sociology, public participation, civil society activism, new media, sustainability, and technology governance.

**Richard Hindmarsh** is Associate Professor in Griffith University's School of Environment, and Centre for Governance and Public Policy where his research lies in environmental politics and policy, and science, technology and society. He holds a PhD in STS from Griffith University (Australia) and his previous book on this topic is *Nuclear Disaster at Fukushima Daiichi: Social, Political and Environmental Issues* (Routledge 2013).

**Rebecca Priestley** is Senior Lecturer in the Science in Society group at Victoria University of Wellington (New Zealand) where her research covers science communication and the history of science. She holds a PhD in HPS from the University of Canterbury (New Zealand) and her most recent book was *Mad on Radium: New Zealand in the Atomic Age* (Auckland University Press 2012).

# Routledge Studies in Science, Technology and Society

# The Fukushima Effect

A New Geopolitical Terrain

**Edited by Richard Hindmarsh
and Rebecca Priestley**

Routledge
Taylor & Francis Group

LONDON AND NEW YORK

First published 2016 by Routledge

2 Park Square, Milton Park, Abingdon, Oxfordshire OX14 4RN
711 Third Avenue, New York, NY 10017

*Routledge is an imprint of the Taylor & Francis Group, an informa business*

First issued in paperback 2017

*Library of Congress Cataloging-in-Publication Data*
A catalog record has been requested for this book

ISBN: 978-1-138-83078-3 (hbk)
ISBN: 978-0-8153-7056-7 (pbk)

Typeset in Sabon
by Apex CoVantage, LLC

# Contents

# Figures

# Tables

# Acknowledgements

The idea for this book arose when the editors met in Singapore at the Asia Pacific Science, Technology and Society Network (APSTSN) conference in 2013. Under the conference theme of "Knowing, Making, Governing," Richard Hindmarsh and Rebecca Priestley were presenting in a session on States of Risk, with Richard overviewing his recently published *Nuclear Disaster at Fukushima* volume and Rebecca presenting a paper on attitudes to nuclear power in New Zealand through the 20th century. Discussions that began at the conference were continued from the editors' respective universities in Australia and New Zealand, and the proposal for this volume emerged.

That publication of this book is a little over two years after the idea first emerged is thanks to the support and hard work of many people. First and foremost we thank our contributors, a leading group of global scholars—situated in science, technology and society (STS) studies—concerned about the long-term effects of the Fukushima Dai-ichi nuclear disaster. We particularly thank the contributors for whom English was not their first language, and to everyone for their prompt engagement in the rigorous editing process.

We also thank Christopher Rootes for providing the foreword, A'edah Abu Bakar for preparing the figures, and proofreader Wendy Smith for her final critical reading of the manuscript. At Routledge New York, we thank Senior Acquisitions Editor, Political Science Research, Natalja Mortensen, Editorial Assistant Lillian Rand, Project Manager Denise File, the copy editor and, of course, the anonymous reviewers who endorsed our initial proposal.

Completion of any book manuscript demands a lot of work on weekends and evenings, and we thank our respective families for their support.

For funding to travel to the APSTSN 2013 conference in Singapore, without which the editors would not have met, Rebecca acknowledges the Faculty of Science, Victoria University of Wellington, and Richard acknowledges Griffith School of Environment and the Centre for Governance and Public Policy, Griffith University, Brisbane, Australia.

Chapter 4 of this volume is a modified version of "Reconstructing the Public in Old and New Governance: A Korean Case of Nuclear Energy Policy" by Hyomin Kim, *Public Understanding of Science* (2014) 23: 268. Sage Publications is gratefully acknowledged. In turn, Chapter 11 is an adaptation of "From Exception to Norm—and Back Again? France, the Nuclear Revival and the post-Fukushima Landscape" by Joseph Szarka, *Environmental Politics* (2013) 22 (4): 646–663, included by permission of the publisher Taylor & Francis Ltd (www.tandf.co.uk/journals).

# Foreword

We live at a time when the evidence that human activity is producing significant and potentially catastrophic climate change has become overwhelming. Aside from the destruction of ancient forests and native grasslands in the interests of intensive agriculture, arboriculture, mining, and industry, it is the burning of fossil fuels, upon which our industrial civilization has so far depended, that is the principal source of the net increase in greenhouse gas emissions. The biggest single source of those emissions is the combustion of coal to generate electricity. As coal remains the largest component of the fuel mix of the economies of most developed and industrializing countries, it follows that much of the action to mitigate rampant climate change should be focused upon the transformation of energy systems.

At the beginning of the present decade, the conventional wisdom was that, for most industrialized or industrializing countries, some combination of renewables and nuclear fission would replace combustion of coal, unless perhaps carbon capture and storage could quickly be made to work on a commercial scale.

Renewables, it was generally agreed, had potential, but wind and solar were compromised by the temporal and geographical variability of their capacity to sustain production of electricity. In the absence of any great improvement in battery technology, efficient large-scale storage of electricity produced outside the hours of peak demand seemed perpetually elusive. Intercontinental transmission grids such as the long-promised Desertec, which would have fed electricity produced by harnessing solar radiation in North Africa into an expanded European grid, required political cooperation and financing on an unprecedented scale. Neither was forthcoming. Biomass was a possible alternative fuel, but growing crops for direct combustion threatened to withdraw land from agricultural production, and converting grain to ethanol for fuel competed with the demands for food of a globally increasing population.

The problems associated with biofuels and the intermittency of electricity production by means of wind and solar even led many advocates of renewables as the long-term solution to industrialized societies' energy problems to regard nuclear energy either as a stop-gap or, more plausibly, as the means of guaranteeing "baseload" energy supplies, and energy security. That is, at least until the technological obstacles confronting the harnessing of more predictable renewables, such as wave and tidal energy, might be overcome.

Nuclear struggled to shake off its reputation as a risky technology, but the Chernobyl disaster, already a generation earlier, was retreating into history. Countries such as Germany and Sweden, which had committed to phasing out nuclear, were postponing the phase-out as the urgency of mitigating climate change increased. The roll-out of renewables was expensive and slow, and the promise of carbon capture technology that might preserve a role for coal remained unfulfilled. Even countries such as Italy, which had a generation earlier rejected nuclear energy, were re-considering, and the United States and the United Kingdom were authorizing the first new nuclear power plants in three decades. In Asia, South Korea and China embarked on major programs of nuclear expansion. Engineers promised that fourth generation nuclear reactors would be safer, more efficient, and more reliable than their predecessors.

The pro-nuclear industrial lobby was vocal and effective; in a de-carbonized energy system, nuclear, as a mature technology that was proven to be capable of reliably generating electricity on a large-scale without emitting climate-changing greenhouse gases, now had a legitimate place in a post-carbon "energy mix." Even environmental NGOs such as Friends of the Earth, which had made their names as opponents of nuclear energy in the 1970s, were now alarmed by the prospects of climate change. Impressed by consultants' reports on the hidden costs—to human health and the environment—of burning coal they were, however reluctantly, prepared to accept nuclear as the lesser evil, in the event that a demand reduction could not ensure that energy needs could be met by renewables in the short-term.

There were, however, signs that all was not going entirely to plan. The construction of new nuclear power plants in Finland and France was both long overdue and massively over-budget. The idea that nuclear technology might be an "interim solution" to the energy needs of industrialized countries' was looking increasingly implausible. Nuclear power plants that might take 15 years to design, build, and commission required capital investment, which tested the capacities of even sovereign governments. They could clearly not be a short-term solution to any problem, and their adoption began to look like a long-term commitment to a technology whose most basic problems—the environmentally hazardous mining of uranium and the long-term storage of high-level nuclear waste—had nowhere been solved.

As the contributors to this volume make clear, the Fukushima disaster, coupled to these contexts, has drastically changed the environment for and prospects of nuclear energy. The immediate reaction came in countries where opposition to nuclear energy had long been mobilized and entrenched by social movements. Thus, Germany's nuclear shut-down was accelerated, Switzerland and Belgium decided to phase out nuclear, and Taiwan abandoned plans for a new nuclear plant. In the UK, where anti-nuclear movements had been less extensively mobilized and the domestic nuclear industry had a relatively good safety record, there was also an immediate fall in support for renewing the country's ageing nuclear installations. But, it proved short-lived and the UK government, after evaluating the circumstances of the "serious nuclear accident" at Fukushima, resolved to proceed with its nuclear program. Elsewhere, countries that had been considering commencing or renewing nuclear energy programs, such as Italy and Israel, decided not to go ahead. Even France, which, more than any country on the planet, relies on nuclear for the generation of its electricity, was looking to increase the contribution of renewables to its electricity supplies.

In other countries still committed to nuclear energy, such as the US, Russia, and China, safety regimes were reviewed and tightened. Most importantly, in China—where plans to reduce the carbon-intensity of increased industrial production entailed an ambitious program of nuclear power plant construction that would have quadrupled the number of reactors by 2020—local communities, alerted by the news from Japan and informed by citizen activists, began to mobilize against proposed new nuclear developments in allegedly unsuitable places. These protests met with remarkable success; one project, a nuclear waste repository, was scrapped and several nuclear power plant projects were postponed or suspended as part of ongoing policy and rigorous safety reviews following Fukushima.

Everywhere, the media attention to the events at Fukushima made the public more aware of the risks of the nuclear industry. The Japanese disaster also flagged the known but hitherto hypothetical risk that the failure of a nuclear power plant might be triggered by a seismic event. In the event, the effects of an earthquake and tsunami were compounded by yet another failure of socio-technical systems. Nature had found the "faultline" in Japan's notoriously complacent nuclear safety regime, and it raised questions about the inherent safety of nuclear energy everywhere. In those countries where past campaigns had left a legacy of public opinion disposed to severe skepticism about, if not outright opposition to, nuclear energy, politicians had little room to manoeuver, and the nuclear "hot potato" was smartly dropped, even before the nature of the Fukushima disaster had been fully analyzed. Everywhere, nuclear energy was again under severe scrutiny.

Although this tells us a lot about the way political opportunities are created and exploited, and the path dependency of the development of policy-relevant public opinion and political mobilization, it is the implications of Fukushima that are especially interesting.

Not surprisingly, the biggest, and potentially most significant, effect of Fukushima was in Japan itself. Almost overnight, Japan went from being a country aiming to derive half its electricity from nuclear power stations, to one with no functioning nuclear plant and a sharply enhanced interest in developing renewable energy technologies. This raises the prospect that Japan's formidable capacity in research, development, and high-tech manufacturing might now be highly focused upon developing and exporting the infrastructure for renewable energy.

Fukushima has thus, directly and indirectly, dramatically speeded up the search for alternative low-carbon technologies to make up the energy shortfall created by the eventual retirement of fossil fuels and their associated infrastructure. Where it is not politically impossible to build new nuclear plants, it has, in most circumstances, become almost impossible financially. There is, after all, a limit to the largesse of Canadian teachers' pension funds and the surpluses of Chinese state corporations. In addition, governments becoming reluctant or unable to guarantee the returns that investors require on investment in new nuclear plants with a design life exceeding 40 years and a host of contingent liabilities.

Moreover, in the space of just a few years, the unit costs of new renewable energy infrastructure have tumbled, with the result that installation rates have lessened, and economies of scale have been achieved. In some markets, solar and wind are now cost-comparative with coal as sources of electricity, even before taking into account the massive externalities of coal mining and, especially, the air pollution produced by the combustion of coal. More quickly than almost anyone imagined, we have moved beyond the stage where the installation of renewables could only be stimulated by the provision of lavish subsidies in the form of discount vouchers, tax rebates, and feed-in tariffs. Now, it is the subsidies, direct and indirect, to the producers of fossil fuels that are coming under scrutiny. Concomitantly, the epidemiological evidence accumulates of the damage done to human health by carbon particulates and other emissions produced by the burning of coal, oil, and gas.

Even the longstanding bugbear of wind and solar—the fact that they are intermittent in generating electricity—may be about to fall, as the long-awaited revolution in battery technology promises efficient small- and medium-scale storage. The availability of affordable batteries will not only facilitate demand management at the point of consumption, and reduce the need for large-scale peak capacity power stations, but it will also bring closer the feasibility of decentralized energy systems and of living "off-grid" for consumers whose demand for electricity is merely average rather than self-denyingly low.

Thus, the disaster at Fukushima has also bought time for renewables. It has delayed or terminated decisions about new nuclear installations, and it has catalyzed the pace of innovation in alternatives to both nuclear and fossil fuels. If the direct effects of Fukushima are most visible in the transformation of energy policy in Japan, it is the effects on China that, however, appear most consequential. What happens in China matters because, perhaps more than any other country, China has the technological ability, the financial capacity, and the political will to fashion a sustainable, low-carbon economy. China may not entirely abandon nuclear technology, but the fallout from Fukushima ensures that its dependence on nuclear will most likely be lower than envisaged less than a decade ago. More consequentially, China is now committed to the largest and most rapid roll-out of wind and solar energy infrastructure that the world has ever seen.

If, as now seems possible, the December 2015 Conference of the Parties to the United National Framework Convention on Climate Change in Paris produces a far-reaching international agreement to reduce greenhouse gas emissions consistent with containing the increase in global temperatures to something close to 2°C, Fukushima will undoubtedly have played a part. The major environmental NGOs have long argued that investment in new nuclear energy infrastructure has represented a massive diversion of resources, engineering capacity, and research effort away from renewables, which always promised a more sustainable and ultimately cheaper alternative to fossil fuels than nuclear. Fukushima, by rekindling enduring uncertainties about nuclear energy and withdrawing the nuclear card from the table for many countries, has thus had the indirect effect of encouraging investment in renewables. Now that the virtuous circle of increasing renewable energy infrastructure is kicking in, confidence that it is possible to contain climate change is rising accordingly.

The disaster at Fukushima increasingly thus appears to mark a watershed in energy policy internationally, and this volume goes a long way to showing how and why.

Christopher Rootes

# 1 The Fukushima Effect

## Traversing a New Geopolitical Terrain

*Richard Hindmarsh and Rebecca Priestley*

*The 11 March disaster shook the very foundation of Japanese society, shattering the idea of a safe and secure society guaranteed by the authorities. The Japanese public has started to question the level of risk that they were willing to accept and the model of society that they had aspired to prior to the disaster. Though Japan certainly has both the technical and financial capacities to rebuild the towns affected by the disaster, the social tensions and divisions created as a result will probably take much longer to heal. The Japanese experience thus offers many unique lessons . . . in terms of dealing with future disasters.*

(Hasegawa 2013: 7)

The immediate and remarkable effects of the disaster at the Fukushima Daiichi nuclear power plant (NPP) on March 11, 2011 (3/11), now widely referred to as the "Fukushima disaster," or "Fukushima," ranged from radiation leakages from the plant, which saw tens of thousands of citizens displaced from their homes, to the shutdown of Japan's entire nuclear reactor fleet. Indeed, "the then Japanese Prime Minister, Naoto Kan, described the disaster as the worst crisis that Japan has had to face since the Second World War" (see Hasegawa 2013: 15). The effects of the disaster then rippled out internationally, causing unprecedented safety checks of nuclear power plants worldwide. Some countries, such as Germany, Switzerland, and Belgium, even decided to phase out nuclear power, whereas others made decisions against embracing or re-embracing nuclear power. Overall, Fukushima appears to have effected a slowdown in the growth rates of nuclear power.

But, so far, little exploration has occurred of the "Fukushima effect" on global nuclear power development, management, and policy. This book addresses this gap, and like the prior book of first editor Richard Hindmarsh, *Nuclear Disaster at Fukushima Daiichi: Social, Political and Environmental Issues* (Hindmarsh 2013a), it is placed at the forefront of critical reflection on the Fukushima disaster.

In *Nuclear Disaster at Fukushima Daiichi*, which was also informed by an international cast of scholars in science, technology and society

(STS) studies, Hindmarsh discerned five themes of key areas of issues and implications informing the disaster one year out: (i) the social or political shaping and subsequent compromise of nuclear power safety in Japan; (ii) the flawed concentrated or multi-reactor siting of nuclear power plants and accompanying inadequate public participation at the local level in regard to ensuring more legitimate and effective siting decisions; (iii) failed institutional risk communication to citizens involving radiation data and other information post-disaster; (iii) the actual substance of the disaster—for example, to what extent the social realm could be seen as causative alongside the natural; (iv) the future of nuclear energy and the alternatives; and (v) the failings of emergency responses to adequately manage the disaster (see Hindmarsh 2013b: 214).

In all of these themes, the need was identified for enhanced and more responsible and inclusive science, technology, and environmental governance, according to notions of good governance and sustainability (also Martin et al. 2012). Key principles and practices that inform these notions include openness or transparency of decision-making processes and information, (active civic) participation in decision-making and governance, precaution about risk and hazard, accountability, inter- and intra-generational equity, and policy effectiveness and coherence for long-term policy learning and change (e.g. CEC 2001; Dovers 2005). Such principles and practices well-inform new integrated development, planning, and regulatory approaches that closely knit the technical and social infrastructures of sociotechnological systems—here energy systems—to make them more coupled, functional, and legitimate (Hindmarsh 2014; Miller et al. 2008; Star 1999). If they are not, as became apparent with nuclear power in Japan, then energy systems can become decoupled and inefficient, sometimes making *breakdown*, such as that demonstrated by the Fukushima Daiichi disaster, almost inevitable.

In exploring such implications further, *The Fukushima Effect* builds on, extends, and to some degree critically reflects on these themes, as well as on an emergent international literature. More succinctly, the *aim* of *The Fukushima Effect* is to investigate the international effect of the Fukushima disaster three to four years out from the disaster, when an effect can be discerned. The follow-on aim is to determine the extent and scope of this effect on national histories, debates, and policy responses in areas such as the safety of nuclear energy, radiation risk, nuclear waste management, development of nuclear energy, anti-nuclear protest movements, nuclear power representations, media representations of the effect, and any other areas considered relevant by the contributors.

Accordingly, this book makes a valuable contribution to the fields of governance, and the politics and policy of nuclear energy and, more broadly, to energy studies, in traversing what we have detected to be "a new geopolitical terrain." Although Fukushima builds on the legacy of Three Mile Island, Chernobyl, and at least "33 serious incidents and accidents at nuclear power stations since the first recorded one in 1952 at Chalk River in Ontario, Canada" (The Guardian 2011), this new geopolitical terrain reflects the sweeping and unprecedented global impact of Fukushima. The effects we explore in this book include radically changed energy pathways in some countries; the dramatic effect on global populations newly connected via the Internet; the subsequent rise of what seems to be enduring civic unrest and concern over nuclear power at a time when participatory governance trends are increasing worldwide; increasing global environmental concerns, particularly over the safety of nuclear power in the face of the rising frequency and intensity of extreme weather events[1] and the lack of safe geological repositories for the final disposal of high-level radioactive waste; and, finally, the growing and increasingly viable renewable energy sector and energy conservation industries.

The book's findings thus have implications for effective and responsible regulation and good governance of controversial science and technology, or technoscience, and the development, management, and future of nuclear power. A range of significantly different countries are investigated. Developing and developed countries are covered; most are nuclear nations but one nuclear-free country—New Zealand, where much debate has occurred about nuclear issues—is included for an interesting and relevant comparison.

This selection of countries, from Europe, Asia, North America, and Australasia, is geopolitically representative. The volume begins at the geopolitical "epicenter" of the Fukushima effect, in Japan, then follows the ripples of the effect it created outwards to South Korea, China, Taiwan, and India, and then to Germany, Sweden, France, Switzerland, Belgium, post-Soviet countries, Finland, the UK and the US, and finally, to New Zealand. As can be seen, included in this investigation are the "big six" nuclear-generating countries pre-Fukushima, which together generated more than 70% of the world's nuclear electricity: the US, France, Russia, South Korea, Japan, and Germany. However, post-Fukushima, Japan's fleet was paralyzed and Germany is in the process of phasing out nuclear power.

Before we turn to the investigations of the contributors, we first provide an overview of the Fukushima effect in Japan. We then review the broader international literature to provide a contextual background for the contributors' chapters and to facilitate understanding of the post-Fukushima geopolitical terrain. Following that, we set out the specific objectives of the book, along with the field of investigative inquiry, and then introduce the substance of the contributors' work. Finally, we provide an end chapter—*The Effect of the Fukushima Effect: From Strong to Weak*—which summarizes and reflects on the contributors' analyses and findings, and the implications raised for future policy and research.

## Japan and the Fukushima Effect

With the Fukushima Daiichi nuclear power plant having inadequate facilities and safety mechanisms in place to protect it, it was at the mercy of huge tsunami waves. The World Nuclear Association reported the tsunami that disabled the Fukushima Daiichi nuclear power plant was 15 meters (50 feet) high.[2] These waves were triggered by the 2011 Tōhoku earthquake or "Great East Japan Earthquake," which occurred in the north-western Pacific Ocean with its epicenter about 70 kilometers east of the Oshika Peninsula of Tōhoku,[3] the biggest recorded earthquake ever to affect Japan. As well as sweeping away entire towns, and destroying or damaging many fishing ports and industrial and commercial zones (Matanle 2011), the tsunami waves also "breached the protective walls at the Fukushima Daiichi plant located in Okuma and Futaba Towns, and knocked out the mains electricity supply and backup generators that [supplied] the six reactors' cooling systems" (Matanle 2011: 825–826). This breach of the power plant's defenses, in turn, directly contributed to the meltdowns of reactors 1, 2, and 3, and severe damage to reactor 4 and to containment systems, which led to "the uncontrolled leak of radioactive materials beyond the vicinity of the plant" (ibid: 826).

Apart from power shortages and radioactive materials being released into the air, it created tens of thousands of nuclear disaster evacuees in an *ad hoc* and chaotic evacuation process described by Hasegawa (2013: 6) as "an evacuation without warning, preparation or knowledge." One source cites up to 160,000 evacuees: 85,000 in response to government instructions and another 75,000 or so outside the exclusion (no-go) zones who fled voluntarily as "self-evacuees." The Japanese government and the Fukushima Daiichi operator, the Tokyo Electric Power Company (TEPCO), have had to subsequently pay out billions of dollars to provide accommodation for these evacuees and as compensation for "psychological damage" downstream.[4] Another estimate was of 120,000 evacuees, portrayed as "uprooted" and living in "nuclear limbo,"[5] with "reportedly 741 victims of 'disaster-related death', especially old people uprooted from homes and hospitals because of forced evacuation and other nuclear-related measures" (WNA 2014).

A survey by the Fukushima prefectural government conducted in 2014 found half of the households forced to evacuate were living apart, "while almost 70% had relatives suffering from physical and mental health problems" (cited in McCurry 2014). McCurry (2014) seemed to well capture the context of stress and disillusionment in citing one former resident from an abandoned home stating: "The nuclear accident turned everything upside down . . . Even if the evacuation order is lifted, no young people or children will go back. We have asked everyone—the village office, decontamination workers, environment ministry bureaucrats—when it will be safe to return. But no one can give us an answer."

This is because of the release of radioactive material, which continues today in various forms, mostly into Japan's atmosphere and the eastern seaboard of Japan, with consequent highly contaminated fisheries (e.g. Högberg 2013), and at least 30,000 square kilometers of contaminated land across Japan (Starr 2013). According to the report of the Japan Atomic Industrial Forum:

> Around 15,000 terabecquerels of caesium-137 was released from reactor[s] 1–3 at the Fukushima Dai-ichi nuclear power plant, 168.5 times that of the atomic bomb dropped on Hiroshima. Radioactive materials from the Fukushima accident, including iodine-131, caesium-134, and caesium-137, were detected around the world, including in North America and Europe. High levels of radioactive isotopes were also released into the Pacific Ocean. People within a 20-km zone around the Fukushima Dai-ichi nuclear plant had to leave the area . . . The disaster was classified as a Level 7 nuclear accident, the highest level on the International Nuclear Event Scale, equal to that of the Chernobyl nuclear disaster.
>
> (Kim et al. 2013)

A level 7 "major accident" describes an event involving a "major release of radioactive material with widespread health and environmental effects requiring implementation of planned and extended countermeasures" (IAEA 2011). The nuclear fallout of caesium-137 was found by Stohl et al. (2012: 2313) to be "about 43% of the estimated Chernobyl emission," with "radioactive clouds" detected in North America on March 15 and in Europe on March 22, 2011. By April, the fission product xenon-133 was also detected in Darwin, Australia, in the Southern Hemisphere (Stohl et al. 2012: 2314).

The long-term health implications of Fukushima are thus of great concern both in Japan and internationally. The World Nuclear Association, though, reported that no harmful health effects were found in 195,345 Japanese residents in the vicinity of the plant who were screened at the end of May 2011 (about 10 weeks post-disaster). About a year later, however, in July 2012, it stated "a Hirosaki University study reported on I-131 activity in the thyroid of 46 out of the 62 residents and evacuees subject to detailed investigation in April 2011 . . . [but the activity was] much smaller than the mean thyroid dose in the Chernobyl accident . . ." (WNA 2014).

Then, about three years later, on March 11, 2015, the fourth anniversary of Fukushima, Australian Broadcasting Commission journalist Matthew Carney reported that: "Before the disaster, there was just one to two cases of thyroid cancers in a million Japanese children but now Fukushima has more than 100 confirmed or suspected cases, having tested about 300,000 children." It was unclear, however, whether this increase in thyroid cancers was directly related to Fukushima radiation or the result of "simply doing more testing with more sensitive equipment" (Carney 2015; see also Oiwa 2015).

Notably, Stohl et al. (2012: 2314) found: "Altogether, we estimate that 6.4 PBq of 137 Cs [caesium-137], or 18% of the total fallout until 20 April, were deposited over Japanese land areas, while most of the rest fell over the North Pacific Ocean. Only 0.7 PBq, or 1.9% of the total fallout were deposited on land areas other than Japan." Also notable in relation to the nearby ocean contamination was the finding by Emspak (2012): "In some places, researchers from the Woods Hole Oceanographic Institution . . . discovered cesium radiation hundreds to thousands of times higher than would be expected naturally."

Carny (2015) also reported that 100 Fukushima residents were taking the local and central governments to court on the matter, and were conducting their own radiation tests near a local school where they claimed radiation levels were 100 times that of the rate in Tokyo. Such claims most often emanate from many of the "gray zones" (safe to unsafe radiation zones) off Fukushima Daiichi, particularly to the north-west, which lie outside the mandatory evacuation zones, where community groups continue to actively monitor radioactive contamination (Freiner 2013). David Boilley, French nuclear physicist and chairman of ACRO (a French NGO with a nuclear testing laboratory), has been working with Japanese citizens since the beginning of Fukushima, and identified Kashiwa City of Chiba Prefecture—in the metropolitan outskirts of Tokyo, about 200 kilometers to the south of Fukushima Daiichi—as one gray zone suffering from radiation hot spots (as shown in Figure 1.1).[6]

Hotspots are random, high-radiation concentrations, which sometimes are quite hard to detect. For example,

> a small area of about one meter radius, was found in a vacant lot in Kashiwa. Radiation levels of 4.11 microsieverts per hour were detected one meter above the surface of the soil, equivalent to some areas in the evacuation zone around the crippled nuclear power plant. Up to 450,000 becquerels per kilogram of radioactive substances were [also] detected in the soil below the surface, an Environment Ministry official said, Fuji TV reported.
>
> (Japan Today 2011)

In turning again to ocean contamination, apart from marine and fisheries radiation impacts, on November 25, 2014, CBS (the US commercial broadcasting TV and radio network) reported that two US sailors had died from radiation exposure after the US Navy was deployed off Japan's coast to assist with the cleanup of the Fukushima nuclear plant, apparently from the ship's water filtration and ventilation systems being

*Figure 1.1* Radiation Hotspot in Kashiwa
Source: Abasaa, Wikimedia Commons, 2012

contaminated as the USS *Ronald Regan* passed through what CBS called a "radiation cloud"[7] (c.f. Stohl et al. 2012). About 200 Navy sailors and Marines are now planning a lawsuit against TEPCO and other defendants, who they blame for a range of radiation-related complaints. The US Navy, however, has denied that those aboard the USS *Ronald Regan* received harmful levels of radiation (Goldenberg 2014).

Downstream, the disaster has had an unprecedented effect on Japan's energy system. Before 3/11, nuclear power contributed around 30% of Japan's energy mix (Skea et al. 2013). At that time Japan had 54 reactors located at 17 seaboard plants around the country. Of Fukushima's six reactors, four were knocked out in 3/11. In 2012, Japan's 48 remaining viable reactors were shut down for comprehensive safety inspections (Poortinga et al. 2013: 1205), due to pressure from regulatory rules, adverse public opinion, and other factors, including media pressure. For example, in 2011, the second most popular national newspaper, *Asahi Shimbun*, published 18,641 nuclear-related articles, 75% of which were negative (Scalise 2014: 103).

As of April 2014, 60% (or 30) of Japan's 48 remaining reactors, DeWit (2014) reported, had still not been considered for approval to restart by Japan's Nuclear Regulation Agency. At least 13 appeared to be write-offs, DeWit (2014) continued, "due to age, proximity to a seismic fault, and other factors that render them incapable of satisfying the new safety standards of the NRA. For that reason, at present there are only 17 reactors for which restart applications have been filed. Of these, it appears—even to Japanese supporters of nuclear power—that perhaps only 8 will finally get approval and be restarted," which may not be sufficient to constitute a baseload power supply (DeWit 2014). Plans to bolster nuclear energy to 50% of Japan's electricity-generating capacity have thus apparently been abandoned for the time being. In 2012, the government introduced subsidies to boost renewable energy under the Purchase of Renewable Energy Sourced Electricity by Electric Utilities Act 2012 (Hunenteler et al. 2012).

Skea et al. (2013: S41–42) report the Fukushima effect on nuclear power in Japan has involved five major changes: (i) a consensus has emerged about the need for a less nuclear-dependent society; (ii) promotion of decentralized energy, particularly renewable energy; (iii) a strengthened regulatory system, with nuclear safety decisions separated from the Ministry of Economy, Trade, and Industry (METI) and re-institutionalized as an external agency of the Ministry of Environment, to develop a more independent and objective attitude towards NPP safety management; (iv) potential liberalization of electricity markets and restructuring of the management of the electricity grid to break regional utility monopolies; and (iv) progression of energy conservation. Such effects are not that surprising when it is reported that "post-accident management measures, including the decommissioning of the crippled reactors and compensation for the nuclear evacuees, are estimated at a cost of more than €200 billion" (cited in Hasegawa 2013: 22).

Such developments have also been pressured by attitudinal changes in Japanese perspectives about nuclear energy post-Fukushima, as Poortinga et al. (2013: 1205) commented:

> The accident at the Fukushima Dai-ichi nuclear power plant that followed the devastating Tohoku earthquake and tsunami on the 11th of March 2011 has thrown nuclear power as a publicly acceptable energy technology into doubt. Before the accident, public support and trust in the regulation of nuclear power had already been seriously tested following a series of accidents in Japan . . .

Post-Fukushima, widespread public distrust about governmental capacities to ensure nuclear power safety deepened, particularly following disclosures about regulatory incompetence confirmed by the findings of three major investigations in 2012. A key finding was the absence of a nuclear industry safety culture amidst the " 'cozy relationship' between the Japanese nuclear industry and the regulatory agency" (Schmid 2013: 197; also Funabashi and Kitazawa 2012; Hara 2013; Hindmarsh 2013a: 216; Nakamura and Kikuchi 2011; Poortinga et al. 2013; Schmid

2013: 197; Szarka in this volume). The investigation "of the privately-funded Rebuild Japan Initiative Foundation and the NAIIC1 investigation ordered by the National Diet (legislature) of Japan concluded that Fukushima was also a man-made disaster rather than one caused directly by the earthquake and ensuing tsunami" (Poortinga et al. 2013: 1205). One subsequent action was the strengthening of the regulatory system, as mentioned above.

Alternatively, Hindmarsh (2013a: 3) has advanced that the Fukushima disaster represented "a *new* type of major nuclear disaster, which is found at the intersection of a chronic technological disaster and a natural disaster" (author's emphasis):

> On one hand were the natural components of the preceding earthquake and tsunami; on the other hand, the social components that marked it also as a "chronic technological disaster" involving social actors, decisions and policies or lack thereof found in the discursive interplay of the various stakeholders involved and their practices and actions, the synergy of which (inadvertently) invited high vulnerability to natural disaster.
>
> (Hindmarsh 2013a: 42)

The disaster and subsequent public distrust, along with a number of other post-disaster aspects, including the incompetence of the government to handle the ensuing emergency situation (see Funabashi and Kitazawa 2012; also Nakamura and Nishimura in this volume), has bolstered sustained public concerns about nuclear energy (e.g. see Figure 1.2) (DeWit 2014; also Koyama 2013). Nevertheless, in July 2012, according to Hara (2103: 33), in a symbolic event to signal that the resources of the pronuclear power side were still stronger than those of opposing interests, the Japanese government approved the first restart of two nuclear reactors of the Oi nuclear power plant in Oi town, Fukui Prefecture. However, it also had the inadvertent effect of strengthening both local and broader protests against nuclear power.

At the broader scale, for example, one demonstration in Tokyo, held outside the Prime Minister's official residence, reportedly drew up to 200,000 protestors, making it one of the largest anti-nuclear rallies post-Fukushima (BBC 2012). A month after the restart, a government opinion poll on Japan's energy future showed 50% of respondents wanted to end nuclear power generation by 2030, which was a far larger proportion than those supporting more gradual reductions in nuclear power (Mainichi News 2012). Nearly two years on, an opinion poll by *Asahi Shimbun*, published on May 18, 2014, found 59% of those polled opposed nuclear restarts.[8]

A compounding factor informing public distrust and opposition has been the continued inability of the Tokyo Electric Power Company (TEPCO),

*Figure 1.2*  Anti-nuclear Power Plant Rally on September 19, 2011 at Meiji Shrine
Outer Garden, Shibuya, Tokyo

Source: Author 2011, Wikimedia Commons

the owner/operator of the Fukushima Daiichi nuclear power plant, to manage large amounts of unstable radioactive wastes and radioactive polluted water accumulating in the plant since the disaster (Fackler 2014, Kingston 2013, Starr 2013), with some even escaping into the Pacific Ocean (Huff 2015). Such nuclear management incompetence reflects the earlier finding of the investigation by Funabashi and Kitazawa (2012: 11) that "the government and the plant operator, Tokyo Electric Power Company (Tepco), were astonishingly unprepared, at almost all levels, for the complex nuclear disaster that started with an earthquake and a tsunami."

Concomitantly, with the nuclear fleet disabled, Japan has been forced to rely largely on fossil fuel imports as a replacement for nuclear power. Fossil fuel energy generation reached 88% of total energy consumption in 2013, which contributed to Japan's largest trade deficit of 11.5 trillion yen (US$118 billion) (METI 2014). In turn, this consumption has generated government and environmentalist concerns about the country's increase in greenhouse gas emissions (METI 2014). For example, *The Japan Times* (2014) reported:

> Japan's greenhouse gas emissions rose in fiscal 2013 to the equivalent of 1.395 billion tons of carbon dioxide, its worst total since comparable data became available in fiscal 1990, according to the Environment Ministry . . . Thermal power generation, which generates large amounts of carbon dioxide, has increased sharply

since the 2011 Fukushima disaster began . . . To improve the situation, a ministry official said Japan will push more energy-saving steps and maximize the use of renewable energy to reduce the use of hydrochlorofluorocarbons, which emit large amounts of greenhouse gas.

With international pressure now on Japan to reduce its emissions, we turn to the international effect of the disaster.

## The International Effect of Fukushima

In the broader international geopolitical terrain, there has also been a significant Fukushima effect on the publics and nuclear policies of many countries (e.g. Kim et al. 2013). Issues of nuclear safety and emergency response functionality, and social and environmental justice, intergenerational equity, and socio-ecological risk from radioactive pollution all inform significant public distrust of and opposition to nuclear power development in many countries (as the contributors in this volume also report). In this disturbed and increasingly publicly contested nuclear energy terrain, Germany, Switzerland, and Belgium are phasing out nuclear energy in favor of energy conservation and renewable energy. Such a turn could be a long-term goal for other economies as these energy options become more advanced, varied, affordable, and efficient. Furthermore, after "reviewing the post-Fukushima situation some countries have now decided that they will not enter or re-enter the nuclear expansion business (e.g. Taiwan, Chile, Israel, Venezuela) . . ." (Joskow and Parsons 2012: 24).

Other countries, such as the UK, see themselves at the forefront of a nuclear renaissance or "rebuild" (Shim et al. 2015; also Blowers 2010). And some Middle Eastern countries, in response to concerns about the exhaustion of their own natural resources, are now introducing nuclear power (Joskow and Parsons 2012; Kim et al. 2013). The World Nuclear Association reported that in 2015 there were more than 435 commercial nuclear reactors operating in 31 countries (about 15% of the world's countries, providing about 11% of the world's electricity), with 70 more reactors under construction. There were a further 240 research reactors operating in 56 countries, and 180 reactors powering some 140 ships and submarines (WNA 2015). Joskow and Parsons (2012) reported that the majority of forecasted new construction lay in China, Russia and the former states of the former Soviet Union, India, and South Korea—countries all included in this volume's investigation. It is the view of these authors that "the accident at Fukushima will not 'kill' the much discussed renaissance of nuclear power, but it adds one more negative pressure on the rate of growth globally" (ibid: 13), which includes significant economic and technological challenges (e.g. Skea et al. 2013).[9]

Perhaps the most immediate Fukushima effect internationally was on the safety of nuclear power plants. Post-3/11, nuclear regulators in all 31 nuclear nations quickly ordered stress tests for their reactors, "to ensure that the facilities could withstand various external impacts" (Hibbs 2012: 12) including earthquakes, tsunamis, fires, and extreme weather events. In addition, longer-term assessments were made of regulatory procedures and safety criteria (Joskow and Parsons 2012), and some manufacturers began developing safer reactors, such as the Russian third-generation AES-2006 reactors with enhanced meltdown protection (e.g. see Stsiapanau in this volume). However, the World Energy Council, which established a taskforce to improve nuclear governance, found that while there was support for harmonized nuclear safety norms, "there also seems to be comparatively lower support for the international enforcement of safety standards" (WEC 2012: 22). Thus, haphazardly, perhaps incompetently, "while aspects of the EU stress tests were mirrored in the US, Russia and China, no common approach was agreed, given the context of competing reactor designs and secrecy over their operation" (Szarka in this volume, p. 214).

Concomitantly, the Fukushima disaster has created a heightened international scrutiny of energy choices for a sustainable long-term future, in the eyes of publics and many governments. Another major disaster may well signal the death knell for nuclear power, particularly for siting nuclear power plants in close proximity to populations. Issues around investment, cost, insurance, political and public acceptability, and renewable energy options becoming more viable alongside energy efficiency and conservation, are also threatening the viability of nuclear power (see also Rootes in this volume). As Joskow and Parsons (2012: 28) remarked: "While the international nuclear industry appears so far to have dodged being hit square in the head by a bullet from Fukushima, it should not expect that it will get another chance if there is another serious nuclear accident anywhere in the world."

If this were to occur, an age of mass decommissioning might well emerge. In such context, the overall effect of Fukushima is as an ongoing pressure on the nuclear industry to improve its safety performance and consideration of people's and communities' concerns in relation to siting nuclear power plants in their proximity; but also there is the fundamental effect of reshaping the way in which many people, regions, and countries perceive the risks of nuclear power in relation to health, safety, social well-being, participation, localism, good governance, and the natural environment, particularly when considering future and sustainable energy pathways.

A popular indicator of the effect of a nuclear accident (or disaster) is that of measuring associated changes in public attitudes to public acceptance of nuclear energy. Little, though, has been done on a broad international scale in regard to Fukushima. The well-designed study of Kim et al. (2013) is a start. This study examined the effect of the Fukushima disaster on public acceptance of nuclear energy in 42 countries (involving 24,556 respondents). It found 52.7% favored the use of nuclear energy before the accident, a figure that declined to 45.4% after the disaster.

Indeed, the level of public acceptance of nuclear energy declined in 40 countries, with Japan, not surprisingly, experiencing the largest decrease in public acceptance rates (22.8%), followed by Iraq, Egypt, Kenya, Bangladesh, and China. Interestingly, "public concern and fear increased with greater distance from a nuclear plant site." This finding appears to align with the finding of Parkhill et al. (2010) that nuclear accidents and incidents at distant locations, including those in neighboring countries, lead to renewed anxiety and concern about nuclear power facilities everywhere, which further stigmatizes nuclear power as an energy technology (Flynn 2003).

At the same time, "public acceptance of nuclear energy sharply decreased after the accident in countries with a high density of nuclear power reactors," which was seen by Kim et al. (2013: 827) to suggest "the total operating time of nuclear power generation may be seen as a potential risk rather than as creating trust for the public." In addition, positive inclination towards nuclear power post-Fukushima was linked to a reluctant dependency on nuclear energy that was seen as not being easily changed in an existing national energy mix in a short time. Further, the more governments were perceived to be controlling the media post-Fukushima, which tended to be seen as media suppression of adverse nuclear power issues, the lower the public acceptance of nuclear energy. In addition, people in countries experiencing serious earthquakes also reflected a greater decrease in public acceptance of nuclear energy post-Fukushima (Kim et al. 2013: 826–827).

Against such findings, Kim et al. (2013: 828) concluded (among other things):

> Public acceptance of nuclear energy is highly correlated with a government's political decision-making. Identifying the fundamental issues that affect public acceptance of nuclear energy and considering its country's unique technological, industrial, and safety status regarding nuclear energy will help each government establish a better national energy policy. Additionally, governments should provide a convincing nuclear energy policy reflecting the change in public acceptance of nuclear energy after a catastrophe like the Fukushima nuclear accident . . . Additionally . . . it will be also important and interesting if we examine the effect of the Fukushima accident on the preferences of other energy sources. The effect can be heterogeneous on different energy sources which affect the future energy policy of a country. Therefore, future research on the effect of the Fukushima

accident on a comprehensive energy policy and an energy mix at the national level seems to be necessary.

Such overtures provide an apt point to turn to the research objectives of this book, the research approach, and an overview of how its chapters contribute to discerning the scope and extent of a Fukushima effect on international nuclear power development.

## Objectives and Field of Investigative Inquiry

### Objectives

Five objectives variously inform this book and the contributors' chapters. *First*, to investigate the scope and extent of a "Fukushima effect" in terms of its effects on national histories, debates, and policy responses on nuclear power development, in both well-established nuclear nations and emergent ones, with Japan at the "epicenter." *Second*, to critically reflect upon social, health, environmental, technological, and political and policy *effects* of the Fukushima nuclear disaster in a country context, through a diversity of science, technology and society (STS) perspectives and theories, in Japan, the immediate region, and globally.

*Third*, to facilitate dialogue about how we better understand and respond to this highly topical and important event—as one of the most significant events, so far, of the 21st century—for knowledge production, good governance, and policy learning in Japan, its immediate region of the Asia-Pacific, and internationally. *Fourth*, to enhance science, technology, and environmental governance and knowledge about nuclear power development internationally. *Fifth*, to make further contribution to STS scholarship and the field internationally by including critical perspectives from both the region directly affected and the wider world, in best addressing the broad effects of Fukushima.

### Field of Investigative Inquiry

The same research approach as Hindmarsh's prior volume *Nuclear Disaster at Fukushima Daiichi: Social, Political and Environmental Issues* is applied. Accordingly, *The Fukushima Effect* aims to offer a well-balanced, constructive, and critical reflection positioned at the crossroads of STS studies as embedded in a broad theoretical framework informed by political science, as well as sociology, cultural studies, environmental studies, media studies, the history of science, and anthropology—all of which traditionally contribute to STS studies.

In exploring a Fukushima effect on nuclear power development and management internationally, we find that our contributors invariably focus on policy, cultural, and political histories, which some authors describe as technopolitics. Most contributors, though, can be seen to reflect a "'policy in action'" approach, a concept Hindmarsh (2013a: 4–5) conceived in summing up the approach of most contributors to *Nuclear Disaster at Fukushima Daiichi.*

This approach is informed by French sociologist of science and anthropologist Bruno Latour's (1987) approach, called "science in action," which involves following scientists and engineers through society to best understand the practice of what they do in an integrated sense. Applied to a historical sociopolitical and policy investigative context, and in building on political and policy analysis more generally, this policy in action approach maps out the culture, history, politics, and policy development and implementation of nuclear power development pre- and post-Fukushima. The purpose, again, is to ascertain the scope and extent of a Fukushima effect on nuclear power development, particularly in regard to policy responses to various concerns raised by, and implications of, the Fukushima disaster.

To sum up, *The Fukushima Effect: A New Geopolitical Terrain* presents a powerful, insightful, constructive, and sometimes provocative, analysis of the disaster's international effects on nuclear power development.

## The Contributors' Perspectives and Topics

In Chapter 2, on Japan, Akira Nakamura and Wataru Nishimura look at two Fukushima effects: the fragility of national leadership at the time of the crisis and the unstable role and function of local governance in the Fukushima area. On the fragility of national leaders, the authors advance an ideal form of political stewardship under crisis. Their second Fukushima effect addresses concerns about the fate of local communities near the Fukushima plant. While outlining several social and financial reasons for a rural government to remain near the nuclear plant, Nakamura and Nishimura comment that it usually illustrates the exodus by both local government and residents from the region following the accident.

In Chapter 3 on Taiwan, Dung-sheng Chen explores the existence and extent of a Fukushima effect on public attitudes towards nuclear power in Taiwan, on its anti-nuclear movement, and, in turn, on the political and policy contours of nuclear power development. In sum, the Fukushima nuclear accident in nearby Japan, Taiwan's neighbor to the northeast, was seen as a cataclysmic event that led to Taiwan's civil society becoming more active in pushing the Taiwan government to phase out nuclear energy. This occurred to some extent, Chen narrates, with Taiwan deciding not to expand nuclear development in, abandoning the construction of the Fourth Nuclear Power Plant, but with pressure remaining to go all the way to complete phase out.

In Chapter 4 on Korea, Hyomin Kim investigates the Fukushima effect on nuclear energy regulatory policies. Pre-Fukushima, from the late 1980s, South Korea started to change from old governance technocratic-driven policies to more open dialogue. Post-Fukushima, however, the separation between local and general publics in old governance continued, with the local siting of nuclear power plants discounted for the "good" of the broader public. With both technocratic and participatory governance used to influence the general public to support nuclear energy, Kim argues that renewed concerns about nuclear power risk and transparency and civic engagement in regulation are unlikely to overly change the nation's progress to nuclear at this time or in the short-term future.

In Chapter 5, on China, Xiang Fang investigates the Fukushima effect by focusing on four cases—two pre-Fukushima and two post-Fukushima—that show how local people and political stakeholders engaged with the policy decision-making of siting nuclear power projects, particularly in regard to addressing local concerns, both inside and outside the political and policy system. Fang argues that evidences of increased influence post-Fukushima might become a contribution to China's civil nuclear industry's gradual transformation from a top-down governance approach to a negotiated participatory risk governance approach.

In Chapter 6, on India, Anupam Jha investigates the Fukushima effect in social, policy, legal, and institutional contexts of nuclear power governance, with a particular focus on civic concerns around the safety of nuclear power plants in relation to the surrounding environment, land acquisition, and environmental impact assessment. These concerns are examined in regard to willingness to compensate victims in the case of nuclear disaster, the role of public participation in decision-making. In addition, government willingness to be transparent with information on nuclear safety in regard to, for example, seismicity and radioactive pollution. Some shifts in Indian policy are evident, although much more could be done, Jha argues.

In Chapter 7, on countries of the former Soviet Union, Andrei Stsiapanau focuses on how nuclear interests are dealing with the Fukushima effect on nuclear program development. An embedded technopolitical and engineering logic rationalizes that safety aspects of Japan's nuclear development are less relevant in the post-Soviet terrain. Nevertheless, citizen pressure for more safety and less risk sees a revised logic to protect reactors from the "outside": of adverse public opinion, social mobilization, environmental harm, and energy controversies. Such context, at the time of writing, is informing three new nuclear power plant projects: in the Baltic, Belarus, and Lithuania.

In Chapter 8, on the UK and Finland, Susan Molyneux-Hodgson and Marika Hietala address nuclear waste disposal debates in the UK and Finland as case studies to open up the effects of Fukushima on their countries' respective socio-technical imaginations. The countries present interesting contrasts in terms of volumes of waste, public deliberation approaches, and timelines for policy work. Official documents, industry newsletters, and local news media are analyzed to reveal various positions in the waste debates. The role that Fukushima played in shifting, bolstering, or negating different perspectives is investigated, where the repercussions of nuclear disasters are found to interact to varying degrees with local concerns and national imperatives.

In Chapter 9, on Germany, Detlef Jahn and Sebastian Stephan highlight the important effect the Fukushima disaster has had on energy policy of faraway countries with a focus on Germany, which experienced a dramatic turning point. To emphasize the "special" case of Germany, which moved quickly to phase out nuclear energy, Jahn and Stephan compare the situation in Germany with Sweden, and more broadly with Europe, in revealing an enabling political and policy window of opportunity, against a robust history of anti-nuclear activism, to change energy policy in Germany. Although other European countries have also shifted quite dramatically post-Fukushima, the German phase out decision is found to be unique in its scope.

In Chapter 10, on Switzerland, Fabienne Crettaz von Roten investigates Swiss citizen positions on two policy decisions made by the Swiss Government soon after the aftermath of the accident and finds the Fukushima effect had a significant sociopolitical impact in Switzerland, first in the decision to ban and phase out nuclear energy, and second, in the demand for a significant reduction in energy consumption. The results of several public surveys on nuclear power and energy conservation conducted between March 2011 and April 2013 are discussed and analyzed. They reveal convincing citizen agreement with these two decisions, which are part of a broader frame of public concern, distrust, risk perceptions, and knowledge related to nuclear energy.

In Chapter 11, on France, Joseph Szarka analyzes France's promotion of a nuclear renaissance, notes that it stalled, and asks whether the stall was caused by the 2011 Fukushima disaster. Six categories of norms are examined: (1) cognitive, (2) diplomatic, (3) geopolitical, (4) technological, (5) market, and (6) safety and security, which together provide a framework for assessing the global governance of nuclear power. Szarka folds these norms into the three stages of the nuclear revival and argues that whereas Fukushima contributed to the stall by highlighting unmet safety norms, France's bid to lead the revival lay more in difficulties encountered prior to Fukushima, with construction delays and cost overruns revealing the sector's limited economic viability.

In Chapter 12, on the United States, William Kinsella addresses questions regarding the viability of nuclear power, still evolving in the aftermath of the Fukushima disaster, further complicated in the US by a parallel controversy related to the management and disposal of used nuclear fuel. Kinsella argues that the nuclear waste confidence controversy provides a critical window into larger questions of nuclear safety and risk, argumentative strategies of the nuclear industry and its critics, inclusion and exclusion in democratic policy debate, and technology governance in a risk society, as affected by the Fukushima disaster.

In Chapter 13, on New Zealand, Rebecca Priestley looks for a Fukushima effect in New Zealand decades after the country's 1978 rejection of nuclear power as a possible energy source, and the adoption of a "nuclear free" policy in the 1980s. Political, public and media attitudes to nuclear power are canvassed through New Zealand's history, with a focus on pre- and post-Fukushima attitudes, with issues raised about the lack of public scientists available to speak to the media, and the public, in the wake of the Fukushima disaster.

## Conclusion

Following these country-focused chapters, a concluding chapter sums up and comments on the findings of, and the geopolitical and policy implications raised by, the contributors. Three meta-themes are detected that appear across all the chapters: (i) media responses, public opinion, and social mobilization; (ii) energy policy, politics, and governance; and (iii) questioning nuclear renaissance. In the second meta-theme a useful conceptual device—a Fukushima effect spectrum—is applied to illustrate the effect of Fukushima from strong to weak.

Again, the book aims to facilitate wide dialogue to better understand and respond to the Fukushima Daiichi disaster and its effects. We hope the findings of this book will be used for knowledge production, good governance, and social and policy learning in Japan and globally, and to promote and encourage long-term social and environmental well-being and sustainability in relation to energy choices, safety, and futures.

## Notes

1 Extreme weather events that have already threatened nuclear power plants include coastal flooding, extreme heat, drought, and wildfires (see Union of Concerned Scientists 2014), blizzards and snowfalls (see Hirji 2015), and tornadoes (see http://www.nrdc.org/nuclear/fallout/, accessed June 1, 2015). In addition, see, for example, the appraisal of the World Future Council, http://goo.gl/Y9zVYZ, accessed June 1, 2015.
2 See http://goo.gl/1zXVZ0, accessed May 31, 2015.
3 See http://goo.gl/9wc3tK, accessed May 31, 2015.
4 See http://goo.gl/SRMptb, accessed May 15, 2015.
5 Some authors cite 120,000 (McCurry 2014; see also Hasegawa 2013: 22–23).
6 See http://goo.gl/uKBv8x, accessed May 21, 2015.
7 See http://goo.gl/IM17ZO, accessed May 15, 2015.
8 See http://goo.gl/nMm2wb (in Japanese), accessed May 15, 2015.
9 See http://goo.gl/otyjh8, accessed June 1, 2015

# References

Abasaa. "Radiation_hotspot_in_Kashiwa_02," February, 18, 2012, accessed May 20, 2012, http://commons.wikimedia.org/wiki/File:Radiation_hotspot_in_Kashiwa_02.JPG.

BBC. "Japan Switches on Ohi Nuclear Reactor amid Protests," July 1, 2012, accessed June 20, 2012, www.bbc.co.uk/news/world-asia-18662892.

Bell, D., Gray, T., and C. Haggett. "The 'Social Gap' in Wind Farm Siting Decisions: Explanations and Policy Responses," *Environmental Politics* 14 (2005): 460–477.

Bijker, W. "Science and Technology Policies through Policy Dialogue." In *Science and Technology Policy for Development: Dialogues at the Interface*, edited by L. Box, and R. Engelhard, 109–126. London, UK: Anthem Press, 2006.

Blowers, A. "Why Dump on Us? Power, Pragmatism and the Periphery in the Siting of New nuclear reactors in the UK," *Journal of Integrative Environmental Sciences* 7, no. 3 (2010): 157–173.

Carney, M. "Fukushima Disaster: Radiation Levels Posing Cancer Risks on Fourth Anniversary of Earthquake," *ABC News*, March 11, 2015, accessed May 18, 2015, http://goo.gl/5ZXNA0.

CEC (Commission of the European Communities). *European Governance: A White Paper. COM (2001) 428 final*. Brussels: CEC, 2001.

DeWit, A. "Japan's Energy Policy Impasse," *The Asia-Pacific Journal* 12, no. 14.1 (2014), accessed May 15, 2015, www.globalresearch.ca/in-the-wake-of-fukushima-japans-energy-policy-impasse/5376899.

Dovers, S. *Environmental and Sustainability Policy*. Sydney, NSW: Federation Press, 2005.

Emspak, J. "Fukushima Radiation Moving across Pacific Ocean," *Huffington Post*, March 4, 2012, accessed October 30, 2012. http://www.huffingtonpost.com/2012/04/03/fukushima-radiation-pacific-ocean_n_1399843.html.

Fackler, A. "Worst Spill in 6 Months Is Reported at Fukushima." *The New York Times* February 20, 2014, accessed April 10, 2015, www.nytimes.com/.

Freiner, N. "Mobilizing Mothers: The Fukushima Daiichi Nuclear Catastrophe and Environmental in Japan," *ASIA Network Exchange* 21, no. 1 (2013): 1–15.

Funabashi, Y., and K. Kitazama. "Fukushima in Review: A Complex Disaster, a Disastrous Response," *Bulletin of the Atomic Scientists* 68, no. 2 (2012): 9–21.

Flynn, J. "Nuclear Stigma." In *The Social Amplification of Risk*, edited by N. Pidgeon, R. Kasperson, and P. Slovic, 326–352. Cambridge, UK: Cambridge University Press, 2003.

Goldenberg, S. "US Sailors Prepare for Fresh Legal Challenge over Fukushima Radiation," *The Guardian*, August 20, 2014, accessed May 25, 2015 http://goo.gl/Zq926v.

The Guardian. "Nuclear Power Plant Accidents: Listed and Ranked Since 1952." *The Guardian*, March 18, 2011, accessed November 4, 2012. www.guardian.co.uk/news/datablog/2011/mar/14/nuclear-power-plant-accidents-list-rank.

Hara, T. "Social Shaping of Nuclear Safety: Before and After the Disaster." In *Nuclear Disaster at Fukushima Daiichi: Social, Political and Environmental Issues*, edited by R. Hindmarsh, 22–40. New York: Routledge, 2013.

Hasegawa. R. *Disaster Evacuation from Japan's 2011 Tsunami Disaster and the Fukushima Nuclear Accident*. Study No. 05/13 May 2013. Paris, France: Institut du développement durable et des relations internationals (IDDRI).

Hibbs, M. "Nuclear Energy 2011: A Watershed Year," *Bulletin of the Atomic Scientists* 68 (2012): 10–19.

Hindmarsh, R. "Fallout from Fukushima Daiichi." In *Nuclear Disaster at Fukushima Daiichi: Social, Political and Environmental Issues*, edited by R. Hindmarsh, 214–218. New York: Routledge, 2013a.

Hindmarsh, R. "Nuclear Disaster at Fukushima Daiichi: Introducing the Terrain." In *Nuclear Disaster at Fukushima Daiichi: Social, Political and Environmental Issues*, edited by R. Hindmarsh, 1–21. New York: Routledge, 2013b.

Hindmarsh, R. (ed). *Nuclear Disaster at Fukushima Daiichi: Social, Political and Environmental Issues*. New York: Routledge, 2013c.

Hindmarsh, R. 2014. "Hot Air Ablowin': 'Media-Speak', Social Conflict, and the Australian 'Decoupled' Wind Farm Controversy," *Social Studies of Science* 44 (2014): 194–218.

Hirji, Z. "Winter Storm Exposes Vulnerability of Nuclear Power Plants: Shutdown of Pilgrim Facility in Massachusetts Fuels Critics' Challenge," *Inside Climate News*, January 29, 2015, accessed June 1, 2015, http://goo.gl/IgiulC.

Högberg, L. "Root Causes and Impacts of Severe Accidents at Large Nuclear Power Plants," *Ambio* 42, no. 3 (2013), 267–284.

Huenteler, J., Schmidt, T.S., and N. Kanie. "Japan's Post-Fukushima Challenge: Implications from the German Experience on Renewable Energy Policy," *Energy Policy* 45 (2012): 6–11.

Huff, E. "Fukushima Radiation Spikes 7,000% as Contaminated Water Pours Into the Ocean," *Natural News*, February 26, 2015, accessed May 21, 2015, http://goo.gl/uyrzjk.

IAEA (International Atomic Energy Agency). "IAEA Update on Fukushima Nuclear Accident," Fukushima Nuclear Accident Update Log, April 12, 2011, 4:45, accessed November 3, 2012, http://goo.gl/Cr8n3X.

Japan Times. "Japan's Fiscal '13 Greenhouse Gas Emissions Worst on Record," *The Japan Times*, December 5, 2014, accessed May 20, 2015, http://goo.gl/UF1ObK.

Japan Today. "Radiation Hotspot in Chiba Linked to Fukushima," *Japan Today*, November 29, 2011, accessed February 19, 2012. http://goo.gl/L8iLoQ.

Joskow, P.L., and J.E. Parsons. *The Future of Nuclear Power after Fukushima*. USA: MIT Center for Energy and Environmental Policy Research, 2012.

Kim, Y., Kim, M., and W. Kim. "Effect of the Fukushima Nuclear Disaster on Global Public Acceptance of Nuclear Energy," *Energy Policy* 61 (2013): 822–826.

Kingston, J. "Nuclear Power Politics in Japan, 2011–2013," *Asian Perspective* 37, no. 4 (2013): 501–521.

Koyama, K. "Japan's Post-Fukushima Energy Policy Challenges," *Asian Economic Policy Review* 8, no. 2 (2013): 274–293.

Latour, B. *Science in Action: How to Follow Scientists and Engineers through Society*. Cambridge, MA: Harvard University Press, 1987.

Mainichi News. "Gov't Poll Shows about 50% Want to End Japan's Nuclear Reliance," *Mainichi News*, August 22, 2012, accessed June 19, 2013, http://goo.gl/ZDukkP.

Martin, P., Li, Z., and T. Qin. *Environmental Governance and Sustainability*. Cheltenham, GBR: Edward Elgar Publishing, 2012.

Matanle, P. "The Great East Japan Earthquake, Tsunami, and Nuclear Melt-down: Towards the (Re)construction of a Safe, Sustainable, and Compassionate Society in Japan's Shrinking Regions," *Local Environment: The International Journal of Justice and Sustainability* 9 (2011): 823–847.

McCurry, J. "Fukushima Nuclear Disaster: Three Years on 120,000 Evacuees Remain Uprooted," *The Guardian* (Australian Edition), September 11, 2014, accessed May 18, 2015, http://goo.gl/zeX0IF.

METI (The Ministry of Economic, Trade and Industry). *Annual Energy Report 2013*, 2014, accessed May 15, 2015, www.enecho.meti.go.jp/.

Miller, C., Sarewitz, D., and A. Light. *Science, Technology, and Sustainability: Building a Research Agenda*. A Report for the National Science Foundation, Arlington, VA, 2008.

Nakamura, A., and M. Kikuchi. "What We Know, and What We Have Not Yet Learned: Triple Disasters and the Fukushima Nuclear Fiasco in Japan," *Public Administration Review* November/December (2011): 893–899.

Oiwa, Y. "Fukushima Finds 16 New Cases of Thyroid Cancer in Young People," *Asahi Shimbun* May 19, 2015, accessed May 21, 2015, http://goo.gl/dQGONz.

Own Work. "Anti-nuclear power plant rally on 19 September 2011 at Meiji Shrine outer garden," September 19, 2011, accessed May 27, 2015, http://commons.wikimedia.org/wiki/File:Anti-Nuclear_Power_Plant_Rally_on_19_September_2011_at_Meiji_Shrine_Outer_Garden_03.JPG.

Parkhill, K., et al. "From the Familiar to the Extraordinary: Local Residents' Perceptions of Risk When Living with Nuclear Power in the UK," *Transactions of the Institute of British Geographers* NS 35 (2010): 39–58.

Poortinga, W., Aoyagi, M., and N. Pidgeon. "Public Perceptions of Climate Change and Energy Futures Before and After the Fukushima accident: A Comparison between Britain and Japan," *Energy Policy* 62 (2013): 1204–1211.

Scalise, P.J. "Who Controls Whom? Constraints, Challenges and Rival Policy Images in Japan's Post-war Energy Restructuring." In *Critical Issues in Contemporary Japan*, edited by J. Kingston, 92–106. New York: Routledge, 2014.

Schmid, S. "Nuclear Emergency Response: Atomic Priests or an International SWAT Team?" In *Nuclear Disaster at Fukushima Daiichi: Social, Political and Environmental Issues*, edited by R. Hindmarsh, 194–213. New York: Routledge, 2013.

Shim, J., Park, C., and M. Wilding. "Identifying Policy Frames through Semantic Network Analysis: An Examination of Nuclear Energy Policy across Six Countries," *Policy Sciences* 48 (2015): 51–83.

Skea, J., Lechtenböhmer, S., and J. Asuka. "Climate Policies after Fukushima: Three Views," *Climate Policy* 13 (2013): DOI: 10.1080/14693062.2013.756670

Star, S.L. "The Ethnography of Infrastructure," *American Behavioral Scientist* 43, no. 3 (1999): 377–391.

Starr, S. "Costs and Consequences of the Fukushima Daiichi Disaster," Environmental Health Policy Institute, 2013, accessed May 15, 2015, http://goo.gl/lwFFLt.

Stohl, A., et al. "Xenon-133 and Caesium-137 Releases into the Atmosphere from the Fukushima Dai-ichi Nuclear Power Plant: Determination of the Source Term, Atmospheric Dispersion, and Deposition," *Atmospheric Chemistry and Physics* 12 (2012): 2313–2343.

Union of Concerned Scientists. "How Climate Change Puts Our Electricity at Risk (2014)," April 2014, accessed June 1, 2015, http://goo.gl/PjBsvm.

WEC (World Energy Council). *Nuclear Energy One Year after Fukushima*, 2012, accessed May 16, 2015, http://tinyurl.com/knqnw64.

WNA (World Nuclear Association). *Fukushima: Radiation Exposure*, September 2014, accessed May 18, 2015, http://goo.gl/NfDNkb.

WNA (World Nuclear Association). *Nuclear Power in the World Today*, February 2015, accessed May 5, 2015, http://goo.gl/RGh2Az.

# 2 The Fukushima Effect in Japan

## Reflections on Political Leadership and Local Governance

*Akira Nakamura and Wataru Nishimura*

The 2011 failure of the nuclear power plant in Fukushima had several major repercussions for Japan's social and political configurations. These repercussions can be investigated under a key theme of science, technology and society studies called "science, technology, and governance," which looks at the political and policy implications of a topic under investigation; here, the "Fukushima effect" in regard to political leadership and to local communities and government.

The Fukushima effect has been marked especially by debate on the quality of political leadership at a time of national crisis. Central to this polemic has been defining what abilities government leadership should possess to meet horrific challenges such as the Fukushima accident. Both academics and mass media in Japan frequently note that, as the spearhead of government, competent national leaders should have several intrinsic attributes (e.g. Masciulli et al. 2009), which are indispensable for the leaders to steer and salvage the country following such an emergency.

This attitude has surfaced because, at the time of the Fukushima accident, government leaders appeared overwhelmed and failed to provide clear policy directions for the country. Instead, much of the national leadership, including the prime minister, often seemed confused, and frequently exposed a lack of knowledge and experience in dealing with the breakdown of the atomic plant. Subsequent mayhem and pandemonium in the central government fueled public distrust of government and expanded confusion about the government policy-making process.

In addition to revealing fragility in national leadership, the accident at Fukushima brought serious disruption to adjacent communities. Atomic fallout contaminated several towns and villages next to the Fukushima Daiichi plant, forcing residents to leave their homes and move to nuclear-free zones. Many decided to move to Tokyo or neighboring metropolitan areas, while others stayed within the Fukushima Prefecture, although in locations remote from the nuclear facilities (see Figure 2.1). Similarly, local governments near the heart of the Fukushima nuclear site were required to relocate their offices.

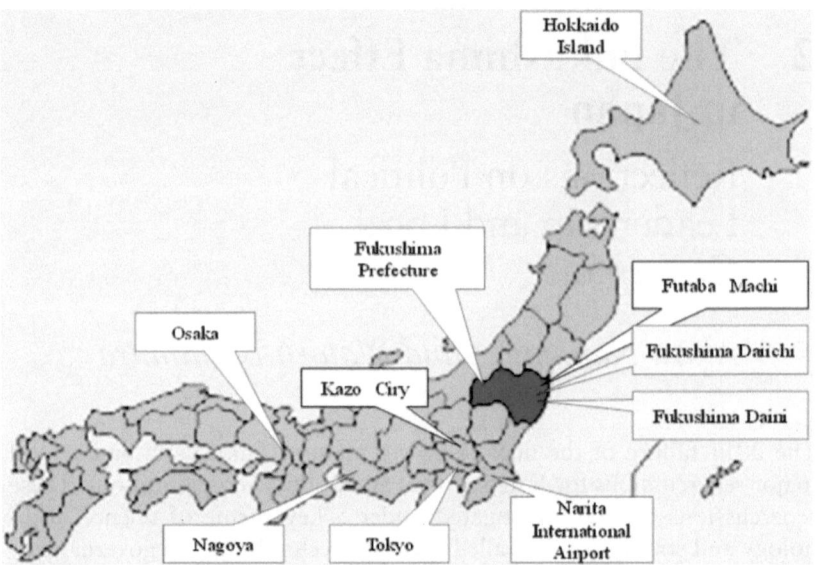

*Figure 2.1* Fukushima Daiichi and Nearby Areas
Source: ACWorks Co., Ltd.

A relevant example is the town of Futaba Machi, the location of the Fukushima Daiichi nuclear plant. The Japanese word "Machi" means both "town" and "government." Futaba Machi, therefore, often connotes both local government and community in Futaba town. Both government and citizens were originally relocated to a neighboring community, then moved to Saitama City and later to Kazo City in Saitama Prefecture, approximately 260 kilometers from Fukushima, although close to Tokyo (Working Group 2012: 5–6).

The city office in the new area extended various local government services to the people from Futaba Machi, including certificates of residency, which allowed them to claim short-term government financial aid or damage compensation from the Tokyo Electricity Power Company (TEPCO), the electricity power utility for Fukushima Prefecture, which operated all of the power plants there. This new arrangement, however, had one odd aspect: although the office provided regular services, only a small portion of the townspeople stayed in the new location. Many former residents were scattered throughout the country, most away from the Fukushima region.

On June 17, 2013, the town moved its office to Iwaki City in Fukushima Prefecture, a location close to the original site of Futaba Machi. This relocation could possibly help to alleviate the anxieties of both city officials and former residents. It appears to be the first instance anywhere in Japan of a local government extending services and functions to people, when some live beyond the town's administrative jurisdiction. Even at this date, the town regularly holds town council elections, has its own annual budget, and receives various national subsides similar to any other local government in Japan (Futaba Machi 2014: 1–10). Given the traditional mode of Japanese administrative practice, the current situation of the Futaba Machi government operating from Iwaki City appears bewildering and unprecedented to many professional observers.

A primary impetus for this chapter comes from these two Fukushima effects: the fragility of national leadership and the unstable role and function of local governance. In exploring the first effect in regard to the fragility of national leadership, an ideal form of political leadership is delineated. In the view of this chapter's authors, national leaders, especially at a time of national crisis, ought to have five essential attributes. They must have perspective, persistence, and perseverance; they must be prescriptive, and must have persuasion. Not many leaders possess these five essentials. This chapter tries to highlight how the party-centered arrangement of government in Japan failed to generate effective national leadership at the critical phase of the Fukushima nuclear disaster.

The second section of the chapter explores the Fukushima effect on the fate of communities located next to the Fukushima plant. Focusing on Futaba Machi, the site of the Fukushima plant, it first outlines several social and financial reasons why this rural community allowed the lethal power generators to be built in its vicinity. This is followed by a brief description of the exodus by both local government and residents from these towns following the accident. The section concludes with a brief discussion of the prospects of the Futaba Machi government to be reestablished in its former location. Concluding remarks then sum up the findings of both sections.

## The Fukushima Effect on Japan's National Leadership

### Five Qualities of Leadership

The Fukushima effect has shed light on the importance of national leadership in Japan's political environment. Traditionally, Japanese national leaders are weaker and less powerful than those in the US or other preeminent democracies. One of the reasons for this unsettling leadership profile comes from a cultural legacy. Unlike countries where individualism is highly valued, group cohesion has been a critical social norm in Japanese society. Japanese specialists in the US and the UK frequently note that decisions in Japan are often made by a group, for the sake of the group. In their views, "Groupism" stands out as one of the most conspicuous features in the country's policy process (see, e.g. Stockwin 2008: 29–47). Unfortunately, this group orientation proved ineffective during a national emergency: in fact, it became a liability when the country faced the disastrous effects of the Fukushima nuclear plant accident in March 2011. The traditional norm impeded quick decisions, as it called for prolonged discussions to reach group consensus.

This cultural tradition was augmented by the political philosophy of the governing party at that time, the Democratic Party of Japan (DPJ). The party had taken control of the government, mainly because of its popular slogan: "Reducing bureaucratic control and enlarging party power." Remaining firm in this doctrine, the DPJ did not trust public officials, even in the aftermath of the Fukushima disaster. The party failed to seek professional advice and opinions in dealing with the nuclear accident. Although they were non-professional amateurs, party members thought they could surmount the terrible nuclear crisis (Independent Committee 2012: 95–119). This was a gross mistake, as the subsequent narrative indicates.

From the view of the authors of this chapter, a national leader should possess several attributes to be a spearhead of government. A leader should be able to examine an issue from a broad perspective and avoid a narrowly focused point of view (the essential attribute of Perspective). The Fukushima effect made this clear. When the nuclear plant in Fukushima began leaking polluted water into the sea, neighboring countries immediately had serious concerns. Unfortunately, Japan's national leadership was too immersed in domestic concerns to acknowledge international apprehensions. Both China and the Republic of Korea were extremely alarmed by the lack of crisis information from the Japanese government. This issue could have been avoided had the prime minister been able to view the issue from a broad perspective.

Similarly, a competent leader should have a philosophy or principle that guides political conduct, and must remain constant to it. The leader's code of conduct ought not to fluctuate over time: it must be consistent and clear to followers (the essential attribute of Persistence). Unfortunately, some officials in the DPJ government at the time of the Fukushima accident appeared capricious. As detailed in the following section, the chief cabinet secretary was one of these officials: his TV announcement about the dangers of formula feeding arising from possible radioactive contamination of tap water confused the public and triggered an increasing distrust of government (see in more detail below).

A leader should not try to tackle a host of critical issues alone, but should delegate or devolve authority to deputies. While delegating responsibilities, the leader should remain patient and allow deputies to manage important policy mandates (the essential attribute of Perseverance). In this regard, Prime Minister Kan's behavior had an important impact. Not only did he take an impulsive helicopter flight to check the Fukushima accident site, but he also abruptly entered the headquarters of the Tokyo Electric and Power Company, shouted at the executives of the firm, and demanded immediate action. This erratic behavior caused mayhem and delayed responses to the disaster (Yoshida 2012: 2).

A proactive rather than a reactive stance is another important leadership quality. A good leader should stay put in office, considering alternative options in the event that the ongoing program proves unproductive. A proactive leader always remains ready to suggest another course should the previous one be ineffective (the essential attribute of being Prescriptive). Finally, an ideal leader requires rhetorical or oratorical skill to convince others (the essential attribute of Persuasion). Of these five essential attributes, the power of persuasion might be one of the most important qualities for a leader in any organization. Unless the leader can convince subordinates, they may not produce the results necessary to help alleviate the crisis situation.

### Institutional Effect on the Governing Party

The Democratic Party of Japan (DPJ) took control of the government in September 2009, a major accomplishment in the Japanese political environment. Since 1955, the conservative Liberal Democratic Party (LDP) had monopolized government and managed the political affairs of the country. In these LDP years, the government had increasingly depended on elite bureaucrats to govern the country. The close politic-administrative relationship eventually became entrenched, growing into one of the most salient features of Japan's conservative rule. In legislative proceedings, for instance, a high-level bureaucrat, rather than a cabinet minister, would often answer policy interrogations put by opposition parties. Similarly, many fast track public officials would change their career path and run for the national legislature under the conservative party label. In the eyes of much of the electorate, the LDP administration frequently looked as if a group of elite bureaucrats was dominating the government (Nakamura 2001: 169–201).

The DPJ challenged the LDP style of government, asserting its intention to replace a bureaucrat-centered administration with a party-oriented government. In the view of the DPJ leadership, a Westminster model of government, in which party members and not public officials would play the key role in policy-making, looked ideal. The DPJ's campaign slogan, to reduce the power of bureaucracy, appealed to many Japanese voters, who seemed displeased with the longstanding mandarin- (or bureaucrat-) dominated LDP rule. In the September 2009 election, the DPJ overwhelmed the LDP in popular votes and, for the first time in many decades, become the governing party. For several months after the election, many experts believed the landslide victory helped solidify the party's political fortune to the extent that the DPJ would stay in government for many years to come (Yamaguchi 2012).

Unfortunately, however, the popular party had to face one of the most serious crises in the country's post-war history. It was under DPJ Prime Minister Kan Naoto that the triple disaster—of earthquake, tsunami, and nuclear accident—hit Japan on March 11, 2011. The triple disaster revealed major problems with DPJ rule. Although the party was inexperienced in managing the country, the DPJ would not actively seek help and support from government bureaucrats in general and technocrats in particular. Steadfastly maintaining a party-centered model, the few non-professional politicos and amateur party members in the Kan office had to meet the enormous challenges head-on.

### Inconsistency in Government Direction

The consequences were obvious. The leading members of DPJ were often reluctant to make critical decisions. They also delayed providing instructions and directives to government officials and local authorities. The end result of these issues was disorder: the different policy directions the DPJ members produced were often contradictory and perplexing, especially to the public. One such example was an announcement by the chief cabinet secretary, Edano Yukio, made on television 12 days after the disaster. Edano, a young lawyer and a rising star in the DPJ hierarchy, suggested that parents should avoid using tap water to make up an infant's formula, thus implying that tap water, especially in the Tokyo metropolitan region, might be contaminated by radioactive fallout. He recommended that families with infants use bottled water to mix formula. Immediately after this announcement, bottled water disappeared from all shelves of stores in the metropolitan region. Families with infants became extremely alarmed and frantic in their quests to get safe water, which was no longer available anywhere in the metropolitan region. Becoming aware of the mounting panic, the secretary once again appeared on television and said: "Any family who is unable to obtain bottled water should use tap water to make formula for the baby. There should be no harm to the infant at least for the time being" (Kyodo News 2011).

The remark was mind-boggling for many Japanese, as it was completely opposite to his original statement. Edano's inconsistent instruction naturally aroused public discontent; many families with babies were agitated and angry. This contributed to a sharp decline of public trust in the DPJ government. (A personal note is in order at this point. A few weeks before the accident, the first author's new grandson was born, while the second author had a new baby girl two months before the accident. Naturally, after listening to the government announcement, both were worried about how to secure safe water for the newborn babies, as stores and shops were sold out of supplies of bottled water. In desperation, one of the authors called friends in Osaka and asked them to ship boxes of purified water. By the time these boxes finally arrived in Tokyo, both authors had been exposed to the new and contradictory announcement from the chief cabinet secretary.)

By way of mass media, a large number of Japanese began to direct harsh criticisms toward the DPJ government and its policies. This was clearly reflected in a survey taken right after the disaster. The result showed that fewer than 2% of Japanese trusted the incumbent government (Research Group 2011). In the view of the authors of this chapter, this marked the beginning of the end of the DPJ rule, which finally occurred in December 2012 when the Liberal Democratic Party was voted back into power.

### The Prime Minister's Dubious Behavior

Another indication of the DPJ's inexperience was displayed in the prime minister's erratic behavior in the immediate aftermath of the disaster. The day after the nuclear accident, Prime Minister Kan Naoto decided to take a helicopter to inspect the accident site. Prime ministers, however, are not ordinary people. They have to bring with them an entourage of subordinates and security personnel, and their departure should be well-planned, -timed, and -scheduled. However, this prime minister overruled all procedural issues and suddenly decided to take off to Fukushima by himself. This abrupt decision caused havoc among his subordinates as well as those working in the frontline of the nuclear accident. Many technical personnel who were frantically working at the Fukushima plant were forced to halt their work and instead prepare for the arrival of the prime minister.

A few days later, in the early morning of March 15, 2011, the prime minister decided to initiate another unprecedented action—an unscheduled visit to TEPCO headquarters. Japanese longstanding political tradition holds: "public revered and private despised," connoting that government officials are the object of popular awe, while those in business are inferior in social status. This being the case, the government would usually summon the president of a company to a ministry, should some issues arise. The public officials would then order or command the firm to take appropriate steps to resolve the problems. If a prime minister should visit a private corporation, the occasion would be social or ceremonial, and would never occur at a time of national emergency.

Kan's unexpected call on TEPCO was unheard of, and considered extremely unusual behavior. According to the media account, the prime minister delivered a fuming speech to TEPCO board members, sternly demanding that their head office make every effort to restore the broken nuclear reactors. This surprise visit disrupted ongoing rescue efforts in the TEPCO Tokyo office and brought chaos to the frontline rehabilitation in the Fukushima plant (Kokkai Jikocho 2012: 285–326).

In September 2014, the LDP government made public the outcome of committee hearings involving the questioning of several DPJ leaders at the time of the Fukushima debacle. These leaders included former Prime Minister Kan and Chief Cabinet Secretary Edano. The records indicate that, even within the governing party, opinions were divided over the prime minister's unprecedented and abrupt decisions to visit the accident site and TEPCO. The committee report once again revisited the evaluation and discussion of the appropriateness of the DPJ's approach to control the nuclear accident (Shi 2014: 1–3).

## Lack of Government Coordination

The Fukushima effect also impacted local governance. Local governments adjacent to the Fukushima plant were in similar chaos; their chief executives were all responsible for the safety and security of their communities. However, none of them knew exactly what steps they should initiate, because little or no information and policy directives appeared forthcoming from central government. According to the Special Law for the Preparedness of Nuclear Power Accidents (Genshiryoku Saigai Taisaku Tokubetsu Sochi Ho), enacted in 1999, atomic energy suppliers were required to submit to central government the nature and extent of a nuclear accident when it happened. The national authority would then relay that information with a policy recommendation to concerned local units of government.

A number of post-disaster studies indicate that this legal procedure was not properly followed, due to the magnitude of the Fukushima Dai-ichi disaster. TEPCO stated that the company did inform the national nuclear safety agency of the breakdown of the facility in compliance with the law. The central government similarly indicated that they issued the evacuation directive to local governments immediately after the accident. Nevertheless, the local chief executives in the affected zones all noted that they did not receive any instruction from either TEPCO or the central government. The town and village chiefs depended, instead, on television news for both information and guidelines (Working Group 2012: 4–28).

During the critical hours in the aftermath of the disaster, a sharp schism developed between party leaders and public officials. This was a direct effect of the institutional arrangement of government that the DPJ espoused, that is, reducing the power of bureaucrats and strengthening the position of party members. A small group of leading party members assembled first on the second floor and later on the fifth floor of the prime minister's office. At the same time, a small group of bureaucrats from different agencies gathered in the basement of the same building, as prescribed by the crisis management manual. One of the unfortunate consequences was the lack of communications between the party members and the public officials, again a direct consequence of the DPJ's political agenda. In remaining true to their campaign agenda, the DPJ had insisted on not putting trust in government bureaucrats.

The group of bureaucrats in the basement, therefore, did not or could not initiate any policy measure responsive to the calamities. They had to sit and wait for instructions to come from the fifth floor of the building. In the meantime, the party government leaders began to organize emergency headquarters in different parts of the central authority. Each day following the nuclear disaster, the government hastily formed several new decision-making offices. The mushrooming headquarters naturally increased confusion and conflict about decision-making at the center (Nishimura 2012: 77–98).

One of Japan's leading bureaucrats confided to one of the authors of this chapter in the immediate aftermath of the disaster that many prolonged meetings and arduous discussions were held in the prime minister's office, but unfortunately no decisions were forthcoming. In fact, it has since been noted that the public officials in the basement stored much vital information critical to both the evacuation and rescue of suffering residents. However, the information was not shared with the party members, but remained in the control of the bureaucrats in the basement of the prime minister's office; consequently, DPJ members had to explore solutions to the nuclear issues by themselves, without relevant data and information (Asahi Shimbun 2012: 60–79).

### Enlarging Evacuation Zones

According to the post-disaster investigation, Prime Minister Kan, in consultation with his inner circle of colleagues, issued the first evacuation order at 21:23 on March 11, which failed to reach several local governments located close to the Fukushima plant. Nevertheless, the document contained two instructions: people who lived in a three kilometer radius from the damaged plant should immediately evacuate the region, while those who resided in the three to 10 kilometer range should stay at home. Thirty minutes earlier, the Fukushima governor had independently issued a different order asking citizens living within two kilometers of the accident site to flee the hazardous zone. These two different directives confused victimized residents; however, these directives were not the end of the ordeal. Complications continued to multiply. Prime Minister Kan amended the range of the evacuation zone several times in the next few days.

By the end of the day on March 15, the prime minister had enlarged the radius of the evacuation district to 30 kilometers; by April 22, he had designated three different zones within this area. The first, a "No Go Zone," covered the area within 20 kilometers of the power plant. Being that close to the nuclear plant, the region was potentially contaminated; therefore, nobody was allowed within that zone, and violators were subject to penalties. The second, labeled "Evacuation Zone," comprised areas lying between 20 and 30 kilometers from the accident site. This directive was not legally binding; the government encouraged residents to leave the area. Some residents stayed in this zone, while others followed the government instructions. The third zone was called the "Stand Ready Zone." Although outside the 30 kilometer confines, the area was vulnerable to possible nuclear particle fallout. Should this happen, residents would be required to leave the district (Asahi Shimbun 2012: 260–264).

With the three zones designated, the question remained: who would enforce these area classifications? Because party members lacked the leadership to control government subordinates, the Kan government left this responsibility in the hands of the local chief executives, which contributed to the fragmentation of decision-making in the center. There was an additional politically practical reason for this delegation of evacuation enforcement: evacuation orders would be highly unpopular. Many residents would protest, because they would hate leaving their home towns. Politically shrewd, the incumbent Democratic Party of Japan tried to avoid being the target of public blame. Instead, in the name of decentralization, the DPJ government delegated the difficult decision of evacuations, along with its potential to arouse public anger, to local governments.

In the post-disaster investigation, many local government chiefs stated that they received little or no direction from either central or prefectural governments. They all testified that, amid pandemonium, they had to make the hard decision to order people to leave their home towns, although they knew this would cause physical inconvenience and financial hardship for residents (Working Group 2012: 5–33). In fact, the people in nuclear-tainted regions had to make a long exodus before they found a semi-permanent location in an urban center in Fukushima. The ordeal does not yet seem to be over: many of these displaced residents have been living in provisional housing units and yearn for long-lasting accommodation (Futaba Machi 2014: 41–55).

### Effective Leadership at Crisis

The above examples provide several important lessons regarding the role of leadership in a crisis. Any government needs a competent leader to deal with unexpected situations. As noted earlier, in the opinion of the authors, the leader ought to possess several intrinsic qualities to be an effective spearhead of government, including the capacity to examine critical situations from a broad perspective. A grave issue does not always remain only a domestic problem; it can go beyond national boundaries and develop an international agenda.

Unfortunately, the DPJ leadership seemed to lack this necessary perspective; the vision and focus of Prime Minister Kan and his associates was narrow and short-sighted. When the nuclear accident occurred, Chinese and South Korean leaders expressed deep concerns about the contaminated seawater from Fukushima eventually flowing to their shores.

(One of the authors of this chapter actually encountered this response at a number of international conferences held immediately after the accident.) The US government also regarded Fukushima as a serious issue of global significance and therefore extended an offer of help to the Japanese government. These are a few of the clear indications that the accident had important global implications (Hindmarsh 2013: 1–21). Nevertheless, Japan's prime minister considered the incident only from a domestic perspective and did not spare time to think about its international connotations.

In addition, the authors of this chapter hold that the prime minister, being the leader of the government, must have strong personal convictions that dictate political behavior. The prime minister's guiding principle should not vacillate; political belief should be adhered to even at the time of crisis. "Chops and changes" would be the worst scenario anyone could expect from one's leader during a nuclear accident. Instead of initiating an unexpected trip, the prime minister should have stayed in his office and given the right orders to his deputies at the right moment. Several incidents have exemplified that Prime Minister Kan was too impatient as a leader to keep coherent and consistent political manners. He should not have travelled to the accident site, nor should he have made a surprise visit to the TEPCO headquarters. His unpredictable behavior only worsened the turmoil.

As the top leader of the government, the prime minister should have exhibited untiring perseverance. Not only did his decisions need to be consistent, but he also had to value the work of his subordinates. Government officials at the time of the nuclear fiasco had been laboring hard to produce effective policy and action platforms. Under the circumstances, the prime minister should have paid attention to essential conditions of modern government: "Let a manager manage." The role of a party leader in an emergency should be that of a strategist, extending encouragement and support to his deputies and government officials. Instead, Prime Minister Kan was tied up with his political dictate that the DPJ should neither rely on bureaucrats, nor work closely with government officials. The party-centered government approach Kan had strongly advocated appeared to backfire in this fiasco (Kokkai Jikocho 2012: 285–325).

## The Fukushima Effect on Local Governance

### *The Fukushima Effect on the Depopulated Fukushima Region*

Another Fukushima effect has left its mark on local governance in the northern part of Japan. A brief description of Japan's local system of government is in order here. Japan maintains a unitary system of government. Under the central government, the country is divided into 47 prefectures (equivalent to states or provinces elsewhere). One of them is

Tokyo. Being the capital of the nation, the area has a distinctive designation, called the Tokyo Metropolitan Region (TMG). The TMG has 23 wards in the center and 26 cities on the periphery. Other prefectures are further divided into cities, towns, or villages, depending on the size of the population. As of April 2014, Japan had 1718 local units of government: 47 prefectures, 790 cities, 745 towns, and 183 villages. They are all incorporated and, theoretically at least, autonomous entities, although they receive both block and categorical grants from the central government. Each has its own elected chief executive with an elected local legislature, and a sizable number of local officials recruited by open, fair, and competitive civil service examinations.

Several towns and villages are neighbors to the Fukushima plants. Because they are mostly rice-farming communities, many members of the younger generations have migrated to urban areas in general, and to the Tokyo Metropolitan Region in particular, to secure more lucrative and exciting professions than toiling on a farm. Other young people have moved to the Tokyo capital region, which includes the TMG as well as three neighboring prefectures, to seek higher education, and would likely stay in Tokyo even after graduation from college. Rural areas in many parts of the country have been facing a loss of residents. Depopulated rural communities have, therefore, tended to welcome industry to their regions to help create jobs and keep young people in the region.

Towns and villages around the Fukushima plant had chosen construction of nuclear power plants in their communities as a safeguard against depopulation. Several factors persuaded them to take this option. Long before the 2011 disaster, the central government enacted legislation that extended subsidies to rural areas willing to accept nuclear power facilities. The government scheme posited that a community consenting to adopt nuclear generators would be entitled to receive ¥520 million (equivalent to US$5.2 million) up front for three years even before the start of the project. During the next six or seven years, while the construction was undergoing, the community could collect an additional ¥7.5 billion (US$75 million) from the national government. Various grants over these 10 years would total more than ¥44.49 billion (US$449 million) (Agency for Natural Resources and Energy 2010: 6). For a small, depopulated town, these government appropriations looked extraordinary and difficult to refuse. Consequently, many desolate rural areas were eager to take government incentives and allow power suppliers to build nuclear plants in the regions (Shigen Enerugi Cho 2010; also Hara 2013).

In addition, these potentially dangerous facilities would bring large property taxes to rural government, often augmented by various forms of financial aid that TEPCO and other major power suppliers provided to the communities. All these financial aids contributed hugely to improving the fiscal and social health of nuclear towns and villages. Futaba

Machi is a typical example of a well-subsidized community. The town government accumulated approximately ¥207 million (US$2.07 million) each year from the central authority as soon as the power plant began to operate.

Aside from the government's direct financial support, the nuclear power plant brought other indirect benefits to the town. Employment opportunities increased, while many shops and small businesses thrived. The different funds made available to the town government were used to, among other things, construct new community centers and heated swimming pools, and improve school facilities (Shigen Enerugi Cho 2010). At some point in the past, both government officials and citizens of Futaba Machi probably felt pleased to have the nuclear power plant, because the town had received monetary windfalls from different sources.

The fate of Futaba Machi was totally altered on March 11, 2011. The town previously had 6932 residents, forming approximately 2393 households. The Futaba Machi government had 104 officials, in addition to an elected chief executive and an eight-member town council, which is smaller than an average Japanese community legislature. The town housed six nuclear reactors at the Fukushima Daiichi plant. On the day following the disaster, the local government decided to evacuate the residents to an area at least 10 kilometers away from the plant. The population moved to six shelters in an adjacent town, Kawamata Machi. A week later, on March 19, they were once again required to move, to a giant commercial multi-purpose gymnasium, Saitama Arena, in Saitama Prefecture. This facility was 266 kilometers from Futaba Machi, next to the Tokyo Metropolitan government (see Figure 2.2 below). The local government prepared 40 buses to help evacuate the residents to the dome. Unfortunately, after a few weeks, the evacuees once again had to be moved, and eventually settled in one of the former high schools of Kazo City in the same prefecture.

In Naraha Machi, which is the site of the Fukushima Daini (the second Fukushima Prefecture) plant, 12 kilometers south of the Fukushima Daiichi (the first) plant, the destiny of the residents looked similarly exhausting and miserable. As soon as information about the nuclear plant breakdown reached the town, the chief executive called a nearby big city, Iwaki, and requested that the city mayor accommodate the townspeople. Then, the chief executive utilized mini-buses to shuttle the people. The journey to Iwaki City would normally take only 20 minutes by public transportation; however, extremely heavy traffic congestion impeded the transfer, and the shuttle buses took more than three hours to reach the receiving municipality. This had an important effect for both central and local governments in Japan, as they recognized the critical need for several free and open roads in an emergency. At a time of turmoil, these highways should be utilized to transport crisis personnel and food as well as victimized citizens to safe locations (Working Group 2012: 5–55).

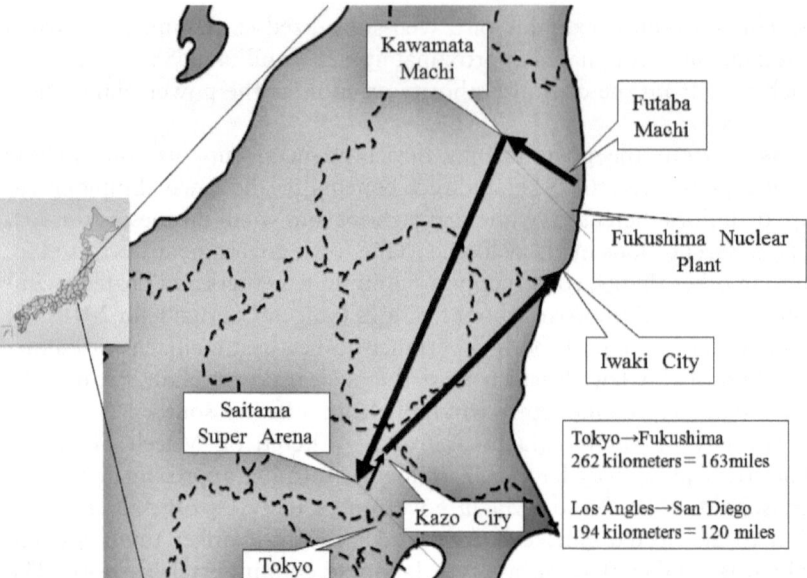

*Figure* 2.2 The Routes that Futaba Machi Took Before the Town Settled in
Iwaki City

Source: ACWorks Co., Ltd.

### Displaced Local Governments and Residents

The current record indicates that more than 3780 residents from Futaba
Machi were relocated to other areas within Fukushima prefecture, while
approximately the same number (3135) moved out of Fukushima and
have been living further away from their original homes. The Futaba
Machi government similarly moved locations several times until it
finally took root in Iwaki City, 85 kilometers south of Futaba Machi, in
June 2013. In the same year, Fukushima Prefecture planned to provide
764 new housing units for the displaced residents. Of that number, only
379 homes were made available by the end of 2014. The discrepancy
between the planned and actual number of available housing units has
been caused not by a financial shortage, but by a lack of appropriate
space.

All in all, the national government has earmarked ¥19 trillion
(US$190 billion) to rehabilitate the victimized regions in the northern
part of the country. Nevertheless, open space for permanent housing
units has been hard to find. For temporary units, most often schoolyards
are used, but the grounds will need to be vacated at most within two
years. Residents in the temporary units are allowed to stay there for only
a limited duration, most probably up to two to three years. During this
time, they must arrange their permanent domiciles.

The time restriction notwithstanding, building permanent housing units for the evacuees is difficult. The regions most affected by the tsunami tend to be long, flat stretches of land. Ideally, new units ought to be built on the hillside rather than in the level basin area. This situation has been critical for the victimized districts, as so much of the land in the tsunami-stricken regions is inappropriate for building permanent housing units. This terrain problem has contributed to the delay in preparing permanent homes for those victims who lost their properties.

This problem has been compounded by Tokyo's decision to host the 2020 Olympics. Both labor forces and materials in the construction industry have developed into a "bubble" economy, and the cost of building homes has tended to skyrocket. Although local governments plan to build permanent units for evacuees, not many firms are willing to participate in open bidding. Local governments completed only 18% of building contracts with private firms in 2013 (Yomiuri Shimbun, Morning Edition 2014: 1). Consequently, *Time* magazine reported on September 1, 2014 that even three years after the nuclear debacle, many victims still remained in unstable and temporary housing units. Their article notes that these facilities are by no means satisfactory from the perspective of both the physical and mental health of the residents (Time 2014: 22–25).

Objectively speaking, residents who lived close to the nuclear plant will not be able to move back to their hometown for a long period of time, if at all. In fact, the vast area has remained classified as a "No Go Zone." It seems unlikely that the nuclear radiation-exposed communities such as Futaba Machi or Okuma Machi will become safe enough for the evacuees to move back to live. Nevertheless, a majority of the former residents have long hoped to return to their homes someday soon. This public sentiment is embedded in Japan's historical rural culture and tradition. Japanese tradition holds that one's property, including farmland, is a fortune that has descended from one's ancestors over many generations.

This social norm remains especially strong among the aging population in the rural sectors, who maintain an affinity with and attachment to their hometowns or places of birth. Many of them, therefore, would not give up their homeland, probably considering such an action a serious offense to their family name and history. Similarly, rural residents also hold their families' graveyards extremely dear. A family tombstone is seen as their mental and physical roots and should not easily be disposed of or removed to other areas. The desecration of their family graves has been a social inhibition in the rural sector of Japan. In such traditional environments, even if the residents in the nuclear-tainted area are forced to move away from their old addresses, their souls and minds most likely stay in their original homes. These misplaced peoples have been extremely unhappy, longing for any information about their former homes, friends, and neighbors.

In 2013, the Futaba Machi government conducted a survey involving 6915 town residents. As previously indicated, of those residents, 3780 remained in Fukushima Prefecture. Others reside in different districts, including Tokyo and Saitama. In Japan, the law requires that a citizen must notify local government within two weeks whenever one changes one's address. In the case of Futaba Machi, however, many who live away from their hometown have not changed their residency registration. Although they have been living outside of the town, they still consider themselves to be Futaba Machi citizens. Many of the 3780 people who remain in Fukushima have chosen nearby Iwaki City, where the town government keeps its office, as their temporary address. Others have sought refuge in different locations throughout the country; nevertheless, many have kept their residency record and prefer to continue to be labeled as citizens of the Futaba Machi populace. This stems from two different considerations. As described above, one comes from the traditional mental bond with their hometown. Another is derived from a practical reason: many evacuees would not change their residency registration because they fear that a new address would disqualify them from receiving damage compensation from both government and TEPCO at a later date. The total effect of this resident behavior has been that the Futaba Machi government operates from within another city, Iwaki City, with an extremely limited number of the citizens actually living there (Working Group 2013:19–42).

The same Futaba Machi survey posited another question to the townspeople: whether they would move back to their hometown. Of more than 6000 respondents, 38.7% indicated their hopes to transfer back to Futaba Machi in the immediate future. Another 30.4% noted that the town was overwhelmed by atomic residue and said they would not like to return to their old address. The final 26.9% remained undecided at the time of the survey. Despite these results, close to half of the residents (45.5%) dreamed of returning to live with their old neighbors in their former pre-accident communities (Working Group 2012: 131).

### A New Futaba Machi Government

The Fukushima effect has introduced a new and perplexing dimension to the role and function of local governments in the atomic-tainted area. The current situation in Futaba Machi offers an interesting example: after a long, wandering journey, the government has finally settled in another political entity, Iwaki City, a leading municipality in Fukushima Prefecture. In this new administrative setting, the Futaba Machi government performs various local government functions similar to other local units. It issues residency certificates and processes records of marriages and deaths. The government has also been providing various forms of financial assistance to the needy Futaba Machi residents. The only difference from other entities is that Futaba Machi is no longer at its old address, and has been functioning in another local unit. In the present situation, it appears as if the Futaba Machi government has been renting free space from Iwaki City on a temporary basis.

Similar problems exist among the residents. Many of the town citizens have been living in Iwaki City, some accommodated in temporary housing prepared by the prefectural government. However, they are not registered Iwaki City residents. Being residents of their former town, although they live in the Iwaki municipality, their administrative transactions must be processed by the Futaba Machi government. In regard to this situation, on March 11, 2014, the third anniversary of the disasters, the chief executive of Futaba Machi made a number of important and bewildering policy proposals. In his address to the local legislature, he first noted the reopening of the Futaba Machi kindergarten, elementary, and junior high schools in April 2014. The government chief also indicated that the town would start building public housing units with a Futaba Machi label. Similarly, he proposed to construct a community center with lodging facilities. One objective of this building is to allow for former residents relocated to other regions to be able to visit the new Futaba Machi and have neighborhood reunions in the center. There is one critical issue: it appears that all of these projects would take place within another independent unit of government. As noted, these policies and programs are to be initiated under the eyes of the Iwaki City jurisdiction (Chocho 2014: 1–7).

The chief of the Futaba Machi government stated in the same speech that the local tax revenue for fiscal year 2014 was ¥1182 million (US$11.82 million), a decrease of ¥67 million. Block grants (Local Allocation Tax) to the town amounted to ¥839 million (US$8.39 million), while categorical grants (National Treasury Disbursement) totaled more than ¥2830 million (US$28.3 million). According to the chief executive, these revenues should be appropriated to various programs such as town legislature, education, social welfare, agriculture and fisheries, the aging population, and so on, identical to any other local entities. From an administrative perspective, the development many Japanese have been witnessing in and around Futaba Machi has been unprecedented and very unusual. No one has ever heard of a town government keeping an office in another local entity and performing various local service functions to a subset of residents residing in a different political unit.

There seem to be several driving motives for Futaba Machi to maintain its identity and extend different provisions to its citizens. The town government, along with a majority of the residents, believes, ostensibly at least, that they will be able to move back to their old hometown in the not-too-distant future and once again enjoy the traditional pattern of daily lives. Various reports that the Futaba Machi government has made public have frequently utilized the expression *Kari no Machi* (transit or temporary town). This, perhaps, reflects their ardent desire to reestablish themselves in the old town (see, for instance, Working Group 2012: 41–42). Making Futaba Machi free of nuclear residue would, however, require tremendous time and effort, in addition to horrendous financial costs. For all intents and purposes, therefore, before both government and citizens can return to Futaba Machi, they should expect to remain several decades, if not more, in Iwaki City.

## Concluding Remarks

This chapter has touched on two dimensions of the Fukushima effect: the political leadership in a crisis situation, and the growing new role of local governments in the Fukushima Daiichi region. As far as the leadership issue is concerned, the Fukushima effect has disclosed many errors that the prime minster of the country made at the time of the accident. Prime Minister Kan Naoto and his party, the Democratic Party of Japan, were too hasty in incorporating a party-centered model of government to Japan's political environment. Kan and his cohort attempted to reduce the power of bureaucrats, while also wanting to enhance the status of party members. Members of the DPJ, therefore, tended to undercut the role of technocrats and tried to initiate various measures on their own. From this vantage point, the DPJ seems to have become a victim of its own conviction and agenda. When the nuclear accident occurred, serious schisms and fissures between the party leadership and leading bureaucrats were exposed. The party members found that they had to meet the challenges of the nuclear accident by themselves. Subsequently, with little or no support from professionals, they often introduced conflicting policy and action platforms and confusing programs. The confusion and mayhem at the central government level contributed to the sharp decline of public trust in government. This is a major Fukushima effect that must be remembered by Japan's leaders for many years to come.

The second notable Fukushima effect considered is the fate of the local governments at the heart of the Fukushima Daiichi area. In particular, Futaba Machi housed six reactors and became one of the major victims of the disaster. A large portion of the town, even to this date, remains a "No Go Zone" due to atomic residues that persist in the area. In the aftermath of the accident, both local government and residents of the town began to take long, wandering journeys and to seek safe refuge in regions away from the epicenter of the disaster. After several relocations, the town government came to be relocated in one of the urban centers in Fukushima. Over 3000 residents followed the government and found a temporary nest in Iwaki City, 70 kilometers from Fukushima Daiichi.

The town government leader may believe that there is no or little possibility for Futaba Machi to return to its original location. No one can tell his real motives at this time. Nevertheless, the fact remains that the chief executive of Futaba Machi has been trying to rebuild a town and local government within another major urban district. The growth of an independent political entity within the boundaries of a neighboring city government is a new and an interesting experiment. Certainly, many stumbling blocks will be encountered before a renewed Futaba Machi takes root in a different political milieu. However, if the attempt succeeds, the outcome would represent another important Fukushima effect, one that Japanese officials and academics should consider as a possible option for a new role and function for local units of government in the aftermath of future natural and/or human-made disasters.

## References

ACWorks. Map of Japan, accessed February 8, 2015, www.map-ac.com/main/list?q=%e6%97%a5%e6%9c%ac

Agency for Natural Resources and Energy (Ministry of Economy, Trade and Industry 電源立地制度の概要 [*The outline of the support system for communities building energy sources*], 2010.

Asahi Shimbun, Tokubetsu Hodo Han [Asahi Newspaper Special Task Force] (eds.), *Purometeusu no Wana [Promethean pitfall]*, Tokyo: Gakken, 2012.

Fukushima Genpatsu Jiko Dokuritsu Kensho Iinkai [The Independent Committee for the Investigation of the Fukushima Nuclear Accident] [Abbreviated as 'Independent Committee in the text above]. *Chosa Kensho Hokokusho [Report of the research and investigation]*, 2012.

Futaba Machi. *Futaba Machi Fukko Machizukuri Keikakku [The rehabilitation program of Futaba Machi]*. Futaba Machi Government, 2014.

Futaba Machi Gikai Teireikai [Futaba Machi Town Council Regular Meeting]. Chocho Shisei Houshin [Future policy direction of the chief executive], 2014.

Genshiryoku Saigai Kento Waking Gurupu Hokokusho [Report of the Working Group Exploring the Nuclear Power Plant Accident]. Abbreviated as 'Working Group in the text above]. *Fukushima Daiichi Genshiryoku Jiko ni yoru Genshiryoku Hisai Jichitai to Chosa Kekka [The research report of the Fukushima Daiichi Nuclear Plant and its effects on local governments in the region]*. Zenkoku Genshiryoku Hatsuden Shozai Shichoson Kyogikai and Genshiryoku Saigai Kento Waking Gurupu, 2014.

Gotham, K. "Disaster Inc.: Privatization and Post-Katrina Rebuilding in New Orleans," *Perspective Politics* 10, no. 2 (2012): 633–646.

Hara, T. "Social Shaping of Nuclear Safety: Before and after the Disaster." In *Nuclear Disaster at Fukushima Daiichi: Social, Political and Environmental Issues*, edited by R. Hindmarsh, 22–40. New York: Routledge, 2013.

Hindmarsh, R. "Nuclear Disaster at Fukushima Daiichi: Introducing the Terrain." In *Nuclear Disaster at Fukushima Daiichi: Social, Political and Environmental Issues*, edited by R. Hindmarsh, 1–21. New York: Routledge, 2013.

Kokkai Jikocho (*The Diet Independent Commission for the Investigation of the Fukushima Nuclear Accident) (Tokyo Denryoku Fukushima Genshiryoku Hatsudensho Jikochosa Iinkai [The report of the Diet Committee for the investigation of TEPCO's nuclear plant accident in Fukushima]*) [Abbreviated as Kokkai Jikocho in the above text].Tokyo: Tokuma Shoten, 2012.

Kyodo News. 東京の水道水から検出　乳児基準超えるヨウ素 [*Iodine over the standard for the infant was detected from drinking water in Tokyo*] March 23, 2011, accessed May 15, 2014, www.47news.jp/CN/201103/CN20110 32301000502.html.

Masciulli, J., Molchanov, M.A., and W.A. Knight. "Political Leadership in Context." In *The Ashgate Research Companion to Political Leadership*, edited by J. Masciulli, M.A. Molchanov and W.A. Knight, 3–27. Surrey, UK: Ashgate, 2009.

Nakamura, A. "Party Members, Elite Bureaucrats and Government Reform in Japan's Changing Political Landscape." In *Politicians, Bureaucrats and Administrative Reform*, edited by G. Peters and J. Pierre, 169–201. New York: Routledge, 2001.

Nakamura, A., and K. Masao. "What We Know, What We Have Not Yet Learned: Triple Disasters and the Fukushima Nuclear Fiasco in Japan," *Public Administration Review* 71, no. 6 (2011): 893–899.

Nishimura, W. Kuni to Chiho no Kikikanri–Gyosei kara mita Kadai [Intergovernmental relationships at the time of crisis: From the public administration perspective]. In *Seiji Gyosei eno Shinrai to Kikikanri* [*Public trust in government and crisis management*], edited by A. Nakamura and K. Ushiyama, 77–98. Tokyo: Ashi Shobo, 2012.

Research Group on "Public Trust in Government." "Survey Research on Public Trust in Government" (Unpublished Report). The research headed by Akira Nakamura and conducted under the auspices of the Japanese Society for the Promotion of Sciences, the Ministry of Education and Sciences for 2011 [*Abbreviated as Research Group in the above text*], 2011.

Shi, K. "Genpatsu Shisatsu o Kyoko [Prime Minister Kan went ahead with the visit to the Fukushima plant]," *Yomiuri Shimbun*, September 12, 2014, Morning Edition: 1–3

Shigen Enerugi Cho [Agency for Natural Resources and Energy]. *Dengen Rittchi Seido no Gaiyo–Chiki no Yume wo Okiku Sodateru [An outline of the program for building energy supplies: Making local dreams true]*. Shigen Enerugi Cho, 2010.

Stockwin, J. *Governing Japan: Divided Politics in a Resurgent Economy*. Oxford: Blackwell, 2008.

Takagi, J. *Genpatsu Jiko ha Naze Kurikaesunoka [Why have accidents in nuclear plants kept repeating]*. Tokyo: Iwanami Shinsho, 2011.

Time Magazine. "The World's Most Dangerous Room," *Time Magazine* 184, no. 8 (2014): 22–25.

Yamaguchi, J. *Seiken Kotai toha Nanidattanoka [A change of government: What has it meant?]*. Tokyo: Iwanami Shinsho, 2012.

Yomiuri Shimbun. "Nyusatsu Fucho Sinsai Mae no Yonbai [Unsuccessful open bids increased by four times since the Great East Japan Earthquake]," *Yomiuri Shimbun*, September 11, 2014, Morning Edition: 1.

Yoshida, R. "Kan Hero, or Irate Meddler," *Japan Times*, March 17, 2012: 2.

# 3 Taiwan's Civil Society in Action

## Anti-nuclear Movements Pre- and Post-Fukushima

*Dung-sheng Chen*

This chapter explores the extent of any Fukushima effect in Taiwan, with a focus on public attitudes towards nuclear power, the anti-nuclear movement, and the political and policy contours of nuclear power development. The Fukushima nuclear accident in Japan, Taiwan's neighbor to the northeast, was seen as a cataclysmic event, one which led to Taiwan's civil society becoming more active in pushing the government to phase out nuclear energy. In investigating the possible effect of the Fukushima disaster on the politics of technology—here, nuclear power development—and democracy in Taiwan, I align with the objective of this book to critically reflect upon social and political effects in a national-level context as raised by the Fukushima nuclear disaster.

I situate this inquiry within a perspective of science, technology, and governance (Miller et al. 2008: 3), a core pursuit of science, technology and society (STS) studies. This chapter's main focus is on how elitist political power can be materialized in design and implementation processes and then realized in terms of artifacts such as nuclear power plants. It also investigates how various social actors, especially ordinary citizens, respond to unequal political relationships by way of human-manufactured products. In this context, my analytical framework is the theory of politics and technology (Nahuis and van Lente 2008; Winner 1980), which focuses on the design of technologies or application of technologies involving power relationships among artifacts, the state, experts, civil organizations, and the general public.

Nuclear energy technology is a highly complicated system, operating on large-scale sites and by a centralized organization. Therefore, it is usually compatible with an authoritarian power structure. This is especially the case in Taiwan, which has a long history of authoritarian rule, particularly before 1987. The Taiwan Power Company established in that political atmosphere, is the operator of the three nuclear power plants in Taiwan. It is a nationalized company that monopolizes the electricity market. The first nuclear power plant, located in Taipei County, began operation in 1978, followed by the second, also in Taipei County, in 1981 (both in the north of Taiwan). Finally, in 1984, the third plant, located in Kaoshiung County (in the south) (see Figure 3.1), began operations. The Fourth Nuclear Power Plant (FNPP), in Gongliao District, Taipei County, was planned to start in 1980 but the government delayed the project because of powerful protests by local residents. In 1992, President Lee Teng-hui approved the FNPP project, and in 1996 it was granted a construction license by the government, but the project was not completed, as I detail below. In 2013, the proportion of nuclear energy of total electricity supply was 12.5% (Taiwan Power Company 2014).

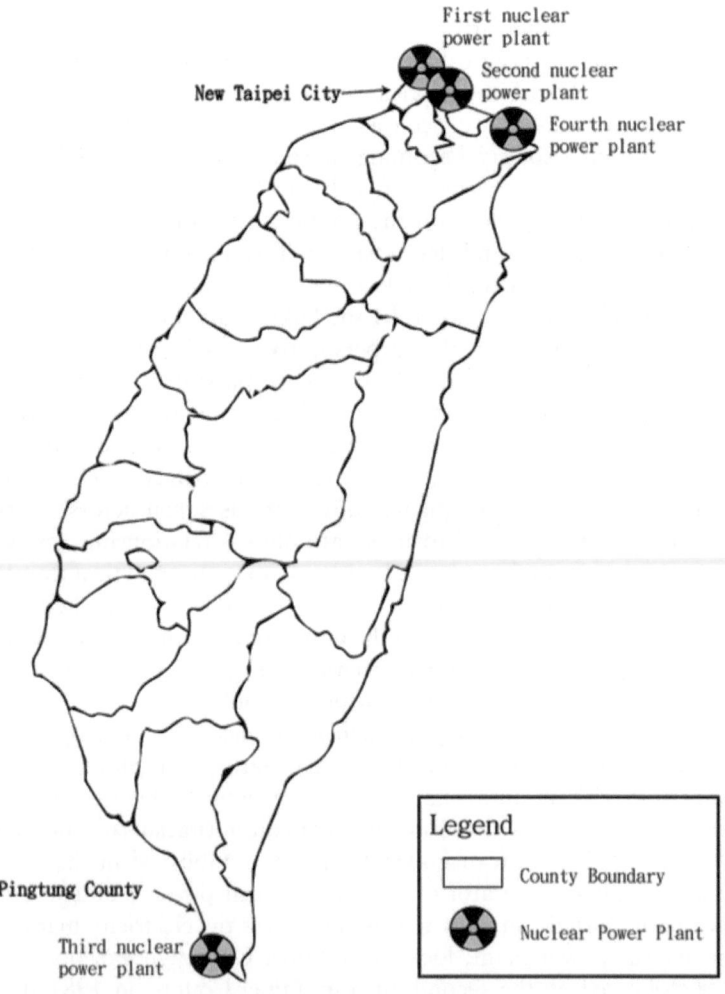

*Figure 3.1* Taiwan Nuclear Power Plant Locations

After the lifting of martial law in 1987, a democratic political system was introduced in Taiwan. General elections were held to elect the president, representatives of the Legislative Yuan (the Parliament), and city mayors, who were formerly appointed by the ruling Nationalist Party (the Kuomintang). As a result of this change in the political system, the relationship between the application of nuclear energy technology and politics gradually changed.

These dynamics, which involved significant policy changes toward nuclear technology application, are mapped out below. First, I explain my analytical approach to politics and technology, then set out the national context of nuclear power development in Taiwan, and then describe the formation of continuous, anti-nuclear social movements after the Fukushima disaster. Finally, I provide a comparative analysis of the suspension of the Fourth Nuclear Power Plant in 2000 and its termination in 2014 after the Fukushima disaster, based upon a number of considerations: (i) how civil organizations changed their capabilities with regard to commencing social movements before and after the accident; (ii) why some of the general public altered their attitudes from desiring a convenient lifestyle and stable economic growth supported by nuclear power to a deep concern about environmental sustainability and safety; and (iii) how governments and legislators responded to a demand from these social movements for "no nukes," and how their reaction influenced subsequent activities of civil society, both in 2000, after the first political regime rotation, and in 2014, after the Fukushima disaster.

## Politics and Technology

Since all nuclear power plants are controlled by the nationalized Taiwan Power Company, nuclear policy is planned by the central government, and nuclear safety is regulated by the National Nuclear Council. Concomitantly, applications of nuclear technology obviously involve diversified political processes, such as interactions between the administrative branch and the legislative branch, political conflict among different political parties, and social protests against the government by civil organizations. Thus, it is useful to employ a modified theory of political opportunity structures to emphasize the interactions between governments, parliament, social movement organizations, and the general public, so as to elaborate the social construction process of nuclear technology (Koopmans 1999; McAdam 1982; Polletta 1999; Tilly 1978).

Fundamentally, "political opportunity structures" refers to variables derived from a political system which affect the possibilities or options that collective political actors in society can use to increase pressure for change; here, in regard to anti-nuclear social movements pressing for no nuclear energy in Taiwan. These structures include institutional arrangements to enable political opportunity, historical precedents for social protest, and specific configurations of various resources for social mobilization (Giugni 2011; Kitschelt 1986; Koopmans 1999).

First, such "institutional arrangements" are reflected by the degree of state autonomy from the influence of interest groups, administrative effectiveness, openness of decision-making to civil society, independence of the judicial system, patterns of party-political competition, and stability of political support from the general public (Constain 1992; Ho 2006; McAdam 1982; Oberschall 1996). As such, the stronger state autonomy and administrative effectiveness are, and the weaker openness is, the fewer the political opportunities for anti-nuclear groups.

Second, "historical precedents" are those previous events—for instance, previously successful social movements, rotation of political regimes by elections, or specific historical conjunctures such as natural disasters—that can affect social mobilization, the development of social movements, and the success of social protests in applying pressure and forcing shifts in policy. With regard to anti-nuclear protests in Taiwan, the change from an authoritarian regime to a democratic one in 1987 was a significant historical precedent in weakening the power and legitimacy of the state on nuclear policy decision-making. This provided good opportunity for anti-nuclear groups to take action. In addition, three major nuclear disasters in the last four decades have been crucial historical events in increasing public awareness internationally of the risks of nuclear technology.

Finally, these "specific configurations" are the macro variables that either facilitate or restrain civil organizations in mobilizing monetary resources, normative support, and human resources. For instance, strong, long-term alliances between civil organizations can enhance mobilization of resources, while strong competition between such organizations, especially between social movements and counter-social movements, can constrain their acquirement of resources. In the case of anti-nuclear protests, stable, inter-organizational connections between various social movement organizations have been established and maintained in Taiwan since 1987. Anti-nuclear groups have been in the vanguard as a platform to distribute information, diffuse new social-movement strategies, and coordinate various resources for collective action.

In summary, this approach emphasizes the investigation of interactions between external (institutional) structures, civil organizations, and public attitudes in making political opportunities for social movements, in this case, the ability of Taiwanese anti-nuclear movements to shape policies on nuclear power. However, political opportunities for anti-nuclear organizations are not only created by the state, but also by the activities of civil organizations, continuous exchanges between anti-nuclear groups and the state, and changes in public attitudes. Even one significant event can generate opportunities for anti-nuclear action (McAdam et al. 2001). In this chapter, I investigate the political structure opportunity provided by the Fukushima disaster and how anti-nuclear groups employed this opportunity to influence nuclear power politics in Taiwan. In the next section, I outline the national and historical context of nuclear power development in Taiwan.

## The National Context of Nuclear Power Development Pre-Fukushima

Anti-nuclear opinion in Taiwan was first publicly voiced by professors and experts in physics and the environmental sciences, and in a number of anti-nuclear articles in the mass media, which increased from 1979 onwards, after the Three Mile Island nuclear disaster in the US. Thereafter, accidents at nuclear power plants constantly triggered waves of anti-nuclear opinion in Taiwan. On July 7, 1985, a fire at Taiwan's Third Nuclear Power Plant led to a group of professors and experts forcefully lobbying legislators to propose a resolution to postpone the Fourth Nuclear Power Plant (FNPP) construction project.

This same group of anti-nuclear intellectuals established the *New Environment Magazine* to monitor governmental nuclear and environmental policies, and to promote anti-nuclear positions. One year later, the Chernobyl disaster saw a high tide of anti-nuclear opinion among the public, yet various anti-nuclear protests made no significant impact on energy policy because of strong political control being exerted by the authoritarian regime (Ho 2006: 44–45).

After the lifting of martial law in 1987, many new environmental and anti-nuclear organizations were founded. In 1994, the Alliance of Anti-nuclear Action was formed from more than 80 civil organizations, in order to coordinate an annual anti-nuclear demonstration. In November 2000, the No Nukes Action Alliance was established by more than 170 non-governmental organizations (Ho 2006: 197, 290). Clearly, the anti-nuclear movement has a very solid social and organizational foundation.

The goal of the Taiwan anti-nuclear movement is a society with no nuclear energy. Since it was very unlikely to close three operating nuclear power plants immediately, the short-term goal of the movement was to suspend the ongoing construction of the FNPP, which began in 1996. Most anti-nuclear actions had this project as their target. The FNPP had been initiated by the government in 1980. Its budget, however, had been frozen by the Legislative Yuan in 1986 after Chernobyl and declining national energy demands, but the project resumed in 1992 (Chang 1997).

The first public opinion poll on nuclear energy was conducted by a major newspaper, the *United Daily*, on March 11, 1988. It was quite likely that the Chernobyl disaster in 1986 strongly influenced public opinion; while 28% of respondents agreed that the FNPP construction should go ahead, 44% opposed the project. In 1988, the Taiwan Environmental Protection Union held anti-nuclear workshops across several townships in Taipei

County near the three nuclear power plants, initiated anti-nuclear social protests, and invited 500 professors to sign an anti-nuclear petition. At the same time, residents of Taipei County's Gongliao District, where the FNPP was to be built, founded a local residents' anti-nuclear organization and started a long-term anti-nuclear protest.

As can be seen from Figure 3.2, public support for the FNPP construction project broke the 50% threshold yet still fluctuated across some 15 percentage points between 1990 and late 1996. In 1991, a serious conflict occurred between the Gongliao anti-nuclear association and the police. One policeman died when members of the association forced their way into the construction site, and severe clashes occurred with the police force based there. However, the government resumed the FNPP construction project the following year after a highly positive evaluation of nuclear safety from nuclear energy experts and a rapid increase in electricity demand (Hu 1995).

On May 22, 1994, Gongliao residents took part in a non-legally binding referendum on the FNPP's construction, organized by the local anti-nuclear association: 96% of those who voted were opposed to the project. That same year, Lin Yi-Hsiung, a key opposition politician, began a hunger strike demanding a national referendum on the FNPP in July, and in September commenced a 1000-mile march to promote his agenda. In September 1995, the annual anti-nuclear demonstration recruited more than 30,000 new participants. Although a petition for a national referendum did not elicit any response from the Kuomintang (KMT) government, Mr. Chen Shui-Bian, the Democratic Progressive Party (DPP) nominee, won the Taipei City mayoral election and held a non-legally binding referendum in the city in March of 1996. In a turnout of 58% of eligible voters, 46% voted in favor of the FNPP and 53% against. Two months later, in May, the Legislative Yuan agreed to halt the project.

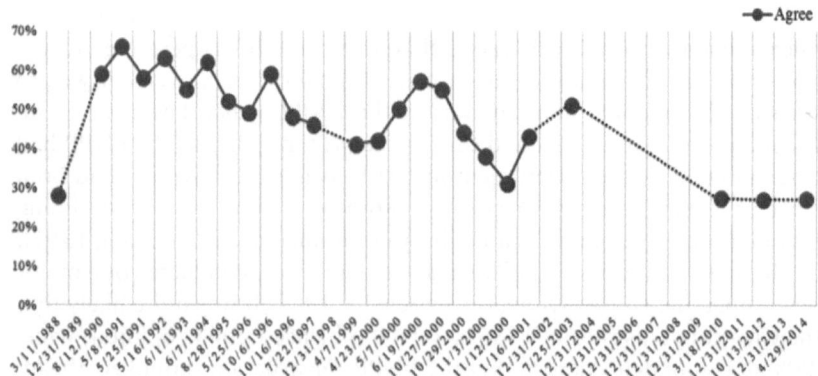

*Figure 3.2* Public Support for the Fourth Nuclear Power Plant Construction Project

However, apart from that win (although temporary), anti-nuclear movements did not significantly change public attitudes or alter government policy.

Soon thereafter, in 1996, a bid for the construction of the FNPP was approved and a license was granted by the National Nuclear Council. In response, in October 1996, members of the Gongliao anti-nuclear association broke through a security cordon and ran onto the site to stage a sit-in. Public opinion gradually built against nuclear power. In 1998, I-Lan County (close to Taipei County) held another non-legally binding referendum on the FNPP project. From a turnout of 44% of voters, 36% voted in favor and 64% against. In 1999, at an anti-nuclear rally in Taipei, the Democratic Progressive Party (DPP) chairperson Lin Yi-Hsiung announced publicly that Taiwan could not afford a nuclear accident. Such events would impact public attitudes toward nuclear energy. From the end of 1996 to April 2000, support for the FNPP construction project stood at less than 50% for most of the time.

May 2000 was a crucial turning point for both the anti-nuclear movement and for democratic development in Taiwan. In that month, the DPP won a presidential election for the first time and promised to implement a "no nukes" policy. From May to October, although the proportion of those who supported the FNPP project was generally more than 50%, the new DPP government was against the project. On May 13, 2000, the Taiwan Environmental Protection Union and other civil organizations initiated an anti-nuclear march to support the suspension of the FNPP construction as proposed by president-elect Chen Shui-Bian. On October 3, Prime Minister Tang Hui, a firm supporter of the FNPP, resigned his post. The new Prime Minister, Chang Chun-Hsiung, announced a suspension of the project on October 27, after intensive public consultation over the previous six months (Ho 2006). After the suspension was announced, public support for the FNPP declined from 50% to just 31% by November 12. It was thus clear that action taken by the government, opposition parties, and/or civil organizations during this period had the effect of changing public anti-nuclear attitudes (Chiu 2001).

The decision to suspend the project, however, led to political turmoil, as this political action occurred immediately after President Chen—who supported the "no nukes" policy—had met with chairperson Lien Chan of Kuomintang (KMT), which was now the largest opposition party. The ensuing decision to suspend was interpreted by the opposition parties and many citizens as an intentional humiliation of their political opponents by the DPP government. Subsequently, an alliance of the three opposition parties initiated a petition for the president's recall in the Legislative Yuan, and on November 4, 2000, with the signatures of 143 members, it almost passed the two-thirds quorum needed to become a motion, and in the process, seriously weakened the legitimacy of the DPP government in the eyes of many onlookers.

To conciliate such serious challenges as this from the opposition alliance, the DPP invited representatives from different societal groups to attend a national affairs conference. A request was also sent by the government to the Constitutional Court in November for clarification on the authority of the prime minister to suspend construction of the FNPP project. On November 11, anti-nuclear organizations held a "No Nukes Homeland" march to support the suspension. After intensive meetings, on January 15, 2001, the Constitutional Court delivered its verdict on the government's suspension of the FNPP's construction. The verdict highlighted that the administrative branch had not completed the necessary procedures for such an important decision, which required that the prime minister report to the legislative branch and concede to the majority opinion of the Legislative Yuan.

The verdict also indirectly revealed that the government's decision to suspend the FNPP lacked legitimacy, which created an excellent political opportunity for opposition parties while also enhancing public support for the resumption of the project. According to a survey conducted by the *China Times* (a major newspaper in Taiwan) on January 16, 2001, the proportion of citizens who agreed with the project's continuation, compared to the previous November, increased from 31% to 38%, and shortly thereafter rose to 43% (China Times 2001).

Availing themselves of such good political opportunity to challenge the DPP government, the opposition alliance proposed an agenda for the continuation of construction of the FNPP, which was approved in the Legislative Yuan on January 31, 2001. (As an aside, on several other occasions the DPP government could not gain support from the parliament to fulfill major policies.) This crucial political action forced the government to announce resumption of the FNPP construction on February 14, 2001, despite vehement protests from some DPP leaders and anti-nuclear organizations. The whole affair lasted 110 days and represented a significant setback for Taiwan's anti-nuclear movement (Ho 2006). In sum, the overall effect of the political turmoil surrounding the DPP's suspension of the FNPP project was to decrease public support for anti-nuclear organizations and quiet public debate on nuclear safety. Subsequently, during 2002–2009, there was only one public opinion survey on the FNPP construction, with more than 50% of citizens supporting construction.

After this major setback for the anti-nuclear movement in 2001, the Green Citizens' Action Alliance began to enhance public renewed support for the anti-nuclear movement by means of island-wide documentary tours, no-nukes music concerts at Gongliao, and community mobilization (Tsui 2011). Subsequently, public support for the FNPP's construction had significantly declined by 2010. After the Fukushima disaster in 2011, public support for the FNPP shifted further, to the lowest ever support for nuclear power. Apart from this Fukushima effect, did

Fukushima actually influence any change in nuclear power development in Taiwan, and if so, to what extent?

## Investigating the Fukushima Effect

Immediately after the Fukushima disaster of March 11, 2011, the *China Times* conducted a nationwide survey of public opinion on Fukushima and nuclear safety: 60% of respondents were worried about radiation from Japan threatening Taiwan, and 75% were worried about nuclear safety in Taiwan. In addition, 84% of respondents were afraid of an earthquake with magnitude 9, a similar scale to that of the Fukushima disaster, hitting nuclear power plants in Taiwan. Some 57% had no confidence in the government's ability to deal with nuclear disasters and emergencies. Only 30% of respondents were optimistic and confident about the emergency response capabilities of the KMT government (China Times 2011). This survey made it clear that the Fukushima disaster enhanced risk perception among the general public and increased distrust in the government's pro-nuclear position. These findings could thus be seen to align with the "cultural opportunity proposition," a variation of the political opportunity theory that suggests sudden disasters tend to render public attitudes toward political organizations more negative, spur public opposition, and finally, possibly, increase the vulnerability of political systems to change (McAdam 1994; Polleta 1999).

In general, the preferences of any given citizen about life in general are not consistent, but, rather, contradictory. In my 2009 study (Chen 2011), I found that

> an extremely high proportion of respondents (88.6%) agreed that nuclear waste would pollute living spaces for a long time, destroy human habitats for future generations. Conversely, 54.7% still considered nuclear power a safe technology. A total of 56.4% of respondents supported the continuous development of nuclear technology, and approximately 40% did not rate environmental protection as more important than economic development.
>
> (Chen 2011: 571)

Another survey on FNPP construction, conducted in April of 2014, showed that more than 50% of the respondents who supported the termination of the project would not, however, accept an electricity price hike of 40% (Common Wealth 2014). It appeared that some wished to maintain a comfortable lifestyle at a relatively cheaper cost while at the same time avoiding the potential for nuclear accidents and disasters. Based on these two studies, it might be suggested that the public were desirous of both quality of life from nuclear power and environmental protection, two possibly contradictory concepts.

In the wake of Chernobyl and Fukushima, for instance, environmental protection and nuclear safety in the public mind became more prominent than material gain, as many citizens perceived considerable risks and hazards from major nuclear disasters. Safety or sustainable environment were rated more highly than pecuniary considerations (Finucane and Holup 2005: 1604). Disasters that involve radioactive pollution have been found to trigger negative emotions such as dread, anxiety, fear, frustration, or sorrow, which tend to amplify risk perceptions of nuclear power and subsequently alter attitudes toward negative perceptions of nuclear power (Hindmarsh 2013; Slovic 1999, 2000). Flexible and contradictory personal attitudinal structures often result in significant modification of public opinion under the influence of such major events. Such occurrences also support the hypothesis proposed by framing theories in political and science communication that people will change their preference weightings and attitudes as a result of continuous exposure to media priming of disasters (Chong and Druckman 2007).

The change in public attitude post-Fukushima was highlighted by a 2013 survey. After a 200,000-participant anti-nuclear event on March 9, 2013, the Research and Evaluation Council (2013)—the governmental organization responsible for policy research and evaluation—presented the findings of a nationwide survey of public opinions on nuclear power plants. First, about 39% of the public agreed that the FNPP project should be terminated even if it could be demonstrated to be safe. Second, about 43% supported a shutdown of the project even if Taiwan were punished by international environmental organizations for exceeding its $CO_2$ emissions cap by its use of fossil-fuel power stations. Third, and in contrast, 14% agreed that the FNPP should operate even if nuclear safety approval was not granted by the regulatory agency. Lastly, only 26% supported the anti-nuclear position of closing all three extant nuclear power plants and terminating the FNPP project.

According to these findings, it can be argued that about 25% of the general public took the strongest anti-nuclear position in desiring a non-nuclear status immediately and at any cost. In contrast, 14% adopted the most favorable stance possible on nuclear energy. The remaining 61% of the general public might then on some occasions change their attitudes toward nuclear energy under the influence of a nuclear disaster, anti-nuclear social movements, governmental promotion, and/or some other relevant event. Public attitudes have also been found to be dependent on the degree of public trust in public policy actors (Kasperson et al. 2003). In other words, social construction of public attitudes towards nuclear technology is a dynamic process. However, while more than 50% of the general public supported the suspension of the FNPP project, the government's nuclear energy policy did not change. In other words, there needed to be crucial factors apart from public opinion to prompt change.

As mentioned in the previous section, the DPP government announced the suspension of the FNPP project in 2000, a decision that led to relentless opposition from the parliamentary opposition parties. This initiative inadvertently became a major setback for both the DPP and the anti-nuclear alliance. From 2001–2011, no anti-nuclear demonstrations were mobilized, and an anti-nuclear policy was not even listed on the DPP's presidential manifesto in 2004 (Ho 2006: 314–320). The KMT resumed power in 2008 and further strengthened its pro-nuclear position. In other words, the voice of the anti-nuclear movement became quiet, but the anti-nuclear civic alliance maintained its regular operation (although with scarce resources), waiting for appropriate political opportunities to emerge.

These opportunities occurred with the Fukushima disaster. Similar to previous nuclear disasters, Fukushima triggered public awareness of the deadly radioactive threat to life and the environment, and led to ongoing civic protests. As such, it re-activated the anti-nuclear organizations in Taiwan, and facilitated subsequent anti-nuclear social demonstrations.

One of the most important anti-nuclear civic organizations in Taiwan is the Green Citizens' Action Alliance (GCAA), which was founded in June 2000. The organization states its mission as the promotion of social justice for victims of environmental disasters and the maintenance of a sustainable environment. Post-Fukushima, the GCAA put the majority of its resources into the study of nuclear power and other energy issues, the mobilization of civil society in expressing social concerns over nuclear issues, and the surveillance of the Taiwan Power Company and government policy on nuclear power and energy (GCAA 2014).

To expand public support of anti-nuclear movements, the Green Citizens' Action Alliance adopted a number of innovative mobilization strategies. The organization distributed "No Nukes, No More Fukushima" flags to coffee shops, convenience stores, independent bookstores, bakeries, and other local shops across the nation. In addition, it successfully facilitated the establishment of the Anti-Nuclear Fathers Front and the Alliance of Mothers Watching Nuclear Power Plants. Strong support for the anti-nuclear movement has also come from artists, writers, filmmakers, singers, designers, and other creatives. From November 22 to December 1, 2013, the GCAA collaborated with cultural and artistic groups to organize anti-nuclear exhibitions in Taipei including films, posters, and photography.

After a large-scale anti-nuclear demonstration on March 9, 2013, the GCAA started a new project—the "No Fourth Nuclear Power Plant Friday Night Gathering"—with numerous cultural and artistic groups. This project continued every weekend in Taipei's Freedom Plaza, in order to cultivate public support for both the anti-nuclear movement and suspension of the FNPP project. A number of famous artists and independent creative artists became enthusiastically involved in these anti-nuclear activities and helped to increase public support for the anti-nuclear movement.

More broadly, following the Fukushima disaster, anti-nuclear groups resumed demonstrations to pressure the government to implement a non-nuclear policy. For example, on March 20, 2011, the Environmental Protection Coalition, along with other civil organizations, launched its "No Nuclear Disaster" movement, which mobilized up to 5000 people. On March 11, 2012, the "Rethinking Zero Nuclear Power" movement was reignited, with 10,000 participants across the country. On March 9, 2013, the numbers increased significantly to around 200,000 people in Taipei, Taichung, and Kaohsiung, as estimated by the Green Citizens' Action Alliance, who joined the now annual "post-Fukushima" anti-nuclear demonstration as the largest anti-nuclear protest in three years (which Figure 3.3 shows in part). Yet "people-power," although resilient, remained unable to change the KMT government's pro-nuclear policy.

Instead, the government reassured, arguably rhetorically, the public of the safety of nuclear power in Taiwan. In the immediate wake of the Japanese catastrophe, and in response to the media's reportage of public distrust in the crisis management capability of the government, they

*Figure 3.3* Taiwan Anti-nuclear Protest, March 8, 2013
Source: Gong and Catenacci

provided a nuclear accident reaction plan. On March 21, 2011, the Taiwan Power Company also responded to public concerns by stating that assured safety was the only way forward for nuclear energy in Taiwan, which, however, given the circumstances, also suggested an attempt to "close controversy" (see Beder 1991).

The Ministry of Economic Affairs followed immediately afterwards (on March 22, 2011), announcing that if all three extant nuclear power plants were closed, then between a quarter and a third of Taiwan's factories would be shut down. Almost in tandem, the day after this announcement, the Taiwan Power Company made another point to highlight the indispensability of nuclear energy to Taiwan—that nuclear was cheaper than coal, gas, wind, or any other form of electricity generation. A concerted effort by Taiwan's political-industrial nuclear complex was obviously underway, seeking to minimize the political damage of the Fukushima disaster.

Nevertheless, only two months later, on May 20, 2011, as the enormity of the Fukushima disaster became clearer (e.g. Hindmarsh 2013), Taiwan's President, Ma Ying-Jeou, departed completely from the hardline pro-nuclear position in his annual speech. He declared a nuclear energy reduction plan including the phase-out of all three extant nuclear power plants at the end of their design life, no new nuclear reactors in those plants, and no new nuclear power plants. Six months later, however, the president repeated his plan but with significant modification: the FNPP would still come online in 2016, which would pose a sufficient condition for the early retirement of the First Nuclear Power Plant. Subsequently, on June 14, 2011, the annual budget for the FNPP was approved by the Legislative Yuan, as controlled by the KMT.

This policy revision, which it seems was initially made to appease public concerns, immediately lost legitimacy because it side-stepped the anti-nuclear position of more than half of the general public. Civic demonstrations resumed and targeted the budget approval for the FNPP, adding to the now regular anti-nuclear demonstration held annually in March. In response, the government used the same economic and safety arguments to defend its position on nuclear energy as it had immediately after Fukushima.

Consequently, in the 2014 annual march, the largely student-initiated "Sunflower Movement" which, as revealed below, was part of an anti-nuclear strategy, occupied the meeting hall of the Legislative Yuan for one whole month to demand no Taiwan-China trade agreement before a law was passed on monitoring and regulating any agreement. The movement gained strong support from citizens, finally mobilizing about half a million participants to show up at an anti-trade-treaty march. This was also because the KMT government itself had generated anxiety and suspicion amongst the general public by implementing its China-Taiwan policies, which catalyzed constant anti-government demonstrations.

The Sunflower Movement was part of these continuous protests, which provided a "master framework" of connecting to everyday citizens, to better legitimate its part in organizing collective action for subsequent anti-nuclear movements (c.f. McAdam 1994). The trade agreement challenge from the Sunflower Movement heavily weakened the authority of the KMT government. According to Kitschelt (1986), such actions in maintaining the momentum of social movements create exceptional political opportunity for other protests, which, in this case, worked toward the anti-nuclear interest of the Sunflower Movement.

At the same time, an important anti-nuclear leader, Lin Yi-Hsiung, had originally intended in late March, 2014, to go on an indefinite hunger strike, but this action was delayed because of the Sunflower Movement's action in occupying the meeting hall of the Legislative Yuan. On April 14, 2014, Lin wrote an open letter concerning the "suspension of the FNPP construction and realization of democratic procedures in holding a referendum on nuclear energy," and announced his indefinite hunger strike to start on April 22. On that day, the DPP chairperson, the prime minister, and major political leaders went to visit Lin, who's act increased public support of anti-nuclear actions.

Indeed, on the third day of the hunger strike, professionals in art and culture initiated a "1000-person sit-in" anti-nuclear protest in support of Lin. The next day, more than 400 professors and researchers signed an anti-nuclear petition and DPP legislators sat in front of the Presidential Hall to protest against the KMT government's nuclear energy policy. These actions were followed, on April 27, by 50,000 citizens attending an anti-nuclear rally and occupying two main roads around the Taipei railway station. In a hasty response, on April 28, the prime minister announced that the first and second reactors at the Fourth Nuclear Power Plant construction site had been sealed and construction suspended. In other words, under tremendous political pressure, it seemed the government had eventually accepted the demands made by anti-nuclear groups and Lin, and had finally compromised with the civil society majority against nuclear power.

### The Policy Significance of the Fukushima Effect

To highlight the policy significance of the Fukushima effect in changing the nuclear policy of the Taiwan government, I now compare and summarize its effect with the failed case led by the DPP government in 2000. The main characteristics are shown in Table 3.1. The 2014 suspension of the FNPP's construction and operation was disaster-driven, while the 2000 suspension was only politically driven. In addition, in 2014, the major counter-governmental actors were anti-nuclear civil organizations, while in 2000 they were opposition political parties. In 2014, the arena for counter-governmental nuclear protest was on the streets, while in 2000 it was at the meeting hall in the Legislative Yuan and the Constitution Court.

*Table 3.1* Comparison of the Government's Reactions to Challenges from Varied Groups on the Issues of Nuclear Energy

| Event | DPP government suspension of FNPP construction in 2000 | KMT government suspension of FNPP construction and operation in 2014 |
|---|---|---|
| Nuclear disaster? | No | Fukushima |
| Political context | The first rotation of the political regime in Taiwan | Continuous mobilization of anti-nuclear movements after Fukushima |
| Governmental position | No-nuke policy and suspension of FNPP | Substitute FNPP for old plants |
| Public attitudes toward nuclear power plant | From more than half of the general public against FNPP to less than half | More than half of the general public against FNPP |
| Counter-governmental forces | Alliances of opposition political parties | Alliances of anti-nuclear and other civil organizations |
| Political crisis weakening governmental legitimacy | Initiation of presidential recall by alliance of opposition parties | Sunflower Movement against Taiwan-China trade treaty |
| Immediate crucial political incident | Approval of FNPP construction budget by Legislative Yuan after a Constitutional Court ruling | Hunger strike of a respected anti-nuclear leader |
| Outcome | Decision on the suspension of FNPP reversed | FNPP policy abandoned and the project frozen |

From 1998, negative public attitudes toward the FNPP exceeded 50%, but such public opinion did not directly affect the government's nuclear policies. However, both the presidential recall organized by an alliance of opposition parties in 2000, and the Sunflower Movement in 2014, significantly weakened governmental authority and created strong political opportunities for oppositional groups to challenge the government. At the same time, public opinion was affected by these political events, and civil society came under governmental rhetorical closure or consensus, to some extent, on the FNPP project one way or the other (see Pinch and Bijker 1984). Anti-nuclear public opinion, although mounting, could not change government policy, but it nevertheless functioned as a powerful foundation upon which counter-governmental groups challenged ruling power. Subsequently, political action in the wake of Fukushima was able to overturn pro-nuclear energy policy.

In summary, these two cases demonstrate the complex social construction process of nuclear energy policy. Both resulted in a reversion of governmental decisions on the FNPP when the government was continuously challenged by either opposition parties or anti-nuclear movement organizations and in the process began to rapidly lose legitimacy. When we contrast the two cases, one before Fukushima and one after, the 2000 case was a politics-driven process, which activated the formation of an alliance of opposition parties. In turn, the 2014 case was a disaster-driven process, which catalyzed a very strong anti-nuclear mobilization of civil society. The former was primarily a political conflict between political parties, the outcome of which actually had the effect of inadvertently suppressing the anti-nuclear momentum of civil organizations. In contrast, the 2014 case was principally a conflict over the application of nuclear technology waged between the state and civil society. Continuous negotiation over nuclear policy sustained the anti-nuclear momentum within civil society, and led to significant policy change and the abandonment of the Fourth Nuclear Power Plant, although the three other nuclear reactors remain in action at this stage.

## Conclusion

This investigation on the effect of Fukushima on the politics and policy of nuclear power development in Taiwan has revealed that adequate political and cultural structures, and the facilitative circumstances or opportunities posed by the Fukushima disaster, enabled anti-nuclear groups to increase public support for anti-nuclear policy and push through changes in nuclear energy policy by way of massive social demonstration. This was the key effect of Fukushima: generating opportunities for change to the extent that nuclear power now appears to be in real decline in Taiwan.

The struggle also showed that public attitudes toward nuclear energy in Taiwan reflected a complex terrain difficult to negotiate. More than 50% of citizens held no constant position in regard to nuclear energy and could be swayed one way or another depending on external influences. The Fukushima catastrophe changed that by strengthening public opinion against nuclear energy, which led to greater public concern about nuclear power risks and hazards, which civic organizations drew on in subsequent anti-nuclear actions.

In sum, the Fukushima disaster bolstered anti-nuclear movements and the expansion of anti-nuclear organizational networks in Taiwan. Prior to Fukushima, the first large-scale anti-nuclear demonstration was organized in 1997 by a number of environmental protection organizations. Thereafter, the strength of the anti-nuclear movement declined because of the failed attempt to suspend the Fourth Nuclear Power Plant project by the Democratic Progressive Party government in 2000. However, from that time onward, environmental and anti-nuclear organizations determinedly sustained their operations with limited resources, to renew and reinvigorate after Fukushima. A key maneuver used by these organizations to gain support was to adopt an "open-door" strategy to mobilize public support across social class, gender, and occupation, to expand organizational networks and build strong capacity to pressure for closure of the Fourth Nuclear Power Plant development, which occurred. The focus now, again bolstered by the political opportunity of the Fukushima effect, and subsequently the Sunflower Movement, is to shut down Taiwan's remaining three nuclear power plants.

# References

Beck, U. *Risk Society: Towards a New Modernity*. London: Sage, 1992.

Beder, S. "Controversy and Closure: Sydney's Beaches in Crisis," *Social Studies of Science* 21 (1991): 223–256.

Chang, Y.-S. "The Brief History of the Fourth Nuclear Power Plant Construction Project," 1997, accessed May 5, 2015, http://tinyurl.com/q4onswt.

Chen, D.-S. "Taiwan's Antinuclear Movement in the Wake of Fukushima Disaster, Viewed from an STS Perspective," *East Asian Science, Technology and Society: An International Journal* 5 (2011): 567–572.

China Times. "Public Opinion on the Nuclear Power Plant Construction Project after the Constitution Explication," *China Times*, January 16, 2001, accessed May 5, 2015, http://tinyurl.com/lme65fj.

China Times. "The Survey Report on Nuclear Safety," *China Times*, March 17, 2011, accessed May 5, 2015, http://tinyurl.com/mmy8x2f.

Chiu, H.-Y. "Public Opinion and the Fourth Nuclear Power Plant Construction." Unpublished Manuscript, Institute of Sociology, Academia Sinica, Taipei, 2001.

Chong, D., and J.N. Druckman. "Framing Theory," *Annual Review of Political Science* 10 (2007): 103–126.

Common Wealth. "60% of the General Public Supported Suspension of the FNPP Project but Half of Them Don't Accept Increase in Electricity Price," *Common Wealth*, April 29, 2014, http://tinyurl.com/qdx77gy.

Constain, A.N. *Inviting Women's Rebellion: A Political Process Interpretation of the Women's Movement*. Baltimore and London: Johns Hopkins University Press, 1992.

Finucane, M., and J. Holup. "Psychological and Cultural Factors Affecting the Perceived Risk of Genetically Modified Food: An Overview of the Literature," *Social Science & Medicine* 60 (2005): 1603–1612.

Giugni, M. "Political Opportunity: Still a Useful Concept?" In *Contention and Trust in Cities and States*, edited by M. Hanagan, and C. Tilly, 271–283. New York: Springer, 2011.

Gong, S., and R. Catenacci. "Taiwan Anti-Nuclear Protest," March 8, 2013, accessed December 11, 2014, licensed under Attribution 2.0 Generic (CC BY 2.0), http://preview.tinyurl.com/nxkjb6n.

Green Citizens' Action Alliance. "About Green Citizens' Action Alliance," n.d., accessed July 14, 2014, www.gcaa.org.tw/about.php.

Hindmarsh, R. (ed.). *Nuclear Disaster at Fukushima Daiichi: Social, Political and Environmental Issues*. New York: Routledge Studies in Science, Technology and Society, 2013.

Ho, M.-S. *Green Democracy: A Study on Taiwan's Environmental Movement.* Taipei: Socio, 2006.

Hu, H.-S. *Nuclear Engineering Experts versus Antinuclear Experts.* Taipei: Vanguard Press, 1995.

Kasperson, J., et al. "The Social Amplification of Risk: Assessing Fifteen Years of Research and Theory." In *The Social Amplification of Risk*, edited by N. Pidgeon, R. Kasperson, and P. Slovic, 13–46. Cambridge: Cambridge University Press, 2003.

Kitschelt, H. "Political Opportunity Structures and Political Protest: Anti-Nuclear Movements in Four Democracies," *British Journal of Political Science* 16 (1986): 57–85.

Koopmans, R. "Political, Opportunity, Structure. Some Splitting to Balance the Lumping," *Sociological Forum* 14 (1999): 93–105.

McAdam, D. *Political Process and the Development of Black Insurgency, 1930–1970.* Chicago: University of Chicago Press, 1982.

McAdam, D. "Culture and Social Movements." In *New Social Movements: From Ideology to Identity*, edited by E. Larana, H. Johnston, and J.R. Gusfield, 36–57. Philadelphia: Temple University Press, 1994.

McAdam, D., Tarrow, S., and C. Tilly. *Dynamics of Contention.* Cambridge: Cambridge University Press, 2001.

Miller, C., Sarewitz, D., and A. Light. *Science, Technology, and Sustainability: Building a Research Agenda.* A Report for the National Science Foundation, Arlington, VA, 2008.

Nahuis, R., and U. van Lente. "Where Are the Politics: Perspectives on Democracy and Technology," *Science, Technology, & Human Values* 33 (2008): 559–581.

Obseschall, A. "Opportunities and Framing in the Eastern European Revolts of 1989." In *Comparative Perspectives on Social Movements: Political Opportunities, Mobilizing Structures, and Cultural Framing*, edited by D. McAdam, J.D. McCarthy, and M.N. Zald, 93–121. Cambridge: Cambridge University Press, 1996.

Pinch, T., and W. Bijker. "The Social Construction of Facts and Artefacts: Or How the Sociology of Science and the Sociology of Technology Might Benefit Each Other," *Social Studies of Science* 14 (1984): 399–441.

Polletta, F. "Snarls, Quacks, and Quarrels: Culture and Structure in Political Process Theory," *Sociological Forum* 14 (1999): 63–70.

Research and Evaluation Council. "The Survey Report on the referendum of the Fourth Nuclear Power Plant." *Research and Evaluation Council*, April 13, 2013, accessed April 6, 2015, http://tinyurl.com/kll7g5c.

Slovic, P. "Trust, Emotion, Sex, Politics, and Science: Surveying the Risk-Assessment Battlefield," *Risk Analysis* 19 (1999): 689–701.

Slovic, P. *The Perception of Risk.* London: Earthscan, 2000.

The Taiwan Power Company. "Basic Data of Nuclear Power Plants," 2014, accessed May 2, 2015, http://tinyurl.com/mrs3lwy.

Tilly, C. *From Mobilization to Revolution.* Reading: Addison-Wesley, 1978.

Tsui, Su-Hsin. "History and Strategies of Taiwan's Antinuclear Movement," *Taiwan Association of Human Right Pas* (Autumn) (2011): 20–22.

Winner, L. "Do Artifacts Have Politics?" *Daedalus* 109 (1980): 121–136.

# 4 The Korean Case of Nuclear Energy Policy Pre- and Post-Fukushima

*Hyomin Kim*

Since the late 1980s, the technocratic nuclear energy regulatory policies of South Korea (hereafter referred to as Korea, as South Korea is commonly referred to) have been relaxed to allow for more open dialogue and public communication. This chapter documents these developments with a special emphasis on the Fukushima disaster to assess the extent of any effect on nuclear power development and management in regard to participatory governance. This aligns with the broader question of how Korean nuclear energy development and the social assumptions of the Korean public have interacted with changing policy environments.

Chung (2012) argues that the decision-makers' safety discourse, which was based on scientific and technical expertise, began to crack in Korean nuclear governance after the nation's political democratization in 1987, when the liberal democracy of the Sixth Republic was established (Kim 2000). Subsequently, anti-nuclear environmental organizations and scientists voicing safety and cost concerns gained public awareness. This led to a new politics of expertise, one no longer monopolized by technical experts associated with the government.

However, after 2005, Korea's politics of expertise experienced another shift as Korea started to catch up with participatory science and technology trends that had emerged in western nations in the early 2000s. Since then, new public engagement practices associated with evolving Korean nuclear regulatory policies show a blend of old and new governance. The assumption that local residents oppose nuclear energy, whereas the majority of the Korean public acknowledges the necessity of it, supported the nation's old technocratic nuclear policy. New initiatives, such as local referendums and open dialogues, while attempting to include public consultation as a step in decision-making processes, have also depended on the positions of general and local publics, which appear to have changed since Fukushima.

The practice of engaging the public in decision-making processes has arisen primarily as a means to ameliorate the public mistrust that has undermined science and technology development plans in recent times (Hagendijk and Irwin 2006). Public involvement is being increasingly institutionalized as part of science and technology policies internationally (Rowe and Frewer 2005; Sclove 2010). Science, technology and society (STS) literature has extensively discussed public participation in decision-making for the governance of controversial science and technology (Irwin 2001; Irwin and Wynne 1996; Jasanoff 2003).

Yet, evidently, a "new" mood for dialogue cannot ignore the history of "old" dependence on technical experts. Technocratic discourses of policies based upon sound science, institutional control, and administrative rationality still co-exist with the new participatory terms of openness, transparency, and mutual learning between lay and expert knowledge (Irwin 2006; Wynne 2006). What is analyzed in the STS literature from technocratic and public engagement initiatives is not simply what is new and old, but what new policy has become an integral part of decision-making processes and what is an optional add-on. The traditional STS question of how the boundary between science and society is (re)constructed is continuously probed, with increasing attention given to two-way dialogues in science technology policies (Barben et al. 2008; Jasanoff 2007). In this chapter, I explore these changing boundaries in Korea, particularly in relation to a Fukushima effect. The next section outlines my methods and materials of exploration.

## Methods and Materials

This chapter has two parts: pre-Fukushima and post-Fukushima. In the first section of the pre-Fukushima part, I review previous literature on Korean nuclear energy and present results of annual surveys on public opinion on nuclear power conducted from 1993 to 2010 by the Korea Nuclear Energy Foundation (KNEF), a government organization for public information on nuclear energy. In the second section, I discuss a 2005 local referendum (the Gyeongju local referendum) to select a nuclear waste repository site as the starting point of "new" nuclear governance. I analyze news editorials and official remarks by KNEF senior researchers on the referendum, along with previous Korean sociological literature. These examples demonstrate that the Korean government, and major newspapers with a pro-nuclear stance, constructed the "local," where nuclear power was sited, as a separate and "isolated" place from the general public.

In the first section of the post-Fukushima part of the chapter, I analyze data obtained from interviews conducted between October and November 2011 in the local region of Kori. The Kori site—in Gori, a suburban village of Busan, the second largest city in Korea, located on the southeastern-most tip of the Korean peninsula—has five nuclear reactors in operation and three under construction. The oldest reactor, Kori-1 (which began in 1978), has been at the center of public debates since the Fukushima disaster in March 2011. I interviewed two local residents, one local council member, one environmental activist, one lawyer participating in a social movement, and one manager at the Kori site. I asked interviewees about their experiences with, and opinions on, increasing demands for transparent nuclear governance.

In the second and third sections, I discuss the results of interviews I conducted with a nuclear regulatory policy researcher and selected interviews in major Korean newspapers covering the accident at the Kori site in February 2012, then present an analysis of selected articles and interviews from Korean newspapers, covering the lifetime extension of the Wolsong-1 reactor in February 2015. I selected these materials to analyze multiple perspectives on the increasing institutional transparency in Korean nuclear governance, and to identify who the "public" are in public engagement with Korean nuclear governance, post-Fukushima.

## Pre-Fukushima: The Local versus the General Public

### Historical Background

Anti-nuclear activism in Korea has, until recently, been considered as a NIMBY (not-in-my-backyard) phenomenon rather than a serious social movement supported by the general public (Park 1992; Valentine 2010). Survey results after the political democratization of Korea in the late 1980s show that a continuously high percentage of respondents understand nuclear energy as necessary and approve of nuclear power plants.

Currently in Korea, 21 nuclear reactors are in operation, with seven more under construction and plans for four new reactors. The four planned reactors will each have a capacity of more than 1000 megawatts and be constructed and operated by domestic companies. The National Basic Energy Plan, announced by the government in 2008, includes programs for increasing nuclear energy to 59% of Korean electricity production by 2030, building 12 more reactors by 2022, and exporting 80 reactors by 2030. In 2010, nuclear power accounted for 34% of electricity generated in Korea.

Three years later, on November 23, 2011, the Nu-Tech 2030 plan announced by Korea Hydro & Nuclear Power (KHNP) and the Korean Ministry of Knowledge Economy included similar proposals to export nuclear reactors developed using wholly indigenous technology. This would make Korea one of the world's top three nuclear energy technology producers, positioned after the US and France. Although this plan was announced after the Fukushima disaster, it reflects much technological optimism for a safer Korean power reactor with a newly developed system to cool reactors for three days in the event of a complete plant blackout.

The general public's support for nuclear energy has been cultivated since the beginning of Korean nuclear technology development. Jasanoff and Kim (2009) discussed nuclear energy in Korea as being portrayed as necessary to build the nation as a modern, economically developed country after liberation from Japanese colonialism following the atomic bombings of Hiroshima and Nagasaki. The history of Korean nuclear energy reflects a history of technical and economic development.

In 1967, the Korean government made a plan to construct three nuclear reactors of 600 MW. At that time, overall Korean energy consumption was only about 1000 MW. The government expected nuclear power to provide the majority of energy needed for the Korean economy. With such expectations, the country's first commercial nuclear power reactor, Kori-1, began operating in 1978. The Kori-2 and Wolseong-1 reactors came on-line in 1983. The Wolseong power plant is located on the south-eastern coast of Korea. These reactors were all purchased from West-inghouse and installed fully complete by government-owned company Korea Electric (currently Korea Electric Power Corporation or KEPCO).

With the increase in nuclear power plants, the government set a goal of building new reactors using indigenous technology to ensure self-reliance in nuclear technology. In 1987, KEPCO began constructing the Youngkwang-3 and 4 reactors (in South Jeolla Province, southwest-ern Korea) as Korean standard nuclear power plants. The US firm Com-bustion Engineering won the contract by agreeing to a full technology transfer. Since the late 1980s, nuclear plants in South Korea have been built using 95% or more indigenous technology. By 1989, nine nuclear units were in operation, providing more than 45% of Korean energy consumption. In 2009, KEPCO signed a US$20 billion contract to export a reactor to the United Arab Emirates, making Korea the world's sixth biggest exporter of nuclear power plants, behind the US, France, Japan, Russia, and Canada (Kim 2011).

So how did major nuclear accidents and disasters affect Korea's nuclear policy? Unlike many other nuclear nations (e.g. see Jahn and Stephan this volume, and Chen this volume), Three Mile Island (1979) and Chernobyl (1986) did not lead to massive anti-nuclear activities in Korea, partly because they occurred during the time of the Korean mili-tary regime and before the advent of the (democratic) Sixth Republic in 1987. Yet, even post-democratization, the Korean government continued its "decide-announce-defend" approach to nuclear policy, taking advan-tage of the general public's support for national economic growth.

In a 1991 survey, 81% of Koreans supported further nuclear energy production, and 85% responded that nuclear power was necessary. In a 2010 survey undertaken after KEPCO's contract with United Arab Emir-ates, public support remained relatively stable: 88% thought that nuclear power was necessary and 61% supported additional construction.[1] The expectation that Korea was set to become a major exporter of nuclear technology was also widely reported and supported by the media. Nota-bly, on matters of science, policy, and the public, the media plays an important part in public translation because "of an increasingly indus-trialized and individualized society and the many attendant technicalities and complexities of controversial science and issues of technology" (Hindmarsh 2014: 197).

The decide-announce-defend approach was, however, not without several anti-nuclear protests. In November 1990, the Korean government announced a plan to construct a nuclear waste repository on a small island, Anmyeon (a western seaside area), with approximately 17,000 residents. Activists and residents organized a large rally with about 20,000 participants, which led to the government canceling the Anmyeon plan in 1993 (Lim and Tang 2002). Similar conflicts around the building of a repository were repeated on Guleop Island in 1994 and Wi Island in 2003.[2]

In all these cases, although the government had to withdraw its construction plans, it could continue to assume that the *general public* supported nuclear power. In a 1996 Korea Nuclear Energy Foundation survey, after the protests in Anmyeon and Guleop Islands, 85% of general public respondents thought nuclear power was necessary, and 66% agreed with plans to construct more plants. The KNEF, while announcing their survey results through press releases, portrayed that local residents' NIMBYist-style protests were an obstacle to realizing the benefits of nuclear technology as agreed to by the majority.

In addition to the KNEF surveys, critical sociologists also noted the Korean general public's willingness to see nuclear power as an inevitable necessity. Lee (1999) pointed out that even though the majority of the public regarded nuclear power as unsafe, they still saw it as indispensable. According to Lee (1999), the high support for nuclear energy seemed to be based upon the public's (mis)understanding that dangerous nuclear radioactivity could be contained to a small region. The government's selection of three small islands as candidate sites for a nuclear waste repository fitted with such popular perceptions of a separation between the "general" public and the "local" public.[3]

The separation between the general and the local public was continuously reconstructed during the development of Korean nuclear energy up to 2012. As the following sections will further discuss, while old technocratic nuclear policy crisscrossed and conflicted with "new" participatory governance before and after the Fukushima disaster, the difference between the local and the general public was socially reproduced in the attempt to downplay nuclear safety concerns to the majority. However, before we get to Fukushima, the Gyeongju local referendum was notable in Korean policy shifts towards participatory nuclear governance.

## Gyeongju Local Referendum

After the earlier failed attempts to construct a nuclear waste repository, the Korean government began a "new" approach, moving away from its decide-announce-defend approach towards participatory governance. The Gyeongju local referendum, introduced in 2005 by the government, has received particular attention in this regard. The government announced four new candidate sites for its waste repository and conducted local referendums as a decisive participatory step to select a low/mid-level nuclear waste repository site. The announcement followed the Special Act on Assistance to the Locations of Facilities for Disposal of Low and Intermediate Level Radioactive Wastes, 2005. As an inducement to accept such a site, reminiscent of Japanese practices (and much along the same lines as occurred in Japan: see Hindmarsh 2013a), the region with the highest approval rate was to receive around US$300 million in a special state subsidy. The coastal city of Gyeongju was finally selected, with a citizen approval rate of 89.5%.

Activists and critical academics, however, regard the Gyeongju referendum as a failure in participatory governance, which worsened social conflicts and simply reinforced that the Korean public's support for nuclear power was for economic reasons. Cha and Min (2006) criticized the referendum for its lack of deliberative processes and for the economic inducements used to sway public opinion. Yun (2006) argued that the process did not consider substantive justice in selecting appropriate sites, distributive justice in administering compensation, and procedural justice in operating the referendum.

In turn, Lee (2006) pointed out that because Gyeongju and the three other candidate regions competed for siting and compensation, the referendum was inevitably tarnished by each local government's questionable activities. For example, the rate of absent voting was 20–30% in all the regions, much higher than the usual rate of less than 3% in Korean elections. It was thus considered that the referendum facilitated a PIMFY (please-in-my-front-yard) syndrome, which reflects a pre-meditated policy position, not participatory or deliberative governance. It thus seemed government-led public engagement still had many limitations to constituting genuinely new nuclear governance on the local benefits and risk and hazards of nuclear power plants.

The degree of shift in Korean nuclear policy towards participatory governance is not the only issue that can be analyzed here. We can also ask: in what contexts did this shift interact with the old assumptions behind Korean nuclear policy? According to Irwin (2006), although transparency and openness can be seen as "mere talk," "new" scientific governance needs to be analyzed as symptomatic of emergent science-society relations rather than simply dismissed as unsubstantial. It is thus notable that the emergent nuclear-society relations in Korea were being constructed and reconstructed by the perceived difference between the general public and the local public.

For example, on September 12, 2005, *Chosun Ilbo*, the leading national daily newspaper in South Korea, known for its conservative orientation, published an editorial column on the 2005 referendum:

> Nuclear waste repository siting used to be a symbol of social conflict in Korea for the past twenty years . . . In case of the Wi Island

in 2003, mob violence continued for two months turning Buan County into a battleground for two months. Things have greatly changed in two years. New attempts to overcome local selfishness are worth examining. Before, questions around economic compensation and procedural legitimacy intensified NIMBY, now, the new policies remove suspicion. First, the Special Act on Assistance to the Locations of Facilities for Disposal of Low and Intermediate Level Radioactive Wastes established in March 2005 stipulated the special subsidy of $300 million and the waste fee of $8.5 million per year. Second, the Act stipulated that local decision-making processes needed to be transparent and democratic. The head of local government [in Gyeongju] organized resident forums and surveys before submitting a proposal [for siting] . . . Transparency and participation have removed conflict.

In the editorial, transparent governance is framed as a means to lessen the NIMBY phenomenon by openly promising economic compensation. It is assumed that local NIMBYism is the key source of social conflict around nuclear energy. The general public, who were supposed to acknowledge the benefits of nuclear power for economic development, were thus not invited to the referendum, as their perceived trust in nuclear technology or regulatory institutes was not seen to be needed for "rebuilding" through public engagement.

Indeed, in a 2005 survey, 95% of respondents answered that nuclear power was necessary, and 51% supported the construction of a nuclear plant in their local regions. The referendum and communication strategies were attempts to win over local residents. The addition of participatory governance to economic incentives as a way to settle the legitimation crisis of the decide-announce-defend approach was contextualized within the general public's continued understanding of nuclear risk being containable in a local region, and the pro-nuclear media's negative framing of anti-nuclear movements as local NIMBYism.

In a symposium on the export of Korean nuclear power plants as the new engine of growth, held in the national assembly in January 2011 and reported on in *Donga Ilbo*, January 3, 2011, Chung Ik-cheol, a KNEF senior researcher, opined that the new governance process in selecting the Gyeongju repository was "a scientific and advanced system to achieve social acceptance of nuclear energy." What he appeared to imply by "social" acceptance was "local" acceptance in addition to general public acceptance. He continued: "The majority of the public acknowledge the importance, necessity and security of nuclear; local acceptance, however, is a different problem as the statistics show," a problem he argued could be met by this new (read manipulative) participatory approach connected to economic inducement.

But of course, and again, economic inducements are far from genuine participatory governance approaches (e.g. Elam and Bertilsson 2003).

Chung also commented on the Korean government's plan to export nuclear reactors to Malaysia and Indonesia. As reported in *DP News* on January 3, 2011, he suggested: "Even though their national long-term policies include the introduction of nuclear energy, they are reluctant to import newly developed Korean reactors because of the public opposition. It will [thus] be useful to export Korean strategies to resolve conflicts, as well as nuclear technology."

It is notable that the Gyeongju referendum was referred to by the Korean government as an attempt to gain nuclear acceptance and an effective response to the Fukushima accident, with its negative impacts upon public perception of nuclear technology. This remarkable claim occurred in 2012 in relation to principles of stakeholder involvement such as transparency and accountability in a session called "Stakeholder Involvement and Public Communication" in a Seoul forum called "Global Nuclear Energy Sustainability: Long-term Prospects for Nuclear Energy in the Post-Fukushima Era." The forum was part of the International Project on Innovative Nuclear Reactors and Fuel Cycles (INPRO) dialogue, organized by the International Atomic Energy Agency (IAEA) and the Korean government and Korea Atomic Energy Research Institute. Some 80 participants attended, representing government, academia, regulators, the nuclear industry, and nuclear operators and research institutions.

## Post-Fukushima Developments

### The Kori Site after the Fukushima Disaster

After the Fukushima accident, it seemed reasonable to expect the Korean public's understanding of radioactive leakages in a small region as likely to change. It was widely reported by the Korean media that the Japanese government initially placed a three-kilometer exclusion zone around the Fukushima Daiichi power plant, which was soon expanded to 30 kilometers. The Kori nuclear power plant in particular received much attention, as a radius of 30 kilometers includes more than four million residents in two highly urbanized regions—Busan, Korea's second largest city, and Ulsan, home to the world's largest automobile plant operated by the Hyundai Motor Company (Greenpeace 2012).

In April 2011, the Busan Bar Association filed a court injunction to have Kori-1, Korea's oldest nuclear reactor, taken out of operation. Kori-1 began operation in 1978. It had suffered a number of minor malfunctions, and was supposed to complete its life span in 2007. Yet, in 2008, it was given a 10-year extension after an International Atomic Energy Agency inspection. Kang Dong-gyu, chairman of the Busan Bar Association's special committee on the environment, opined:

> I didn't have much interest in nuclear power plants before. After the Fukushima incident, I learned that a wide area can be influenced

by such accidents. When KHNP decided to extend the operation of Kori-1, it did not disclose the results of [a] safety assessment. I filed the court injunction not because I am against nuclear energy or I demand the whole Kori site should be shut down. I made a rational demand to stop the Kori-1 reactor temporarily, check its safety and operate it again after examination. Some people raise questions about the 35-year-old Kori-1's safety yet little information is accessible. Even if Korea continues to operate nuclear power plants, it is necessary to publicly raise those questions at least once after the Fukushima incident.

(Interview transcript, No. 1)

Another public interest association is Energy Justice Action, a national-level environment association that has led protests against the extended operation of Kori-1 since 2007. In an interview, Energy Justice Action Chairman Lee Heon-seok stated:

In principle, we want to live in a safe society without nuclear; we are against nuclear power itself of any kind.

(Interview transcript, No. 2)

Lee believes transparency is an important issue to address because "people are interested in accidents and safety issues; safety and transparency go together; you cannot have one without the other." In an official public statement issued in May 2012, Lee wrote:

People are astounded by the exclusiveness of nuclear industries and the government. As all the citizens of the nation (*mo-deun kuk-min*) are already aware, transparency is as important as nuclear safety. Safety assessments conducted in a transparent way will increase not only the level of trust but also of safety.[4]

Korean environmental activists and NGOs expected the public to demand more transparent and participatory nuclear governance after the Fukushima disaster. Unexpected practices of the "local public" were, however, observed. For a safety assessment to be conducted on Kori-1, the Busan Bar Association recruited 97 plaintiffs for its injunction filed in April 2011, although public interest lawsuits are rare in Korea. Association Chairman Kang recalled that "only two or three among them were local residents living near the Kori plant" (Interview transcript, No.1). In the same month, 60 cities and district council members in Busan made a public announcement and demanded that the government shut down Kori-1 and stop turning Busan into a nuclear complex. However, no council members from Kijang, the county where the Kori plant is located, joined the announcement. This begs the question: why were only a few local plaintiffs and no local council members from Kijang represented in these actions?

Some insight was given by Park Hong-bok, a council member of Kijang County. He also serves as a vice-chancellor in a private organization that demands more open dialogue between local residents and Korea Hydro & Nuclear Power. However, he was adamant that other local problems should be dealt with as a priority alongside safety or transparency in nuclear governance, even perhaps as more important. His statement began:

> Members of the Kijang County council did not join the announcement because we represent the local [people]. People in Busan suddenly became anxious about Kori-1 after the Fukushima incident. We are not like them. We are calm and aware of [the] real problems. We have lived with this issue for a long time.
>
> (Interview transcript, No. 4)

The "local" problems to which Park referred included those related to economic compensation for people living in the villages of Gilcheon and Wolnae, located within five kilometers of the Kori plant. Park Kab-yong, the Gilcheon village headman, seemed upset when asked about the local residents' stance on social movements demanding more transparency in nuclear governance:

> We did not join street protests [after the Fukushima accident]. We know about safety, transparency and social acceptability issues. Of course I do not trust what KHNP says about safe nuclear. But more importantly, we are living with the plant. The local economy is terrible because the government sets limits on construction near the plant in case of meltdown. Look at this town [compared to] Busan with its urban glitter. I know young people like you would not even want to visit this town. The fishing community has declined, I think to 20% compared to 30 years ago. We are dying of hunger. Nuclear safety? We have no time to think of that. As people outside talk more about safety, discussion over the local economy becomes neglected. Of course I do not trust what KHNP says about safety. Yet, if we talked about problems about nuclear energy, more people would stop buying our fish. So we stay silent. We are totally different from Busan citizens.
>
> (Interview transcript, No. 5)

Seo Yong-hwa, a local resident whose family has lived in Wolnae village for six generations, agreed with Park:

> So many people moved out because of the worsening local economy. They think we receive massive economic compensation from KHNP.[5] But most of it is spent on construction—roads, sports complexes and

so forth. Local residents cannot decide where to spend the money. It would be really good if people in this village could pay, say, TV and Internet costs of about $30 per month out of it. But we cannot. People in big cities like Seoul and Busan pay attention to the safety of Kori-1 after Fukushima. But they know nothing about the local situation. Their concern is not ours.

(Interview transcript, No. 6)

Conversely, national-level social movements after Fukushima kept their distance from "local" publics hosting nuclear power plants (NPPs), because of their focus on economic compensation for living near NPPs rather than on safety. In recalling the absence of citizen plaintiffs who were local residents living near the Kori site, Kang commented:

After the Fukushima . . . social demands of open discussion about nuclear safety are increasing. Surely, the local public cares more about compensation. I understand their pain but this particular lawsuit cannot be about that.

(Interview transcript, No. 1)

Likewise, Lee Heon-seok, of Energy Justice Action, commented:

It is bitter but [is] the inevitable truth. I admit our goals and [those of] local residents' are different. We cannot join discussions over those [economic] kinds of problems. When some local residents in anti-nuclear movements want to divert us to economic [compensation] issues, we just quit. I think that's the right way.

(Interview transcript, No. 2)

In more detail, the post-Fukushima separation of the local and the public is a continuation of the historically constructed memory of nuclear energy as a *national* issue. Even after the nation's political democratization, and the burgeoning of environmental movements against narrowly-conceived economic development, the media paid little attention to the local voice. Young Gi Kim's (2003) study reveals that between 1987 and 2002, *Chosun Ilbo* and *Joongang Ilbo*, the two most-read Korean dailies, issued only one article about local residents' rights for economic survival in regard to nuclear power issues, compared to 39 articles about nuclear technical safety measures and 53 about the hazardous international movement of radioactive wastes. *Hankoyreh*, the only national daily with a progressive and anti-nuclear editorial orientation, also issued only two articles about local residents' rights, compared to 18 articles about accidents in power plants and 55 about issues of secretive nuclear policies.

Kim (2003) found that both anti- and pro-nuclear newspapers employed mainly political and environmental frames to support their views. For instance, the lack of transparent nuclear governance in Korea was framed either as a call for a national transition to a sustainable energy plan, or as a reason to improve regulatory systems while maintaining the current energy plan centered on low-carbon emitting nuclear energy.

Notably, economic frames were almost exclusively used by pro-nuclear newspapers to emphasize the necessity of nuclear for national economic growth. Only occasionally did progressive newspapers mention economic arguments that, in considering costs for waste management and compensation, nuclear was not cheap. Yet Kim's (2003) analysis showed that such arguments were presented sparingly, without substantial backup material. Indeed, the lack of a local voice in major Korean newspapers was not due to the absence of a media critique against nuclear energy and its social implications; rather, it was due to the major dailies' selection of issues that were seen to appeal to the "general" public rather than the "local." It was thus not just the government and the conservative press that marginalized local residents' rights for local economic survival as a less important issue than nuclear power for economic growth. The progressive media followed suit in appearing to regard the coverage of local residents' economic concerns as a less effective strategy to support anti-nuclear arguments than public appeal for a more democratic and environmentally friendly nation. In both pro- and anti-nuclear media discourse, the nation came first.

Whereas local residents are isolated from national-level media discourses and the focus of social movements on nuclear safety, the strategy of Korea Hydro & Nuclear Power is to dismiss as unsubstantial both local problems and the new demands for transparent and participatory governance after the Fukushima incident. For example, a manager in the external affairs coordination department at the KHNP Kori site regarded trust-building mainly as dealing with local residents' complaints about economic issues:

> It is extremely rare that someone files a so-called civil appeal regarding safety of nuclear power plants. These are the examples of conflicts we deal with [to build harmonious relationships with local communities]. We operate free bus shuttles to a community sports complex as a part of our local support program. Then the commercial bus company files a petition to stop it. It is ridiculous. Well-educated new recruits volunteer to tutor local students. Then we deal with a petition filed by a local resident running a rental study-room business.
>
> (Interview transcript, No. 7)

As the KHNP continues to focus on economic compensation for the "local" as seemingly public relations strategies, its move toward transparency is criticized by citizens as ritualistic, a window-dressing for the still authoritative nuclear governance. Recalling how he began the suit, Kang commented:

> I was reckless to file the injunction. Some nuclear-related information is legally required to be accessible but we could not obtain much information for the suit. The KHNP Kori site gave limited time and access to view information. For example, they gave 30 minutes to read the safety assessment report for Kori-1 in a courtroom. We are not experts in nuclear technology. How are we supposed to understand [the possible problems]?
>
> (Interview transcript, No. 1)

Lee alleged that the seemingly "new" policy toward more openness was a cover-up:

> The Energy Justice Action have, in 2007, 2008 and 2009, continuously demanded that three documents should be accessible to the public—a periodic safety review, a lifetime evaluation report and an environmental impact assessment report, which the KHNP is required to submit to the Ministry of Education, Science and Technology. After Fukushima, the reports are finally disclosed but how? The KHNP told us to come to their office [at the Kori site] without cameras or pencils and just read. The reports were [very] thick. So we held a press conference and stated this was nonsense. It is meaningless. This is what the Korean government calls transparency.
>
> (Interview transcript, No. 2)

The safety assessment report for the Kori plant is 5440 pages long. The KHNP Kori site disclosed the report in May 2011 in response to public safety concerns over the 35-year-old Kori-1 reactor. However, Korea Hydro & Nuclear Power limited the time and location to read the report on the grounds that it contains design plans and other intellectual property. In September 2011, the Busan court rejected the injunction, but an appeal was ongoing at the time of writing.

### The Kori-1 Blackout Scandal

After the Fukushima incident, despite questions about any shifts to participatory governance, the Korean government claimed it made some changes in regard to safety regulation and institutional transformation for new nuclear governance. In October 2011, the Korean government launched the Nuclear Safety and Security Commission (NSSC) as a new regulator, with a chairman of ministerial rank reporting directly to the president. The NSSC, as an independent regulatory body separated from the Ministry of Education, Science, and Technology (MEST), regulates licensing, inspection, enforcement, emergency response, non-proliferation, export/import control, and physical protection. The former regulator, the Korean Institute of Nuclear Safety (KINS), became a technical support organization under the NSSC. Because KINS was formerly under the MEST, which is in charge of promoting nuclear power, the establishment of the NSSC is considered a sign of increasing regulatory independence in Korean nuclear safety regulation. The NSSC is also independent from the Ministry of Knowledge Economy, which controls Korea Hydro & Nuclear Power and is responsible for the construction and operation of nuclear power plants, nuclear fuel supply, and radioactive waste management.

Choi Kwang-sik, a researcher at the Korean Institute of Nuclear Safety, contributed two editorials on transparent and participatory governance to the KINS newsletter in 2007 and 2009. In an interview, Choi emphasized the importance of regulatory independence in new nuclear governance:

> There is little discussion over this [new governance] at KINS. I am personally interested in this as an academic topic. Although more people will demand transparent regulation, KINS will not [adopt more conditions] unless the government strongly drives it. The launch of NSSC is a good attempt toward a new nuclear governance of safety and transparency.
>
> (Interview transcript, No. 8)

However, the government-initiated drive towards new nuclear governance and public trust faced serious obstacles, as raised by the media and activists in 2012. On March 13 (a year after Fukushima occurred), the Nuclear Safety and Security Commission temporarily shut down the Kori-1 reactor after being notified of an accident on February 9 by Korea Hydro & Nuclear Power. During maintenance, a worker broke the power connection and caused a 12-minute blackout at the reactor. Failure of the backup generator resulted in a suspension of the cooling water system, which could lead to a meltdown. For more than a month, the NSSC had no knowledge of the incident, as the manager of the reactor concealed the situation. My search with the keywords "blackout in Kori" in the Korean Integrated News Database System, a Korean equivalent of Lexis-Nexis, found 320 news articles and 35 editorials in major national dailies from March 2012 to September 2012. The head of KHNP resigned in April 2010 and the manager was imprisoned in July 2012.

The Kori-1 blackout scandal not only jeopardized the safety of the nation's oldest reactor, but also placed the reliability of the NSSC as a regulatory organization under serious question. The NSSC was first criticized for failing to monitor the KHNP. Four NSSC officials at the Kori site did not know about the blackout. On July 4, 2012, the Nuclear Safety and Security Commission approved resuming operation of the Kori-1 reactor, but met nationwide opposition. A national survey performed by Representative Kim Je-nam of the National Assembly on July 5 in Busan, and reported in *Hankyoreh* on August 6, 2012, found that 67% of respondents favored shutting down Kori-1, a much higher percentage than 43% from the same survey in May. In addition, 66% of respondents expressed that they could not trust the NSSC's announcement about Kori-1's safety.

Environmental activists also strongly criticized the Nuclear Safety and Security Commission (NSSC). Kim Hye-jeong, from the Korea Federation for Environmental Movement, one of the oldest and largest Korean environmental organizations, commented, as reported in the daily newspaper *Kyunghyang Shinmun*, July 31, 2012: "Although the NSSC was launched to toughen safety regulations, it is only giving indulgences to nuclear industries. It is working for the nuclear mafia and their job security."

The government tried to mitigate the situation through public dialogue. The Minister of Knowledge Economy, Hong Seok-woo, visited the Kori site after the NSSC's decision to restart the reactor. An editorial in *Joongang Ilbo*, on July 5, 2012, commented: "It is a good decision for the Minister to go to the local area (*hyeon-ji*). With as much modesty as possible, he needs to have a dialogue with local residents (*chu-min*) and clear their doubts on safety." While the editorial urged local residents to accept the decision of the NSSC, it proposed the government "have sufficient open dialogue and bring local [concerns] into [decision making] as much as possible," because "no matter how safe the reactor is, it is right and just to delay [restarting it] when there is opposition."

The invitees to the open dialogue exercise with Minister Hong on July 7, 2012, however, did not include activists and Busan citizens but were limited to local residents and council members from the small villages near the Kori plant. A KHNP press release published in the *Edaily Korean News* on July 15, 2012, summarized the local residents' voice as a "petition for good living conditions where local residents can co-exist with safe nuclear as the experts claim," and the "spending of economic compensation for local support programs that the locals really want."

When a joint examination of the materials by the government and local residents was suggested, some local residents requested a time and place where they could read the data without government officials being present. As reported in *Hankook Ilbo*, July 26, 2012, Hong, at a press conference in Seoul, commented:

> More open dialogues would be a good thing. But we should consider the living conditions of all the other people in the nation. We have patiently communicated with local residents near the Kori site yet this [refusal to a joint examination] is moving away from the

direction where we achieve national consensus. We have no choice but to start re-operating the Kori-1 reactor on August 3, at least. We can save about $3 million per day if we do. We are trying to revise the Act on Assistance to Electric Power Plants-Neighboring Areas and give the locals what they really want.

The Korea Hydro & Nuclear Power press release and Hong's remarks indicated a situation of very limited civic dialogue within Korean nuclear policies. Participation is understood by the government as simply a narrow consultative "add-on" process more reflective of a persuasion exercise to get local residents to accept economic compensation, along with the old rhetoric that nuclear is a necessity for national and economic growth, and to marginalize issues of safety and transparent risk governance.

As Chung (2012) noted, although the Korean nuclear policy after the Gyeongju referendum emphasized the importance of residents' social acceptance of nuclear power, in reality it posed as a strategic detour to continue the expansion of what nuclear experts, not the public, see as safe and cost-effective technology. So-called new governance, both before and after the Fukushima disaster, is thus more closely related to supporting technocracy. This differs from laying the groundwork for mutual learning between lay and expert knowledge (e.g. Wynne 2006), which reflects both technical and social concerns to derive more effective policy. What is being done simply represents the pre-Fukushima Japanese policy approach (Hindmarsh 2013b). The Korean government does not appear to have learnt much from the experience of Fukushima.

### Ten Year Extension for Wolsong-1

On February 27, 2015, the Korean Nuclear Safety and Security Commission gave approval for the Wolsong-1 nuclear reactor, which is the second oldest reactor in Korea, to continue operation. Wolsong-1 began its operation in 1983 and was shut down in November 2012 after the expiration of its 30-year lifespan. Yet, similar to Kori-1, which was given a 10-year extension in 2008, Wolsong-1 won a new lease on its operating life until 2022. The decision-making panel comprised two standing NSSC commissioners, three appointed by the Korean administration, two by the ruling party and two by the main opposition party. It was reported by several Korean newspapers that the panel's decision was made when the two members appointed by the opposition party were absent.

The location of Wolsong County, where the Wolsong-1 Power Plant is based, is notable. The county is situated between Gyeongju, a small city with an aging population of less than 300,000, and Ulsan, an industrial city with the highest per capita income in Korea. Known as the heart of Korean heavy and chemical industries, Ulsan has a rich regional history of labor movement protests and anti-military government protests since the 1980s.

Several civic groups and the opposition parties in Korea promptly and strongly protested against the NSSC's decision to grant the license renewal to Wolsong-1. Director of the Korean Federation for Environmental Movement, Yang-yi Won-young, requested the NSSC release the safety evaluation report. In *Business Korea* on March 3, 2015, Won-young was reported as commenting that "the decision [to grant a lifetime extension] completely excluded the intentions of the people and the local residents."

At the local level, the Ulsan Progressive Politics Forum Promotion Committee, which comprises around 50 representatives of civic and labor organizations in Ulsan, also expressed concerns over nuclear safety and its implications for Ulsan citizens. The Ulsan City Branch of a Korean opposition party, New Politics Alliance for Democracy, issued the following statement, which was reported in the same issue of *Business Korea*:

> It is clearly a snatch that the NSSC decided to re-open the Wolsong-1 Nuclear Power Plant without opposition party members [present when the decision was made]. The extension is invalid without definite data presentation about its safety and the investment of the proper expenses and [the] agreements of residents.

The Ulsan City Branch of the Labor Party also criticized the NSSC for making the decision behind closed doors and not releasing relevant information to the public. Lack of public inclusiveness and information transparency, along with nuclear safety, were thus raised as major issues by both national-level and Ulsan-based organizations.

Meanwhile, in Wolsong County, four local committees representing around 17,000 residents requested the relocation of the NPP. Nuclear safety issues thus emerged at the local level, along with economic, social, and emotional ones. Shin, the 63-year-old leader of Na-ah village near Wolsong-1, commented in the *Joongang Daily* of February 27, 2015: "After the Fukushima incident, more people came to regard nuclear power plants as a hate facility. More than 50 percent of shops in our village closed down after the incident." Sharing Shin's concerns, Kim, a 69-year-old resident of Na-ah village, strongly criticized the NSSC's decision to operate Wolsong-1: "Local residents who live right in front of the plant are devastated. What is the point of nuclear safety?" The local population of Na-ah village decreased from 1500 to 830 following the Fukushima incident.

Concerns over radioactivity in Korea have also shattered local tourism and the real estate market in Wolsong. What local residents were expressing in both departing and raising their voices was a mixed sense of fear, frustration, and powerlessness that their economic and social survival was at stake. In such contexts, while local groups occasionally cooperate with national-level civic organizations, local concerns and issues raised in more urbanized areas such as Seoul and Ulsan remain misaligned. On March, 3, 2015, an official in Gyeongju local government was quoted by *Nocut News*, an Internet news agency, as saying: "Each local committee claims that its members want relocation to a safer town and they are concerned about the Wolsong-1 lifetime extension. But some members are [still] more interested in negotiations over economic compensation." Again, safety of the lifetime-extended nuclear reactor and transparent decision-making processes in nuclear governance remain elusive for some local residents: their hometown is already affected by nuclear stigma and long-term residual safety concerns; there are four nuclear reactors other than Wolsong-1, and one more is under construction.

This gap between safety and transparency in decision-making and economic compensation between local and national publics in their nuclear risk awareness thus works in favor of the Korean government's unrepentant commitment to nuclear energy after the Fukushima accident. For example, on February, 28, 2015, *Donga Ilbo* reported Korea Hydro & Nuclear Power as stating that it would establish a cooperative initiative to discuss economic compensation for local residents but mentioned nothing else in its plans to re-start operating the Wolseong-1 reactor during 2015.

## Conclusion

The key aim of this chapter was to investigate what effect the Fukushima disaster had on nuclear power development and management, particularly in regard to participatory governance, and to determine the extent of that effect. This aim aligned with the broader question of how Korean nuclear energy development and its social assumptions of the public interacted with changing policy environments. So, in summary, what did I find, and what are the implications of this Fukushima effect? Overall, it appears from my analysis that the Korean government perceives open dialogue on safety of nuclear power as causing too much delay during processes of decision-making. This is seen in the continued and strategic separation between local and general publics to progress nuclear power. Communicating with the local is seen as strategically important to address the safety debate, but this should not interfere with economic development and the contribution of nuclear power to the well-being of the nation. In this context, dialogue with the local is important to government as an implicit condition that dialogue eventually moves toward the consensus that nuclear energy will continue for the nation.

Since the 1970s, the Korean government has framed nuclear power as something the public must accept for national economic growth. The government's portrayal of the local referendum as "a scientific and advanced method to gain social acceptance of nuclear energy," and of open dialogue as a step towards a fixed conclusion on nuclear safety and cost-effectiveness, is connected to the assumption that concerns over expert decisions can be addressed as local rather than national issues. With increasing social demands for transparency and civic engagement in science and technology decision-making processes, renewed local concerns about the risks of nuclear power—as the main effect of Fukushima in Korea—however, have not overly changed the situation from before Fukushima. Such concerns instead remain in the shadow of nuclear energy building for the nation. Even though "new" regulatory practices focus on increased safety, and trust-building with local residents and their economic support, the social distance between the local and the general public concerned over nuclear safety is likely to be maintained.

## Acknowledgements

This work was supported by the National Research Foundation of Korea Grant funded by the Korean Government (NRF-2013S1A3A2054849).

## Notes

1 In a 2005 survey commissioned by the International Atomic Energy Agency, Korea was listed as the country with the highest support for nuclear power and its expansion among the 18 countries examined, which included the US, UK, France, Japan, and Canada.
2 There were 12 residents of Guleop Island when the repository plan was announced in 1994. Protests were attended by 300 activists and residents of nearby islands.
3 The majority of the public's understanding of nuclear as "necessary" and the local anti-nuclear movements as "irrational" is noted similarly in Taiwan (Fan 2009). Yet unlike the Yami tribe in Taiwan, residents of Guleop Island are not ethnically distinct from other Koreans; the separation between the local island residents and the general public in Korea is related more to their socio-economic differences.
4 See: www.eswn.kr/news/articleView.html?idxno=3421, accessed April 12, 2015.
5 The amount of compensation spent in Gijang County is about US$2 million per year.

## References

Barben, D., et al., "Anticipatory Governance of Nanotechnology: Foresight, Engagement, and Integration." In *The Handbook of Science and Technology Studies*, 3rd edn., edited by E. Hackett, et al., 979–1000. Cambridge, MA: MIT Press, 2008.
Cha, S.-S., and E.-J. Min. "Bang-pye-jang Bu-ji-seon-jeong-eul Dul-leo-ssan Gal-deung-gwa Min-ju-ju-eu, ['The conflict surrounding radiation waste treatment space and democracy']." *ECO* 10, no. 1 (2006): 43–70.
Chung, T.-S. "The Politics of Expertise and the Cracks of the Scientific and Technological Safety Discourse in Site Selection Processes of Radioactive Waste Repository," *Economy and Society* 93 (2012): 72–103 (in Korean).

Elam, M., and M. Bertilsson. "Consuming, Engaging and Confronting Science: The Emerging Dimensions of Scientific Citizenship," *European Journal of Social Theory* 6, no. 2 (2003): 233–251.

Fan, M.-F. "Public Perceptions and the Nuclear Waste Repository on Orchid Island, Taiwan," *Public Understanding of Science* 18, no. 2 (2009): 167–176.

Greenpeace. "Fukushima Must Be a Lesson to Korea's Nuclear Industry— Greenpeace Report," April 4, 2012, accessed May 5, 2015. http://goo.gl/Wquqo1.

Hagendijk, R., and A. Irwin. "Public Deliberation and Governance: Engaging with Science and Technology in Contemporary Europe," *Minerva* 44, no. 2 (2006): 167–184.

Hindmarsh, R. "Megatechnology, Siting, Place and Participation." In *Nuclear Disaster at Fukushima Daiichi: Social, Political and Environmental Issues*, edited by R. Hindmarsh, 57–77. New York: Routledge, 2013a.

Hindmarsh, R. (ed.). *Nuclear Disaster at Fukushima Daiichi: Social, Political and Environmental Issues*. New York: Routledge Studies in Science, Technology and Society, 2013b.

Hindmarsh, R. "Hot Air Ablowin! 'Media-speak', Social Conflict, and the Australian 'Decoupled' Wind Farm Controversy," *Social Studies of Science* 44 (2014): 194–217.

Irwin, A. "Constructing the Scientific Citizen: Science and Democracy in the Biosciences," *Public Understanding of Science* 10, no. 1 (2001): 1–18.

Irwin, A. "The Politics of Talk: Coming to Terms with the 'New' Scientific Governance," *Social Studies of Science* 36, no. 2 (2006): 299–320.

Irwin, A., and B. Wynne. *Misunderstanding Science? The Public Reconstruction of Science and Technology*. Cambridge: Cambridge University Press, 1996.

Irwin, A., and M. Michael. *Science, Social Theory and Public Knowledge*. Maidenhead: Open University Press, 2003.

Jasanoff, S. "Breaking the Waves in Science Studies: Comment on H.M. Collins and Robert Evans, 'The Third Wave of Science Studies'," *Social Studies of Science* 33, no. 3 (2003): 389–400.

Jasanoff, S. *Designs on Nature: Science and Democracy in Europe and the United States*. Princeton, NJ: Princeton University Press, 2005.

Jasanoff, S. "Technologies of Humility: Citizen Participation in Governing Science," *Minerva* 41, no. 3 (2007): 223–244.

Jasanoff, S., and S. Kim. "Containing the Atom: Sociotechnical Imaginaries and Nuclear Power in the United States and South Korea," *Minerva* 47 (2009): 119–146.

Kim, M.-J. *Won-ja-ryeok Dilemma: Won-ja-ryeok Renaissance-eu Mi-rae [The nuclear dilemma]*. Seoul: Science Books, 2011.

Kim, S.-H. *The Politics of Democratization in Korea: The Role of Civil Society*. Pittsburgh: University of Pittsburgh Press, 2000.

Kim, S. "Public Trust in Government in Japan and South Korea: Does the Rise of Critical Citizens Matter?" *Public Administration Review* 70, no. 5 (2010): 801–810.

Kim, Y.-K. "Saeng-tae/Hwan-kyeong Un-dong-gwa Eon-ron, 1987–2002: Ban-haeg-un-dong-eul Jung-shim-eu-ro, ['Ecological/environmental social movements and the press (1987–2002): Framing anti-nuclear movements']." *Journalism Science Research* 3, no. 3 (2003): 51–94.

Lee, H.-M. "Haeg-pye-gi-jang Chu-jin Jeong-chaek-eu Mun-je-jeom, Ji-yoeg-eu Sa-rye Yoen-gu ['Nuclear waste treatment plant siting process: A case study of the Buan']." *Democratic Society and Policy Research* 10 (2006): 78–102.

Lee, P.-L. *Energy Dae-an-eul Chaj-a-seo [In search of energy alternatives].* Changbi: Seoul, 1999.

Lim, J.-H., and S.-Y. Tang. "Democratization and Environmental Policy-making in Korea," *Governance* 15, no. 4 (2002): 561–582.

Park, C.-T. "The Experience of Nuclear Power Development in the Republic of Korea: Growth and Future Challenges," *Energy Policy* 20, no. 8 (1992): 721–734.

Rowe, G., and L. Frewer. "A Typology of Public Engagement Mechanisms," *Science, Technology and Human Values* 30, no. 2 (2005): 251–290.

Sclove, R. *Reinventing Technology Assessment: A 21st Century Model.* Washington, DC: Science and Technology Innovation Program, 2010.

Valentine, S., and B. Sovacool. "The Socio-political Economy of Nuclear Power Development in Japan and South Korea," *Energy Policy* 38 (2010): 7971–7979.

Wynne, B. "Public Engagement as a Means of Restoring Public Trust in Science: Hitting the Notes, but Missing the Music?" *Public Health Genomics* 9, no. 3 (2006): 211–220.

Yun, S.-J. "Ban-haeg-eul Neom-eu Sange-tae Jeon-hwan-gwa Dae-an Energy-ro ['Beyond anti-nuclear movements and toward the ecological turn and alternative energy']" *Environment and Life* 46 (2005): 206–221.

Yun, S.-J. "Looking at the Selection Process of Low and Medium Level Radioactive Waste Disposal from an Environmental Justice Perspective," *ECO* 10, no. 1 (2006): 7–42.

# 5 China's Civil Nuclear Power Development

## Shifts from Government to Risk Governance?

*Xiang Fang*

Over the last three decades, China's civil nuclear industry appears to be gradually decoupling from the military sector and government ministries. This suggests an institutional transformation that signals that civil nuclear power development could be shifting away from the governmental domain of key policy-makers. In its place, a multi-actor risk governance model is emerging that involves, at the top level, experts, policy makers, and authorities in nuclear power companies and, at the bottom level, pressure from local political elites, the media, a range of professional, business and NGO stakeholders, and the general public.

It seems the Fukushima nuclear disaster has sped up this apparent (technology risk) governance process. After the accident, local people in China became more active in participating in anti-nuclear events and pressuring for change, as reported widely in the media. When the Fukushima accident happened, many Chinese people heard about it from all kinds of media, including newspaper reports, and the Internet and TV news. The Chinese media highlighted how incompetent the Japanese government was in risk management post-disaster. News about the possibility of radiation dust from Japan moving to China spread most quickly through Internet media platforms. This kind of news helped shape people's risk awareness of nuclear power projects. Indeed, it appears the accident had a wide effect on reshaping people's perceptions in China of the risks and hazards of nuclear power. To pressure development agendas for nuclear power, local people have turned especially to social media and NGOs, as well as local political elites who represent policy actors *inside* the political system.

The research aim of this chapter is to ask: to what extent did Fukushima speed up policy change and enable a potentially new risk governance approach in China? Four case studies—two pre-Fukushima and two post-Fukushima—are investigated to address this question. I first introduce my conceptual framework and method of investigation; then provide a background of China's civic nuclear power development before Fukushima, including institutional transformation and public participation activities; then investigate the Fukushima effect; finally, I discuss the overall effect of Fukushima, before concluding.

## Conceptual Framework and Investigative Method

Transformation from government to risk governance models in policy decision making is widely seen in many countries as the most practical and effective way to conduct policy development. In particular, it is a way to resolve conflicts around the development of controversial science and technology, including genetically modified products, nuclear power, and nanotechnology (e.g. Beck 1992; Renn 2008). "Governance" is a process that stresses the cooperation of state and civil society, from "top-down" to "bottom-up" in the policy decision-making process, and informs a more equitable distribution of social power (Pierre and Peters 2000). Tait and Lyall (2005) advanced that two levels of transformation occur in shifts from government to governance. First, at the governmental and institutional level, cooperation replaces expert hierarchy. Second, at the public level, non-governmental actors engage more in decision-making processes, which results in more meaningful influence in policy settings (Lyall and Tait 2005).

In regard to controversial technology, risk governance addresses issues of technological risk and hazard in both technoscientific and social-cultural areas, concerning multiple interdisciplinary risk research. The purpose of risk governance is to offer a framework to better understand and manage "an interplay between governmental institutions, economic forces and civil society actors, such as non-governmental organizations (NGOs)," to produce more legitimate and effective policy outcomes (Renn 2008: 8).

Mol and Carter's (2006) research about China's environmental governance offers insightful points in understanding new governance relationships between the state, the market, and civil society. To probe the question of how China deals with environmental threats and risks faced by the country now and in the near future, these authors grouped their analysis into four categories: (i) political transitions, (ii) the role of economic actors and market dynamics, (iii) emerging institutions beyond the state and the market, and (iv) processes of international policy integration (Mol and Carter 2006: 155).

Based on my research on China's civil nuclear industry and social construction of nuclear risk in Chinese society, I suggest the analysis of risk governance of China's civil nuclear power industry would most appropriately involve two categories: (i) institutional transformations from the military and planned economic systems to experts, policy-makers, and authorities in nuclear power companies; accompanied by (ii) social actors' participation and engagement in policy decision-making processes on nuclear development risk and hazard.

In the following section, on the background of China's civil nuclear power development, the first part focuses on institutional transformations, and involves a number of institutional actors and processes. In the second part, on public participation activities pre-Fukushima, two case studies, Han River (Guangdong province) and Silver Beach (Shandong province), are discussed in relation to social actors' participation on nuclear power issues. The next part, on participation activities after the Fukushima accident, discusses the two case studies of Wangjiang County (Anhui province) and Jiangmen City (Guangdong province), to assess to what extent the Fukushima accident affected social actors' participation on nuclear power issues. While this cannot be representative across China, it provides some illustration of what is occurring in this still-early period after Fukushima.

To conduct the case study investigation, interviews (2007–2013) and documentary research were used as data collection methods. Interviewees included staff and experts working for the nuclear industry, and People's Deputies who participated in the inquiry meeting of the inland nuclear power project proposed at Han River in 2007. Documentary research for analysis included published documents about China's nuclear power development process, such as the "Li Peng Nuclear Diary" (Li 2004), official letters between People's Deputies and provincial government departments, and information and coverage about local people's and stakeholders' engagement in nuclear power issues, as found in the media and on Internet webpages and forums.

## Background of China's Civil Nuclear Power Development

Nuclear power has been defined by China's central government as a reliable energy source to be actively developed in the future, as it produces very little carbon dioxide and has high energy efficiency. Currently most of China's electricity is produced from fossil fuels, predominantly from coal, which has caused significant air pollution problems. At the time of the Fukushima accident, China had 13 operating reactors in six different nuclear power stations. There are now 26 reactors located at eight different nuclear power stations with a further 25 under construction, and additional reactors planned to increase capacity threefold by 2020.[1] Despite China's ambitious nuclear power development plans, in 2015 nuclear power comprised only 4% of the country's electricity supply, even with 17 reactors coming online in 2015; this is far less generating capacity than the US, France, and other developed countries (as discussed in many chapters in this volume).

The Chinese nuclear industry faced some resistance to nuclear power before the Fukushima disaster. After Fukushima, the issue was widely reported by the media. As many Chinese people witnessed the terrible results of the nuclear accident they started to pay more attention to the safety issues and potential risks of nuclear power projects (Fang 2012b), and resistance increased.

The accident also prompted the Chinese government to immediately conduct safety checks of all operating plants; review those under construction (completed in October 2011); ban construction of any inland nuclear plants until 2015, citing higher risks in landlocked areas; and to suspend approval processes for new nuclear power plants until a new nuclear safety plan was developed (Li et al. 2011).[2] The original plan put nuclear power projects on hold for a year after Fukushima, but the State Development and Reform Administration only recently reached agreement with nuclear power companies about when to restart the suspended nuclear power projects.[3] In the next section, I begin to explore whether institutional transformations in China's civil nuclear power industry have reflected any shift in risk governance.

## Institutional Transformation

Institutional transformation is seen as the initial stage in the shift in China's civil nuclear power industry from government to governance. From the early 1970s until the end of the 1990s, decision making for nuclear power projects was strictly a top-down process in China. Mol and Carter described the early shift of China's environmental protection system: "Economic transformation towards a market-oriented growth model, decentralisation dynamics, growing openness to and integration in the outside world, and bureaucratic reorganisation process have shifted China's environment governance model away from those common to centrally planned economies" (Mol and Carter 2006: 51–52).

The government and nuclear industry started to implement institutional transformations in the 1980s, which I look at in the following three sub-sections. Since then, the civil nuclear industry has been gradually detached from the military sector and government agencies and has been moving toward cooperation, involving experts, policy-makers, and authorities in nuclear power companies, and a market-oriented style, as well as selectively hearing bottom-up actors.

### The China National Nuclear Corporation

Currently, three state-owned nuclear power corporations hold permits to develop and operate nuclear power plants inside China. They are the China National Nuclear Corporation (CNNC), the China Guangdong Nuclear Power Group (CGNPG), and the China Power Investment Corporation (CPIC). Previously, the CNNC and the CGNPG shared responsibility for developing nuclear power. Then, in 2004, the CPIC was established to also develop nuclear power.

The CNNC, as the largest nuclear power corporation in China, has already seen two stages of institutional transformation. The first stage involved the Second Ministry of Machine Building (SMMB), which was formed in 1958 to develop "nuclear weapons, the nuclear submarine propulsion plant, and all associated industries. The Second Ministry controlled the nuclear industry from prospecting, mining, and processing uranium; processing fuel; constructing nuclear facilities; through to developing and producing all instruments and control (I&C) equipment" (Cappellano-Sarver 2007: 116).

In 1982, the SMMB changed its name to the Ministry of Nuclear Industry (MNI). The MNI organized and owned China's first domestically designed nuclear power station—the Qinshan Nuclear Power Station Phase 1—as a transitional institution in the 1980s. China's nuclear power industry began transferring from military projects to big commercial civil nuclear power projects according to the plan "Retain Military, Transfer to Civilian."

However, the MNI did not receive the same resources and economic support as the military. The central government still offered the best labor resources, as well as funding research and development of new military technology (Li 2004). In 1984, the MNI applied to the State Council for ownership of the Daya Bay nuclear power project—located in Daya Bay in Shenzhen, Guangdong, north east of Hong Kong—but its bid was unsuccessful. The Daya Bay project was owned by the Ministry of Water and Electricity (MWE) before 1990 and afterwards by the China Guangdong Nuclear Power Group.

The necessary technology and equipment (reactors, turbines, and other artefacts and materials) for Daya Bay were imported from France and the UK. With no chance to participate in the design and manufacture of this big project, the 1980s was the most difficult decade for the Ministry of Nuclear Industry (MNI). It lost its leading position in the nuclear industry, a position only achieved during the SMMB period. After the Qinshan Phase 1 Nuclear Power Project, the central government decided that the reactor style of Qinshan 1 was out of date and that the country would no longer build that kind of reactor. The technology for building Qinshan 1 (which was on hold by the MNI) thus lost most of its value.

The second stage of transformation, in 1988, saw the MNI reorganized into the China National Nuclear Corporation (CNNC). However, military programs were still quite evident, as Cappellano-Sarver (2007: 116) noted: "Like the Second Ministry, the CNNC consists of over one hundred subsidiary companies and institutions and still controls the vast majority of the civilian and military nuclear programs." However, according to the CNNC website, the corporation remains

> the main body of the national nuclear technology industry, the core of the national strategic nuclear deterrence and the main force of the national nuclear power development and nuclear power construction, and shoulders the duel historical responsibilities for building the national defence force, [and] increasing the value of state assets and developing society.[4]

The CNNC is thus in charge of both military and civil nuclear technological development, in building and running nuclear power stations, and in economic cooperation with foreign countries and the nuclear import-export business. It is the owner of all nuclear power stations and largely responsible for operating and constructing them. Such developments indicate only a partial transformation of the CNNC from a centrally planned to a public corporation style of management and governance. For the China National Nuclear Corporation to move to a more market-style corporation would be quite a radical shift.

## The China Guangdong Nuclear Power Group

The process of establishing the China Guangdong Nuclear Power Group (CGHPG) reflected the ambivalence of the Chinese government in building a market-oriented modern nuclear power corporation. On one hand, the government wanted to learn from the progressive corporation management style of western countries, to make its business corporation capable of joining the international nuclear industry. On the other hand, the government did not want to lose control of the corporation. The CGNPG was established in September 1994 with a registered capital of RMB 10.2 billion yuan. The China National Nuclear Corporation owns 45% of the shares, the Guangdong provincial government 45%, and the Ministry of Power Industry 10%.

In comparison with its precursor, the Guangdong Nuclear Joint Venture Company (GNJVC), the China Guangdong Nuclear Power Group transformed from a Sino-foreign joint-venture joint-stock company to a state-owned joint-stock corporation. The GNJVC was founded in 1985 as the owner of the Daya Bay Nuclear Power Station, in a joint-venture enterprise with the Hong Kong Chinese Electricity Investment Corporation and the China Guangdong Nuclear Power Investment Company (CGNPI). The Hong Kong Chinese Electricity Investment Corporation owned 25% of its shares and the CGNPI owned 75%. Of the electricity produced by the Daya Bay Nuclear Power Station, 70% was sold to Hong Kong to pay back its commercial loan.

The Chinese government also contributed financially to building the power station. The CGNPI gained the commercial loan from the Bank of China to invest in the Guangdong Nuclear Joint Venture Company. After the latter made money from selling the Daya Bay's electricity, it then started to invest in the Lingao Nuclear Power Project in 1994. When the China Guangdong Nuclear Power Group (GGNPG) was founded in the same year, the GNJVC was integrated into it. The CGNPG then learnt how to run a modern corporation informed by the market-oriented approach. However, after the Daya Bay project was established, the CGNPG came under the control of the CNNC and Chinese government. Nevertheless, the CGNPG enjoys a better reputation than the China National Nuclear

Corporation and the China Power Investment Corporation within the broader civil nuclear industry, as it values the market more than political ideology (Lu 2009).

## The China Atomic Energy Authority and National Nuclear Safety Administration

On October 11, 1983, the International Atomic Energy Authority (IAEA) decided to admit China as a member to its 27th conference. The China Atomic Energy Authority (CAEA) was founded immediately afterwards as an internationally recognized institution for the purpose of best connecting China with the IAEA. Nevertheless, the CAEA still worked for military purposes. For its military purposes, the CAEA has another name, the National Security Technology and Industry Commission (NSTIC).

The CAEA and NSTIC thus share the same staff and work together on both civil and military nuclear power issues.[5] According to the description on the CAEA website, the agency is in charge of creating decision-making policies about, and regulating the peaceful use of, nuclear power. It also cooperates with governmental bodies and international organizations and provides regulations for China's export missiles and missile-related items and technologies, as well as emergency nuclear accident matters. It also participates in and organizes international nuclear conferences, and publishes non-proliferation policy and measures.

Yet another agency, the National Nuclear Safety Administration (NNSA), oversees the safety of the country's nuclear power development, under the supervision of the State Council. It performs independent reviews and surveys nuclear safety for civilian nuclear installations. The NNSA was founded in 1985 as a subordinate department of the Ministry of Science and Technology until 1998, when it merged with the Station Environmental Protection Administration (SEPA) and became a primary department. A number of other agency restructurings around safety occurred in 2002–2003. The National Nuclear Safety Administration is the externally used name for licensing and issuing other regulatory documents.

In sum, this institutional transformation of the civil nuclear industry shows how some parts of the nuclear power industry have begun to detach from the military nuclear industry. It represents an important step in helping the country's civil nuclear power to organize its own system. It is also the foundational step for potential civil nuclear technology governance in the future, in moving to more of a market-style approach coupled to risk governance, with the latter especially emergent after Fukushima. That said the next section discusses how things were in the risk governance area pre-Fukushima, particularly in regard to citizen pressure and participation around nuclear power development and siting.

## Public Participation Activities pre-Fukushima

Until the late 1980s, the public rarely participated in activities related to nuclear issues and risk. The first reported public anti-nuclear activity took place in Hong Kong in 1986, when Hong Kong was still a British colony. The anti-Daya Bay nuclear power project activity (the anti-Daya movement) was similar to anti-nuclear power activities in western countries. It was organized by environmental NGOs and widely publicized in the media. Concerns were expressed about the potential health and environmental risks posed by the Daya Bay project.

In mainland China, before the Fukushima accident, two notable movements contesting nuclear power projects occurred in 2006. The first was initiated from inside the political system. The movement was provoked by news that the first inland nuclear power project might be built alongside the Han River in Guangdong province. People's Deputies in the lower reaches of the Han River were concerned that the nuclear power project would jeopardize the safety of people's drinking water resources. The second instance of public participation in anti-nuclear activities was organized and spread on the Internet, concerning a nuclear power project which might potentially pollute Silver Beach in Shandong province.

### Case Study 1: The Han River, Guangdong Province

When a People's Deputy (respondent SC) from Shantou city (on the eastern coast of Guangdong province immediately north of Hong Kong) first heard that the Guangdong government planned to select sites for an inland nuclear power station in the upper reaches of the Han River, the Deputy realized that the project might cause (radioactive) risk to people's drinking water resources downstream. At the same time, municipal government officers of nearby Chaozhou city read the same news about the nuclear power project in the *Meizhou Daily*. The officers thought it better for People's Deputies to raise the issue at the Provincial People's Congress.[6,7] People's Deputies were considered the "proper" people to raise the issue with the provincial government, as they had political power to raise issues concerning people's daily life and environmental problems (Fang 2012a). Local stakeholders had raised questions based on their local knowledge and life experience of the Han River, but had not been heard by scientists and experts when they conducted the applicability selection test of potential nuclear power sites. When the Provincial People's Congress is convened, People's Deputies have the opportunity to make their voices heard. In addition, the Shantou and Chaozhou municipal governments did not want to appear to confront the provincial government directly on an anti-nuclear activity. Therefore, the Deputies were better placed to raise the issue of potential pollution of drinking water resources and the environment of the Han River basin, which also reflected the people's growing awareness of and involvement in environmental issues.

Subsequently, Deputies from Chaozhou and Shantou cities cooperated to collect information on the inland nuclear power project and its potential harm to drinking water and the environment, and to raise this issue at the Provincial People's Congress meeting on February 4, 2007. At the meeting, the eight Deputies expressed several concerns about the posed risks. The provincial government officials and experts from the China Guangdong Nuclear Power Group answered the People's Deputies' questions and concerns. However, their replies did not satisfy the Deputies, who argued that they were superficial, lacking research data and risk assessment. Nevertheless, government official Li intervened and announced that development of nuclear power was the province's policy. Expert Ma also attempted to convince the audience of the safety of the nuclear power project, but was not seen by the Deputies to present any reliable proof. Arguably, the aim of these officials appeared to be one of marginalizing the Deputies' concerns (Fang 2012a).

Following the meeting, the Deputies summarized their concerns in an internal letter to Zhang Dejiang, the province's prime minister at that time. Their concerns included potential leaking of "heat pollution" (from radioactive materials) into the ecological system of the Han River; potential pollution from the transportation of construction materials; and the social influence of the project on the investment environment of Shantou city and Chaozhou city. The Development and Reform Commission of Guangdong Province (DRCGP) replied to this letter on April 12, 2007. This reply did not satisfy the People's Deputies, and so they sent another letter to the DRCGP. On June 25, 2007, the DRCGP responded by declaring that the potential nuclear power project would be postponed and that nuclear power projects would be restricted to the coastline until 2015.

### Case Study 2: Silver Beach, Shandong Province

In 2006, Shandong's State Development and Reform Administration announced that the Hongyanhe Nuclear Power Station, with four 1000 MW nuclear reactors, was to be built only 4–5 kilometers (or three miles) from Silver Beach, a national travel resort in Rushan city, in the southeast of Shandong province (a coastal province in the East China region). Subsequently, the Silver Beach anti-nuclear Internet campaign began, organized by property owners of retirement and holiday homes who wanted to protect the environment surrounding their homes.

The Silver Beach case, the Han River case, and many other environmental protection activities in China are thus quite similar. Protesters were concerned with protecting their lifestyles and the local environment that supports that lifestyle (Jiang 2003). People involved in the Silver

Beach activity were not permanent residents of Silver Beach, but they had a certain social status that gave them the networks and financial resources to organize and take action. Most had a "middle-class" social status; they were professional people who could both afford a second property for holidays and retirement and finance campaigns to protect the local environment in which their retirement and holiday homes were situated or planned.

Their Internet anti-nuclear movement began in March 2006. Several stakeholders, who had purchased retirement flats in the Silver Beach area, had heard about the planned nuclear power project. These people called themselves "netizens" and began posting articles on the Shandong Property webpage *Shan Fang Wang*. As one organizer, "Jingming," stated: "Silver Beach No Nuclear Power is a self-organizing mass movement. There is no formal organization style. Netizens are spread out everywhere. Most of them have never met each other." From March 2006, netizens started to circulate anti-nuclear power information on the Xinghua Net. The click-through rate of these anti-nuclear power articles was over 10,000.

In April 2006, these netizens drafted a letter to Wen Jiabao (the present premier), mobilized other netizens to sign the letter, and then sent it to Beijing. From that point on, they worked with the broader anti-nuclear Internet movement to resist the Silver Beach project. Their articles and Internet activities drew attention from the local government and from newspaper journalists. On May 25, 2006, for example, The *No.1 Finance* and *Economic Daily* reported news about people's concerns about the Silver Beach project.

However, a month later, in June 2006, the entire Internet forum about Silver Beach in *Shan Fang Wang* was abruptly removed by the website administration. The netizens lost their network for the anti-nuclear movement without any notice. In response, an environmental NGO called "Ocean Environmental Protection Commune" invited the Silver Beach netizens to post anti-nuclear information on its online forum. Throughout this local anti-nuclear power movement, netizen activities were not limited to one online forum, but were found on many. In addition, the Silver Beach netizens wrote letters to the State Environment Protection Administration (SEPA) and to the State Development and Reform Commission (SDRC); organized two "on-beach" anti-nuclear propaganda activities called "Silver Beach No Nuclear," on May 1 and July 27 of 2007; and sent "Silver Beach No Nuclear" promotion materials to provincial and municipal government officials.

The local government, however, welcomed the nuclear power project, even going so far as to organize an exhibition to positively "educate" the public about nuclear technology. They hoped to win over Silver Beach stakeholders. They did not succeed. Netizens were not convinced by the exhibition. The municipal and provincial governments at last realized that the netizens' anti-nuclear movement should not be ignored. On May 28, 2007, the Rushan municipal government announced a public review of the Silver Beach nuclear project on its official website.

On June 12, 2007, the netizens collected 628 copies of these public review forms and handed them to the Silver Beach nuclear power planning office, but they did not receive any feedback from the municipal government. In November 2007, the Civil Nuclear Power Mid-Long Term Development Schedule (2005–2020) was published. In this schedule, the Silver Beach project was not considered an ongoing project. The netizens published and discussed this news in their online forum. Their Internet movement had successfully campaigned against a nuclear power project. But the Civil Nuclear Power Mid-Long Term Development Schedule (2005–2020) declared only that there would not be any nuclear power project built on Silver Beach before 2020, which was more a temporary injunction than a permanent one. Moreover, it was not known whether netizens resisting or other reasons were responsible for suspending the project, as the netizens never received any official reply about their concerns.

As part of my research, I monitored the development of the Silver Beach anti-nuclear Internet movement from 2007. In summary, the movement represented bottom-up, local stakeholder pressure that aimed to influence the policy decision-making process, along with social education to raise public knowledge. During the online anti-nuclear activities, a large amount of information was circulated about nuclear power and its potential to harm the environment and human beings, including accidents that have happened in the history of nuclear power development (e.g. Hindmarsh 2013: 2). Even netizens doing ecological research in the US posted articles on the forum, citing expert knowledge of nuclear radiation and risk. The activity was also successful in drawing attention from the local government which, along with nuclear power companies, sent nuclear experts to communicate with netizens both online and offline in order to bridge the division between local perspectives and expertise on nuclear power. But after these experts communicated with netizens, they became aware that nuclear power was a social as well as a technological issue. Unfortunately, the way in which local government and nuclear power companies dealt with the public review results did not satisfy the netizens, as, again, the results were never published.

The Silver Beach case, however, as a self-organizing Internet anti-nuclear social movement made up of netizens, illustrated well the status of negotiations and interactions occurring during the policy-making process of a potential nuclear power project pre-Fukushima. Citizen stakeholders did find a way to make their voices heard by policy-makers, but in comparison to the power of insider political representatives such as those in the Han River dispute, had little direct meaningful influence on policy outcomes, except for the suspension of a nuclear power plant for some 15 years, which in the short-term might be considered a win. Nevertheless, just having their voices heard was an important step in the risk governance of civil nuclear power in China.

In sum, before Fukushima two things occurred to motivate shifts from government to risk governance. The first was at the central government level, in making nuclear power development more market-oriented rather than solely politically oriented. The second was local political elites and stakeholders aware of the risks from nuclear power projects beginning to wage campaigns to voice their opinions and participate in the decision-making process about the health and environmental issues of nuclear power plant siting.

## Investigating the Fukushima Effect

The Fukushima accident significantly increased people's attention to the risk of nuclear power projects (Fang 2012b; He et al. 2014); motivating them to further question the safety of nuclear power projects. Two notable case studies of post-Fukushima public participation activities in relation to such questioning in suggesting shifts to risk governance are presented in this section to examine the Fukushima effect on public participation and decision-making in regard to civil nuclear power.

### Case Study 3: Wangjiang County, Anhui Province

After the Fukushima accident, the first participation activity was in Wangjiang County, Anhui province, part of the East China region. Wangjiang County, lies across the basins of the Yangtze River and the Huai River, and borders Jiangsu to the east. At the end of 2011, a retired official at the local court (interview respondent AF) learnt of an inland nuclear power station proposed for Pengzhe County, which is next to Wangjiang County. Anxious about adverse risks the nuclear power project might bring to Wangjiang County, the official contacted other retired local government officials to discuss the issue. They collected relevant data from the Internet, visited the location of the potential site, and informally interviewed local people about their opinions of the project (Yue 2014).

After reviewing their data, the ex-officials submitted a request letter to the county government, outlining four issues they considered important in relation to the proposed siting. First, population data for the area near the project was inaccurate and was higher than estimated. Second, the area was vulnerable to seismic activity; building a plant in the area would pose a risk to local people. Third, the project site was very close to an area of concentrated industry, which they considered hazardous in case of an accident. Fourth, a public consultation survey about the proposed siting had not been conducted according to the stated process.

After reading the request letter, in February 2012, the local government of Wangjiang County submitted an official document to the provincial government, raising these issues, and asked the upper level of government to address them. The local government document specifically pointed out the problem of the earthquake zone:

> The nuclear power site will be located at the "Jiujiang–Jingan" fault zone. It is an active earthquake zone. Within the last 10 years there were five earthquakes of magnitudes between 3.2 and 5.7. However, the report provided by the nuclear power company claims that the potential site is situated in a low magnitude earthquake zone.

In issuing this official document, the local government hoped the Anhui provincial government would raise these problems with the central government, and hence influence them to cancel the project. This action thus also represented a bottom-up public participation activity, but one that involved retired government officials who knew how to get the issue onto the policy agenda inside the political system. Because of this local action, and also according to the western media source *The Economist* (2013),[8] so far, the first proposed inland nuclear power station has been delayed (also Yue 2014). However, in September 2104, *Nuclear Monitor* suggested the local citizen and government resistance may still see it stopped, but that might involve existing investment costs being recovered from local taxpayers and the provincial government, which could become a costly burden.[9]

## Case Study 4: Jiangmen City, Guangdong Province

More recently, in July 2013, a mass public participation activity occurred in Jiangmen city, Guangdong province, south China. The city was selected as a potential site by the China National Nuclear Corporation (CNNC) for a nuclear fuel processing plant, with construction due to start at the end of 2013. On July 4, 2013, the Jiangmen City Council announced this proposed project on its website and gave citizens 10 days to express their views about it. Local people who live in and around Jiangmen were strongly against this project, and mobilized with the help of activists who used the Internet as a network to organize public resistance.

On July 12, 2013, an estimated 10,000 local people mobilized to hold a street demonstration that ended at the gate of the city council;[10] one banner included the slogans "OPPOSE nuclear pollution" and "Give us back our green homeland" (The Economist 2013).[11] *The Economist* (2013: 1) stated: "The protest was the first known major public rally against a project involving the nuclear-power industry since China began building nuclear plants in the mid-1980s."

Local people were concerned that the municipal government did not provide any information about the project before making the announcement and that 10 days was insufficient for citizens to evaluate the project.

Local people argued that the municipal government was attempting to avoid their participation. The demonstration was subsequently widely reported by the mainstream media in Guangdong and Hong Kong (World Bulletin 2014).[12]

In an unprecedented response, on the same day as the protest, both the Jiangmen municipal government and the CNNC announced that the US$6 billion project was scrapped. The Jiangmen municipal government pointed out that they cancelled the project because of the people's concerns. This event provoked wide discussion on the Internet. Some nuclear experts, however, argued that the project was not a nuclear power station and thus not as important: it was only a fuel processing plant. They further argued that problems with radioactive leakage and terrorist bombing, which people worried might happen with nuclear power stations, would not happen with a fuel processing plant. One commentator pointed out that the Fukushima accident had made people feel anxious about the safety risk of nuclear power projects (Li 2013). In the opinion of *The Economist* (2013), Fukushima had "changed the public mood," with the role of social media significant in that communication (also Yue 2014).

## The Overall Effect of Fukushima

To reiterate, my research question was to investigate to what extent Fukushima sped up strategies of policy pressure and the enablement of a new risk governance approach in China. In reviewing the four cases studies in China pre- and post-Fukushima, I found three important points that invite further discussion.

First, local political elites, including People's Deputies and local government officials, have informally become positioned inside the political system as a somewhat representative group of local communities opposing nuclear power. These elites are enabled by the political power associated with their positions to express their views to upper levels of government. This is notable in China, as environmental protection NGOs are found to be weak in their ability to pressure government over nuclear power issues (Heet et al. 2014). In contrast, these elites were enabled by their stronger political position to work for local people's interests in critically questioning and resisting nuclear power projects proposed to be built at nearby locations.

In comparison to everyday citizens, political elites have better access to information and upper levels of political leadership to voice their concerns. However, local political elites' participation on nuclear power issues reflects self-interest as well as the social and environmental risk that a nuclear power project might bring to their area. They do not want the nuclear power project to be located anywhere in the county or places near their administration area because of these risks, but they also acknowledge the economic and political benefits from such facility siting.

Prior to Fukushima, in the Han River case, People's Deputies were selected by the municipal government to express concerns about the nuclear project. They used environmental/ecological problems as the main ways to voice these concerns. As People's Deputy Chen pointed out: "We can't use the dangers of nuclear power technology to make our claims. This topic is too sensitive. We only use water pollution to express our concerns . . . issues that [are] relative to people's daily lives" [Interview: August 7, 2007].

Post-Fukushima, matters changed. In the Wangjiang case, local government directly submitted an official document to the provincial government to question the proposal for the nuclear power project. They did not try to avoid the sensitive topic of nuclear power risks. They did not just make an inquiry and show concerns; they pointed out problems directly and asked for the project to be stopped.

Second, among the four cases studied here, only the post-Fukushima Jiangmen city case involved mass participation of everyday citizens. In the other three cases, local people who lived around or very close to the selected location of the nuclear power project remained passive. For example, although the Silver Beach case (pre-Fukushima case study) was organized by citizens, these people were not local residents. Instead, the activists were mostly largely absent property owners who holidayed there or planned to retire to the area. They were concerned about the local environment and had resources to mobilize an Internet campaign to protect it from the proposed nuclear station. However, even with considerable campaign resources, they did not receive any direct response from government regarding their concerns.

The difference between the influence and activism of everyday local citizens pre- and post-Fukushima is an important point to address, as Chen (2007) highlighted of the pre-Fukushima situation:

> Most people are forced to keep silent about environmental hazards. Firstly, living is the most important thing. More people care more about their daily lives than about environmental hazards. Secondly, there is a professional barrier in making claims about environmental hazards. Local lay people lack the scientific knowledge to defend themselves. Thirdly, lay people lack the power to fight companies. Fourthly, lay people normally work individually and the government does not want to see [the emergence of] NGOs. When there is any problem the government is [instead] willing to act as a "father" which helps people work for their benefit.

Another significant reason that local people remained silent on the nuclear power issue before Fukushima was based on perceptions of their lack of agency (power to act) and dependence on the government (Fang 2014). Nuclear power projects are normally planned in economically less-developed rural coastal areas far from cities, as in Japan (Hindmarsh 2013). Politically weak local people in these areas face many kinds of development projects, such as dam and highway construction and controversial industrial facility siting, including waste dumps. They typically accept that they are unable to resist development projects, but can only work for better compensation. However, post-Fukushima, everyday citizens started to become aware of, and highly concerned about, the risks from nuclear power projects and became more active in critically questioning and resisting nuclear power development in close proximity to their residences.

Third, after Fukushima, both the Chinese government and nuclear power companies have become more careful about the development of new nuclear power projects. Instead of silence, they responded to local people's opinions about nuclear power projects in the two case studies investigated. This was especially the situation experienced in the Jiangmen city case, where the municipal government and the CNNC announced the project was cancelled. Nevertheless, it seems the nuclear power station planned for Wangjiang County (the first post-Fukushima case study) has been only delayed, similar to the case of the Han River project (the first pre-Fukushima case study).

In summary, of the four cases presented it appears that only one had any meaningful, long-lasting impact on decision-making at the government level. This was a case study regarding the nuclear fuel processing plant planned for Jiangmen city: a responsive decision to a mass public protest and its subsequent publicity across China. In other words, citizens acted against the usual top-down style of decision-making concerning nuclear power projects. Contestation that the citizens feel is legitimate has thus begun to emerge in various forms. However, it remains to be seen to what extent this "Fukushima effect" will play out over time and influence the government and nuclear power companies to change the typically closed policy style of developing nuclear power projects and waste repositories.

## Conclusion

This chapter set out to explore the extent to which the policy decision-making of China's civil nuclear industry has transformed from a government to a risk governance model, and to what extent a Fukushima effect contributed to that. Nuclear power has always been a sensitive topic in China. As such, Chinese scholars have mainly focused their attention on economic and technology development policy-making in regard to the civil nuclear industry. Research about the social policies of China's nuclear industry is limited both inside and outside China, but is now receiving more attention because of Fukushima. Emergent citizen attention is particularly significant for notions of risk governance with the national civil nuclear industry now in rapid expansion to help fuel China's economic growth and development.

When the Chinese government first embarked on the development of the civil nuclear industry, economic and development considerations were the first priority. It was also believed that nuclear technology was safe and mature and posed little risk of accidents and hazards. With economic growth becoming more a priority over the last three decades, the government has made some moves to detach civil nuclear power from the military sector and government agencies in order to embrace a market-style approach that is seen to best accommodate economic growth.

Nevertheless, the China National Nuclear Corporation, as a state-owned corporation, retains its planned economic characteristics. Although institutional changes have emerged to adjust its monopoly, such as the market-oriented Guangdong Nuclear Joint Venture Company and the Daya Bay Nuclear Power Station, the direction of further development is still to unfold. For example, it is still unknown as to whether nuclear power projects will be opened to other energy corporations (both Chinese and foreign ones), as is the case with wind power.

However, an unforeseen effect of the Fukushima disaster was the civic pressure now being put on China's rapid expansion of nuclear power. At the bottom-up citizen level of society, environmental awareness and negative perceptions of nuclear risk appear to be steadily growing in China, as elsewhere (as the many international case studies in this book illustrate). Visible anxiety about the irreversible harm that nuclear power projects might bring to the local environment is evident; the Jiangmen city experience made that quite clear. People have now openly begun to question the safety of nuclear power projects. So far, the government has made some acknowledgment of the problems but little it seems has changed to reassure society in addressing their concerns, even as difficulties for nuclear power development grow in local and regional areas. Meaningful risk governance remains elusive.

In other words, the effect of Fukushima has been as an emergent *turning point* for government at all levels to give more consideration to the perspectives of local citizens and communities as stakeholders in nuclear power development. My previous research indicated that people's anxiety about nuclear power risk and uncertainty has significantly affected their trust in government decision-making on nuclear issues (e.g. Fang 2012b). A more open and knowledge-partnering decision-making model (e.g. Eversole 2015), engaging multiple actors, is desirable to rebuild trust around controversial facility siting. Recent bottom-up participation on nuclear power issues from inside and outside the political system in China appears a highly useful contribution and motivation to aid the transformation of China's civil nuclear industry from a narrow focus on economics and technicalities to a more open risk governance approach, one that brings in the social infrastructure of development to better plan the siting of nuclear power and reduce social and environmental risks and hazards. While the adoption of safer Russian styled AES-2006 reactors perhaps offers one step,[13] this broader social step promises more, as flawed nuclear power development in Japan has made quite clear. The people's trust and support appears as another key to such transformations, and that probably is the key lesson and effect of the Fukushima disaster, although, and again, meaningful risk governance in China remains elusive.

## Notes

1 See http://goo.gl/xiDLDF, accessed April 23, 2015.
2 Also http://goo.gl/xiDLDF, accessed April 23, 2015.
3 See http://goo.gl/diHH1Q, accessed April 1, 2015.
4 See http://goo.gl/XQoXQr, accessed April 3, 2015.
5 Interview with BL in Beijing (September 2007).
6 Anonymous interviewees: At Chaozhou city with CH and with CW (August 2007).
7 On the People's National Congress and People's Deputies see: http://goo.gl/LOSYmR, accessed April 4, 2015.
8 See http://goo.gl/ybxbBs, accessed April 3, 2015.
9 See www.wiseinternational.org/node/4200, accessed April 24, 2015.
10 See "A Thousand People Marched to Resist Nuclear Power Project in Jiangmen," accessed April 3, 2015, http://goo.gl/7gwJC8.
11 See also http://goo.gl/yrXR7x, accessed March 8, 2015.
12 See also "Jaingmen Nuclear Fuel Project Has Been Cancelled after People's Resistance Activities," accessed 3 April 3, 2015. http://goo.gl/I6Pe1y. Also, "Over a Thousand People 'Walk' to Resist Nuclear Power," accessed April 3, 2015, http://goo.gl/Z6G7JK.
13 See http://goo.gl/5t39HL, accessed April 23, 2015.

## References

Beck, U. *Risk Society: Towards a New Modernity*. London: Sage, 1992.

Cappellano-Sarver, S. "Naval Implications of China's Nuclear Development." In *China's Future Nuclear Submarine Force*, edited by A. Erickson et al., 114–132. Annapolis, MD: Naval Institute Press, 2007.

Chen, A. "An Analysis of Stakeholders in Water Pollution Evens," Beijing International Conference on Environmental Sociology, Beijing. June 30–July 2, 2007, 197.

Eversole, R. *Knowledge Partnering for Community Development*. New York: Routledge, 2015.

Fang, X. "The Growing Power of Risk Construction and Bottom-up Politics in China's Civil Nuclear Development: Participation of Local People's Congress and Deputies," *Social Transformation in the Chinese Society* 8 (2012a): 37–66.

Fang, X. "Woguo dazhong zai heddian fazhan zhong de 'buxinren': jiyu liangge fenxi kuangjia de anli yanjiu ['Interpreting public "distrust" in China's nuclear power development: Based on the empirical research of the controversies of the potential inland nuclear power station in Guangdong']." *Kexue Yu Shehui* 4 (2012b): 63–78.

Fang, X. "Local People's Understanding of Risk from Civil Nuclear Power in the Chinese Context," *Public Understanding of Science* 23, no. 3 (2014): 283–298.

He, G., et al. "Nuclear Power in China after Fukushima: Understanding Public Knowledge, Attitudes and Trust," *Journal of Risk Research* 17, no. 4 (2014): 435–451.

Hindmarsh, R. (ed.). Nuclear *Disaster at Fukushima Daiichi: Social, Political and Environmental Issues.* Book Series in Studies on Science, Technology and Society, New York: Routledge, 2013.

Jing, J. "Environmental Protests in Rural China." In *Chinese Society, Change, Conflict and Resistance,* edited by P. Elizabeth and M. Selden, 197–214. New York: Routledge, 2003.

Li, C. "Jiangmen fanhe 'qirenxiutian ['Resisting activity in Jiangmen about the nuclear power project is not necessary']." *Guangcha,* July 14, 2013, accessed April 3, 2015, http://goo.gl/kewf90.

Li, P. *Li Peng hedian riji [Li Peng nuclear diary].* Beijing: Xinhua Chubanshe, 2004.

Li, W., Wang, Q., and H. Zhang. "Zhongguo hedian 'yuzhen ['Nuclear power shock in China']." *Caijing Magazine,* March 28, 2011: 62–65.

Lu, F. "Beifangzu de 'zhongguozhizao'-Pojie zhongguo hedian miju ['Made in China is in exile, to demystify the puzzle of China's nuclear power']." *Business Watch Magazine,* January 20, 2009: 29–53.

Lyall, C., and J. Tait. "Shifting Policy Debates and the Implications for Governance." In *New Modes of Governance: Developing and Integrated Policy Approach to Science, Technology, Risk and the Environment,* edited by L. Catherine and T. Joyce, 1–18. Surrey, UK: Ashgate, 2005.

Mol, A., and T. Neil. "China's Environmental Governance in Transition," *Environmental Politics* 15, no. 2 (2006): 149–170.

Pierre, J., and B. Peters. *Governance, Politics and the State.* Basingstoke: Macmillan, 2000.

Renn, O. *Risk Governance.* London: Earthscan, 2008.

Rong, J., and H. Lai. "Xiandai qiye zhidu zai Guangdong hedian de diansheng he Fazha ['The birth and development of a modern enterprise system in Guangdong nuclear power']," *Jingji Shehui Tizhi Bijiao* 1 (1999): 2–24.

World Bulletin. "China to Use Russian Nuclear Reactor as Prototype," *World Bulletin,* March 24, 2014, accessed April 3, 2015, http://goo.gl/4IcrEy.

Yue, S. "China's Nuclear Expansion Threatened by Public Unease," *Chinadialogue,* September 23, 2014, accessed March 8, 2015, http://goo.gl/5tXUNf.

Zhang, G. "*Jingyong he Miaobao* ['*Jingyong and Mingbao*']." Beijing: Hubei Renmin Chubanshe, 2007.

# 6 The Fukushima Effect on India's Science, Technology (Nuclear Energy), and Environmental Governance

*Anupam Jha*

Top-level governmental policy-makers traditionally drive technoscientific issues in India. In the wake of the 3/11 Fukushima Daiichi disaster, however, this approach—as it applies to nuclear energy development—has been challenged. In this context, this chapter investigates the scope and extent of a "Fukushima effect" on nuclear power governance in India in relation to social, policy, legal, and institutional aspects.

In particular, I investigate citizens' rising concerns about the safety of nuclear power plants in relation to their surrounding natural habitat, land acquisition policy, norms of environmental impact assessment, and the willingness of the government and nuclear power operators to pay adequate compensation in case of nuclear disaster. In addition, I examine the role of public participation in decision-making processes, and the ability and willingness of government to share adequate information in regard to the safety of people's lives, especially in relation to radioactive pollution—a notable feature of the Fukushima disaster (Hindmarsh 2013a).

My investigation focuses on issues of science, technology, and the environment in relation to democratic governance and its relevance in India in the case of a Fukushima effect. Informed by a policy in action approach (Hindmarsh 2013a: 13), I investigate both policy formation and implementation with respect to nuclear energy and its vulnerability in regard to possible situations of nuclear catastrophe, as occurred at Fukushima Daiichi, and any notable changes post-disaster.

My method is to first investigate *pre-Fukushima* governance relating to large-scale nuclear energy technology development. I focus specifically on how decisions were primarily made on technical criteria, which ignored local social perspectives on acceptability of nuclear siting, particularly in relation to personal safety and possible radiation. I then turn to the *post-Fukushima* situation in investigating the Fukushima effect on India's nuclear governance. First, I provide a background on nuclear energy development and governance in India, then discuss four effects of the Fukushima Daiichi disaster in regard to: (i) policy and legal frameworks;

(ii) livelihood issues of people displaced by land acquisition for nuclear power plants; (iii) environmental governance of nuclear technology; and (iv) safety, disaster management, and issues regarding transparency of decision-making and information, e.g. on the safety of nuclear power plants (NPPs).

## Development of Nuclear Energy in India

Nuclear energy development in India is as old as India's independence in 1947. At the time of independence, total power generation capacity (comprising thermal and hydro) was 1400 MW. Such low capacity to improve on presented a challenge for nuclear scientists, economic planners, and political leaders supportive of nuclear energy for economic development. Jawaharlal Nehru, the first Prime Minister of independent India, stressed the need to develop science and technology for solving national problems (Agarwal 1996).

Concomitantly, Indian nuclear policy became influenced by prevailing international realities of the Cold War, with India's immediate neighbors China and Pakistan aligning with either the USSR or the US. Against this background, India's nuclear policy took shape (Sisodia 1985). Nehru well understood the importance of electricity generation for the reconstruction and rehabilitation of a poor, developing country like India (Pathak 1980). He stressed the peaceful use of nuclear energy, particularly for domestic power and for industry, agriculture, energy, and medicine.

The first legislative action in the field of nuclear energy was the Atomic Energy Act of 1948. Nehru and Homi Jehangir Bhabha, a nuclear scientist, interacted closely to establish the key policy-making body of the Atomic Energy Commission (AEC) in 1948 (Parthasarathy 2008; also Anderson 2010). According to the First Five Year Plan (1951–1956), the AEC sponsored research into nuclear science and subjects relating to the production and deployment of atomic energy for economic and industrial purposes.

The AEC was placed in the Ministry of Natural Resources and Scientific Research but, in 1954, a policy turn boosted the atomic energy program with the establishment of the Department of Atomic Energy (DAE), directly under the Prime Minister (Morehouse 1971). The Department quickly moved to develop and unveil the first nuclear research reactor, called "APSARA," in August 1956, with the help of the US and the Atomic Energy Establishment Trombay (AEET), Mumbai. The AEET was established under the AEC in 1954 to consolidate research and development for nuclear reactors and technology. It was later renamed the Bhabha Atomic Research Centre in 1967. In 1958, Homi Jehangir Bhabha, in consultation with Nehru, had reconstituted the AEC to rapidly enhance nuclear energy development (Sharma 1991). By the early

1960s, one uranium metal plant and another heavy water reactor (called "CIRUS"), with assistance from Canada and the US, started production (Wisconsin 2010).

To facilitate nuclear energy expansion, the Union Parliament (India's supreme legislative body) passed the Atomic Energy Act of 1962 (AERB 2013). The Act empowered the Union government to develop, control, and use atomic energy in relation to the "welfare of the people of India" and for other peaceful purposes. The Union government also established "Indian Rare Earths Limited," responsible for commercial scale processing of monazite to obtain thorium, which is used as a nuclear fuel. In 1963, the Tarapur atomic power station, located in the western state of Maharashtra, started commercial production of electricity with two boiling water reactors and a capacity of 320 MW. In 1973, the Rajasthan Atomic Power Station—a pressurized heavy water atomic power station of 100 MW—was established near Kota in the northern state of Rajasthan.

At the same time, the increasing number of atomic power stations being set up led to the need for a permanent safety committee (Rosen 1977; Sharma 1996). Such need was reinforced in 1979 by the nuclear accident at Three Mile Island in the US (Rees 1994). Subsequently, in 1983, the Department of Atomic Energy established the Atomic Energy Regulatory Board to carry out regulatory and safety functions under the Atomic Energy Act of 1962. Functions included developing safety codes, guides, and standards for both radiation and industrial safety areas in the siting, design, construction, commissioning, operation, and decommissioning of nuclear power plants.

Such policy development was based upon evolving international safety standards and the special local requirements of a nuclear power plant, to factor in risks of high seismicity, prospects of flooding, and so forth. With the establishment of the Atomic Energy Regulatory Board, expansion of nuclear power plants proceeded. In 1987, the Nuclear Power Corporation of India Ltd (NPCIL) was established as a public sector enterprise under the AEC for the generation of nuclear power for electricity. The NPCIL was solely responsible for constructing and operating India's commercial nuclear power plants. With a program of steady development, by 2011, pre-Fukushima, the NPCIL had developed 20 nuclear reactors at seven locations, with an installed capacity of 4780 MW of electricity, which comprised 2% of total energy generated in India (NPCIL 2014).

However, along with development, and as in Japan before Fukushima (Hindmarsh 2013a), a number of minor nuclear accidents, machine failures, leakages, and flooding of nuclear power reactors occurred, which caused some human and environmental health impacts. For example, leakages of heavy water, helium gas, radiation, and tritium were reported at the Madras Atomic Power Station, CIRUS, and at the Rajasthan

Atomic Power Station; and flooding, containment collapse, and valve failure and fire outbreaks also occurred at several other stations.

Nevertheless, political support remained strong for nuclear energy development, given the amount of electricity that was being generated and in regard to national energy self-sufficiency and security. Post-Fukushima, such support has continued and is evidenced by the commissioning of the 1000 MW Kudankulam Nuclear Power Plant in October 2013, with five more reactors of 4800 MW at the Kudankulam, Rajasthan, and Kakrapar NPPs under construction (NPCIL 2014), and many planned for other locations. This continued political support for nuclear development then begs the question: what was the scope and extent of a Fukushima effect?

## Investigating The "Effect" of Fukushima

### Public Activism and Policy Responses

As the aftermath of the Fukushima disaster unfolded through dramatic real-time telecasts broadcasted widely throughout India by television news channels, social media, and newspaper coverage, a huge debate emerged about the pros and cons of nuclear power. Many people had also not forgotten the notorious Bhopal gas leak disaster of 1984, which killed around 2000 people and seriously injured 30,000. Although Union Carbide India was found negligent by the courts, the company did not pay adequate compensation (Baxi 1986). The public had also not forgotten the impact of the 2004 Indian Ocean tsunami on the southeastern coastal belt of India, where two nuclear reactors were being constructed, one at Kudankulam and another at Kovvada, and the resulting damage to them (CNN 2005).

In the shadow of Fukushima, robust public protests erupted at various places where nuclear reactors were located—in Kudankulam, Jaitapur, Mithi Virdi, Kovvada, Haripur, and New Delhi. Protesters demanded denuclearization of energy and immediate safety audits of nuclear power plants, the latter preferably done by a nuclear regulatory authority more independent of the industry than the Atomic Energy Regulatory Board—especially in respect to the threat of tsunamis and earthquakes occurring in the Indian Ocean (Jayaraman 2011). Regulatory independency was also a key public issue in Japan following Fukushima (Hindmarsh 2013a).

In September 2012, in Kudankulam, located in the southern state of Tamil Nadu, public protests intensified against the construction of a nuclear power plant due to go online in 2013. Hundreds of protesters converged on the village of Idinthakarai. They were led by anti-nuclear activist S.P. Udaykumar of the People's Movement against Nuclear Energy, which since 2011 had led protests against this plant by nearby farming and fishing communities. In an effort to halt construction activities, they organized demonstrations and began a relay fast. A number of people went without food in turns, for an indefinite period, to show their commitment to the issue and slow down progress (Bhadra 2012). Nevertheless, construction continued. When the uranium fuel was to be loaded at the plant, the protesters became confrontational. In response, the local police started firing and killed one fisherman.

At Jaitapur in Maharashtra, protesters also turned confrontational, in reaction to police brutality: one protestor was shot dead by the police. That protest was organized in response to the then-Environment Minister's statement that the nuclear power plant at Jaitapur would be built despite the tsunami effects at Fukushima. A press release from the Konkan Vinashkari Prakapl Virodhi Samiti organization announced: "We demand scrapping of all new nuclear power plants and ask the State Government to end the reign of terror in the Jaitapur area" (E.T. 2011). Similar protests erupted in Andhra Pradesh, West Bengal, and Gujarat.

Some political parties, including Shiv Sena in Maharashtra, All India Dravida Munnetra Kazahagam in Tamil Nadu, Trinamool Congress in West Bengal, and Telugu Desam Party in Andhra Pradesh, supported public opinion and organized "bandh" (lock-outs) in many regions (Indian Express 2011). At the same time, many academics, scientists, and journalists formed associations to demand a renewed focus on safety, the democratization of the Atomic Energy Commission, the fixing of risk liability on operators and suppliers, and a new regulatory body that would also include scientists and be independent of the government (Indian Express 2011). At Mumbai, the Tata Institute of Social Sciences prepared a People's Report on the Jaitapur Nuclear Power Plant that laid out public concerns and what should be done in regard to them (Kamble 2010).

Various other civil society organizations voiced concerns and raised public awareness about nuclear power risk and hazard following Fukushima. They included the People's Movement Against Nuclear Energy, Konkan Vinashkari Prakapl Samiti (Konkan Coalition Against Destruction), the Coalition for Nuclear Disarmament and Peace, Lokayat, the National Alliance for People's Movement, Gujarat Anu Urja Mukti Andolan (Gujarat Freedom from Atomic Energy Movement), and Chutka Parmanu Pradushan Sangharsh Samiti (the Chutka Atomic Pollution Protest Committee), and Madhya Pradesh Jan Sangharsh Morcha (Protest Committee).

The Union Government immediately responded to this pressure in two ways. First, it outlawed collective protests. Protesting villagers were subsequently arrested on trumped-up charges of violence, and eminent citizens—including a retired Supreme Court judge and a former admiral of the Indian Navy—were issued orders preventing them from entering contested areas to show their support of the opposition. One retired high-court judge was even detained for five days in violation of legal safeguards against prolonged arrests without charge (Bidwai 2011).

In the ongoing effort to contain protests, several complaints were filed by public authorities against local residents peacefully protesting the Indian government's moves to construct nuclear reactors with foreign help in India. Leaders of these movements were arrested and some were charged with committing grave crimes against the State, that is, sedition (Human Rights Watch 2012). The government has also consistently attempted to discredit and thus delegitimize protesters by negatively stereotyping them as "illiterate," "backward," and "superstitious" (Mukharji 2012).

The government also attempted to close controversy through absorption of civil society concerns at the institutional level, as discussed below. For example, with the safety matter of nuclear power post-Fukushima debated by opposition leaders in the Parliament attracting high public interest, the parliamentary Public Accounts Committee examined the issue. It concluded that failure to have an autonomous and independent nuclear regulator was quite risky in regard to ensuring safety (Doherty 2013). The Committee also followed up with an equally critical assessment of the Atomic Energy Regulatory Board by the highest auditor of the country, the Comptroller and Auditor General (CAG 2012; Ramana 2012).

Both the Public Accounts Committee and the Comptroller and Auditor General reiterated cross-jurisdictional findings detected by Japan's Independent Investigation Commission and the Commission's observation about the need for an independent regulator; Fukushima could have been preventable in the absence of "collusion between the government, the regulators and TEPCO (the Tokyo Electric Power Corporation)" (Nuclear Accident Commission 2012: 16; CAG 2012: 73). The same sort of "cozy" relationship was seen to exist in India between the Atomic Energy Regulatory Board and the Department of Atomic Energy (Raju 2014). In response to these criticisms, the Central Government introduced the Nuclear Safety Regulatory Bill of 2011, according to which a statutory regulatory body would replace the Atomic Energy Regulatory Board.

## Liability for Nuclear Damage

At Fukushima, energy giants Toshiba and Hitachi built reactors based on (US) General Electric's (GE's) Mark I reactor design. But when the reactor could not withstand the tsunami and the consequent meltdown of the reactor, GE, as a manufacturer and supplier of nuclear reactors, was not held liable for the US$235 billion damage (estimated total damage) to Japan (World Bank 2012: 2; also Kobayashi 2011). According to the World Bank, the highest estimated damage to the buildings, public utilities, social infrastructure, and production sectors of agriculture, forests, and fisheries was about 10.5 trillion yen (about US$140 billion), excluding the cost of decommissioning the broken Fukushima Daiichi reactors (World Bank 2012: 2).

Thus, another important effect of the Fukushima Daiichi nuclear disaster—in relation to Japan not being able to get adequate compensation from the nuclear reactor suppliers—was to provide opportunity to the Indian public for more discussion and analysis of India's Civil Liability for Nuclear Damage (CLND) Act 2010, and The Civil Liability for Nuclear Damages Rules 2011 (Devraj 2011). These regulations fixed the maximum liability amount of nuclear power suppliers at 15,000 million Indian rupees (ca. US\$240 million). The Japanese situation also generated a sharp debate internationally, as operators of nuclear reactors are given exceptional rights to seek damages from a supplier under certain circumstances. Suppliers are also subject to tort claims (where a tort is generally considered a wrongful act leading to legal liability), which provide:

> That the provisions in the Act shall be in addition to, and not in derogation of, any other law for the time being in force, and nothing contained herein shall exempt the operator from any proceeding which might, apart from this Act, be instituted against such operator.
> (Section 46, CLND Act 2010)

Interestingly, this provision was hardly debated in Parliament (Ram Mohan 2014). Under the law of torts in India, the compensation principle is based on absolute liability and there is no need to prove negligence or fault on the part of the liable supplier or operator.

International nuclear suppliers, especially those based in the US, have expressed dissatisfaction with these provisions (Jacob 2011). Following the Indo–US 123 Agreement of 2007 (Jha 2008), suppliers have consistently argued that supplier liability be reduced or eliminated. The extent of damages of the Fukushima disaster, the experiences of the Bhopal tragedy, and the current unprecedented level of high adverse public opinion to nuclear power, however, make it difficult for India to accede to suppliers lobbying to ease liability legislation. Political parties, both right and left, oppose any dilution of the liability rules for suppliers (Mukherji 2013).

## Land Acquisition and Nuclear Power Plants

On the issue of land acquisition for the construction of nuclear power plants, policy-makers so far fail to recognize public concerns. The Right to Fair Compensation and Transparency in Land Acquisition, Rehabilitation and Resettlement Act of 2013 requires rehabilitation and resettlement to be provided to people affected by land acquisition (Bedi and Gangwani 2011). Compensation for owners of acquired land is fixed at four times the market value for rural areas and eight times for urban areas. In addition, in the case of the acquisition of land for use by private companies or public private partnerships, a consent clause has been added. This clause, in recognizing the potential sacrifices made by landowners such as farmers, enables landowners to not consent to government acquisition of their land except in situations of national interest.

Before the new government came to power in 2014, the Act was termed regressive because 13 pieces of legislation listed in its fourth schedule, including the Atomic Energy Act, were kept out of its purview (Counterview 2013). As a result, the government could still acquire land in the name of establishing nuclear reactors without addressing issues of rehabilitation and fair compensation. Farmer organizations, including Kisan Sabha, demanded that the fourth schedule be scrapped altogether, as most government land acquisitions took place under this list (Rajalakshmi 2013). In other words, the true intent and purpose of the Act to compensate landowners in a fair manner could not be fully achieved, as these 13 pieces of legislation—covering NPPs, highways, pipelines, and railways—were not covered by The Right to Fair Compensation and Transparency in Land Acquisition, Rehabilitation and Resettlement Act of 2013 (Roy 2014). Many public protests related to land acquisition for construction of nuclear power plants were visible signs of discontent with the new law. The new government amended this law by including the Atomic Energy Act, inter alia, under its purview and removed the consent clause (TNN 2014a). The question of rehabilitation, however, remains to be settled. In the next section we turn to environmental matters.

## Environmental Governance of Nuclear Technology

In India, the environmental governance process begins with an environmental impact assessment (EIA) of a proposed nuclear project by experts, which is followed by advertised public hearings that involve pollution control board officials, administrative officials, citizens, and local representatives. These hearings provide opportunity for affected publics to air their concerns and suggestions about a development proposal and make objections if they so wish. The Ministry of Environment and Forests grants environmental impact assessment certificates.

The Ministry issued an EIA guidance manual for nuclear power plants in 2010, which mandates EIA and public consultation (MEF 2010). However, the methods used in these processes remain questionable in terms of gaining public confidence and adequate risk assessment. In the case of Mithi Virdi (which literally means "sweet well") in Gujarat State, six imported US reactors, each 1000 MW, are to be built by 2020. Some 152 villages are located within a 30 kilometer radius around the power plant. The coastal villages have fertile land and are also dependent on fisheries. The EIA for this plant was submitted in January 2013. It emphasized benefits for land use, citizens' lifestyles, and economic development of the State. On the possibility of a Fukushima type tsunami, the report concluded that the plant would be kept above any estimated flood level caused by tides, tsunamis, and waves.

Local authorities held public hearings. In one hearing, some 5000 villagers (silently) walked out in protest, as they considered the EIA report flawed in ignoring many safety aspects they considered important. In addition, the authorities did not allow villagers to make presentations (oral or otherwise) about procedural issues and their concerns about the NPP. This process was also in violation of the law that a person not living in the proximity of a project site may also participate (through the right-to-ask questions) in the public hearing (Samarth Trust and Others 2010). Subsequent clearance of the Coastal Regulatory Zone of the state government for the proposed siting to go ahead was then given, remarkably, without any site visits or documentation of ground conditions and variables by the regulatory authority (Hindu 2013).

Unsurprisingly, affected communities were highly critical of this process, not only because of a lack of legitimacy and ignorance of environmental siting conditions, but also because they had become aware of possible radiation leaks and inadequate management of nuclear waste (Indian Express 2013). Protest organizations stepped up opposition following the Fukushima disaster. On September 23, 2013, they organized a large rally at Mithi Virdi, around the time the Prime Minister would be in the US, to gain international attention (Statesman 2013). They appealed to non-resident Indians living in the US to hold protest meetings to confront the Indian Prime Minister. The villagers supported the call of these organizations locally. The leaders of 10 villages protested against the procedural lapses made during the public hearings in written submissions to the relevant Ministry of the Government (Krishnakant 2014). Nevertheless, the Ministry of Environment and Forests approved the project without considering these submissions.

In turn, the public hearing process for the Kovvada nuclear reactor in Srikakulum district of Andhra Pradesh was ineffective for citizens. The affected communities in the area, mostly fishers and farmers, expressed their resentment over public hearings being held at nearby Ramchandrapuram and Kotapalem for acquiring some 482 acres for a 6000 MW NPP (Hindu 2013a). A prominent convener of the National Alliance for People's Movement, Medha Patkar, opined that if the plant was set up, radioactive waste stored at the plant could pose grave danger to nearby residents (First Post 2013). Many residents subsequently demanded to express their right of consent or non-consent to proposed land acquisition under the new Land Acquisition and Rehabilitation Act of 2013. This was not only because of the risks of radioactive pollution, but also because some 8000 people were to be displaced by the Kovvada nuclear reactor, along with many livelihoods. Citizens were also alarmed to learn that General Electric, which supplied the reactors to Fukushima Daiichi, would be setting up the proposed nuclear plant at Kovvada.

One protest action organized by the villagers consisted of stage plays to increase public awareness about nuclear radiation; however, the district administration imposed Section 30 of the Police Act in the entire Srikaku-lum division, stating that no activity could occur without the permission of the Deputy Superintendent of Police (Kavula 2013). As a result of this repression, the local people organized a relay hunger strike to demand revocation of the government order to acquire land. Influential politi-cal support was seen when the former Panchayat President of Kovvada, Mylapalli Polisu, said that fishers would lose their livelihoods as restric-tions on fishing activities would follow construction of the nuclear power plant (Hindu 2013a). Support also came from Leftist parties, including the (Marxist) Communist Party of India and the Telugu Desam Party. Retired government servants and scientists also lent their technical knowledge to village protesters (Boora 2013). However, political support from the Tel-ugu Desam Party diminished when it came to power in the State, as the ruling party is dependent on the Union government funding (Hindu 2014).

Yet another example related to the public hearing on the proposed Chutka nuclear power plant, held in the Mandla district of the State of Madhya Pradesh in February 2014; amidst 10,000 paramilitary person-nel. Two public hearings had been cancelled earlier due to fierce pub-lic protests (Times of India 2014; also Trivedi 2013). The proposal of the Nuclear Power Corporation of India Ltd was to build a 1400 MW pressurized heavy water reactor near the banks of the Narmada River. According to the district administration, some 500 people from the region, mostly Gond tribespeople and farmers, were able to participate at the public hearing and express their views. But tribal leaders stated they were not allowed to adequately air their sentiments at the hearing. Subsequently, they complained to the Madhya Pradesh Human Rights Commission. A leader of the Chutka Atomic Pollution Protest Commit-tee opined to a news correspondent: "The jobs the power companies propose are for the educated. We will not even get electricity the plant produces. And if there is a Fukushima-like disaster, we will be the first victims" (Pallavi 2012).

The above examples illustrate that public hearings in relation to envi-ronmental (and social) impact assessment are more of a formal legal necessity foisted on different agencies of the State than an inclusive pro-cess of acquiring local knowledge and perspectives for planning input in the siting of nuclear power plants. This lack of adequate community engagement in some ways mirrors the lack of community engagement over NPP siting in Japan prior to the Fukushima disaster, which Hind-marsh (2013c) argued was a key contribution to why the Fukushima disaster occurred.

Even after the Fukushima disaster, public hearing mechanisms in India failed to be organized in a robust manner so that public misgivings about nuclear technology could be addressed in a fruitful and positive man-ner for more effective planning outcomes. The overarching role of "dis-tant" top-down State authorities remains, which, separated from local contexts, has led to State authorities portraying and marginalizing local concerns as being "unreasonable, unscientific, politically motivated and funded by foreign agencies" (Mukharji 2012).

## Safety, Disaster Management, and Transparency

Because of the severity of the Fukushima disaster and the massive public protests in Jaitapur and Kudankulam, and local actions in many other places over nuclear power development, the Indian Government changed tack and began to comprehensively review and reevaluate the safety readiness of nuclear power plants to deal with extreme natural events. Four task forces were constituted by the Nuclear Power Corporation of India Ltd, covering different types of reactors (NPCIL 2011). These task forces analyzed the Fukushima event based on information available through various international agencies. They then made recommendations to the NPCIL (NPCIL 2011).

Acting independently, the Atomic Energy Regulatory Board also convened a high-level governmental committee with national-level experts to assess the safety of Indian nuclear power plants against severe external events that might exceed the design basis of existing and proposed plants. Subsequently, the committee made several recommendations, which were accepted by the regulatory board (AERB 2011). A summary of the recommendations of the Nuclear Power Corporation of India Ltd and the Atomic Energy Regulatory Board is provided in Table 6.1.

*Table 6.1* Summary of Recommendations of the Nuclear Power Corporation of India Ltd. and the Atomic Energy Regulatory Board

| *Nuclear Power Corporation of India Ltd report* | *Atomic Energy Regulatory Board report* |
|---|---|
| Additional design features are needed, such as providing core cooling with additional means of power source at sites to bring plants to a safe shutdown. | Enhance the severe accident management program, particularly with respect to hydrogen management and reliable containment venting. |
| In the event of sensing seismic activity, automatic shutdown of the reactor must be ensured. | Introduce a seismic trip where it does not exist. |
| Increase the duration of the passive power sources or battery-operated device for monitoring important parameters. | Strengthen the back-up power supply (air cooled generators at higher elevation). Strengthen the provision for monitoring of critical parameters under prolonged loss of power. |

*(Continued)*

*Table 6.1* (Continued)

| Nuclear Power Corporation of India Ltd report | Atomic Energy Regulatory Board report |
|---|---|
| Hook up various instruments, such as cooling water, steam generators, *calandria, calandria* vaults, end shields, and emergency core cooling systems. | Enhance the reliability of cooling through external hook-up points. |
| The water inventory of the plant should be augmented. | Steps for augmentation of onsite water storage are needed wherever required. |
| Emergency operating procedure to be reviewed, with adequate training of the staff. | Training and mock-up exercises for operating personnel. |

Source: Bajaj 2013; NPCIL 2011

The National Disaster Management Authority (NDMA) also reviewed the existing emergency preparedness and response plans of all NPPs to ascertain and evaluate the response and its coordination among different agencies during an offsite nuclear emergency situation. Consultations were also held with state governments on emergency preparedness plans. As the Prime Minister of India is the Chairperson of the NDMA, this institution plays an important role in nuclear disaster preparedness.

New measures, such as installing hi-tech radiation measuring gadgets in 35 cities across India, and creating a disaster response force and procurement of special radiation detection vehicles, were planned (Deccan Herald News 2011). Later in the 2014–2015 budget session question times, the government stated that the implementation of these measures was subject to the availability of funds, and the consent of state governments and other relevant agencies (Kumar 2012; Tharoor 2014). However, despite the best efforts of the National Disaster Management Authority, a preliminary audit report of the Comptroller and Auditor General of India in 2014 revealed that despite a series of natural disasters every year, the authority still needed to install an emergency communication system at its operation center (Yadav 2014).

Emergency preparedness drills around nuclear power plant sites, carried out to test their ability to coordinate emergency responses, have also been found wanting. Although there are national-, state-, and district-level disaster management authorities, comprehensive drills with communities living in the vicinity of nuclear power plants appear not to have been carried out. Subsequently, the Madras High Court expressed concern at the need for emergency preparedness on the part of the Nuclear Power Corporation of India Ltd when it came to the Court's notice that such drills had only been conducted in one village in the vicinity of Kudankulam nuclear power plant in Tamil Nadu State (Sundarrajan 2013). The Court observed such exercises should occur in all villages within a 30 kilometer radius of reactors; and awareness programs needed to infuse confidence in the minds of the local people as to their effectiveness for preparedness.

Apart from inadequate emergency preparedness, transparency measures undertaken by State officials entrusted with the functioning of nuclear technology have also remained weak. With details of past safety audits of reactors remaining publicly obscured, in the wake of the Fukushima accident, on April 26, 2011, the Prime Minister's Office assured the public that "action taken on previous safety reviews would be put in the public domain." Even though, more than three years have elapsed since then, the Nuclear Power Corporation of India Ltd is yet to comply with this assurance (Sarma 2014).

According to a former Union Power (or Ministry of Power) Secretary, Emani Anantha Satyanarayana Sarma, there is an overall lack of transparency on the part of the Department of Atomic Energy and the Nuclear Power Corporation of India Ltd. Significantly, they did not promptly inform the public of the details of accidents at nuclear power plant sites. This malpractice occurred with a series of accidents at the Kakrapar plant in June–July 2012, in earlier episodes, and a more recent accident that took place at Kudankulam in May 2014.

Government scientists driving the nuclear industry in India are also criticized for lack of transparency and inadequate safety oversight (Rahman 2011). The industry still enjoys the old era protection of the Atomic Energy Act of 1962, which allows non-disclosure of certain information under Section 18. Indeed, the Right to Information (RTI) Act of 2005, Section 8(1) (a), contrary to its name, acts to restrict public access to information related to nuclear technology on grounds of scientific, strategic, or security interests of the State. Several residents' requests for information relating to the safety of nuclear power have been either plainly refused or answered with information provided in a "half-baked" fashion.

In response, citizens have organized several protests, which have escalated since the Fukushima disaster and indicate disgruntlement towards the government-controlled industry on this important information issue for local communities around NPPs. In Kudankulam, fishers adjoining the proposed site of a nuclear plant asked to see the disaster management plan, but it was not provided (Dutta 2012). Secrecy also shrouds the fate of site selection committee reports, and the location and management of radioactive spent fuel, and its reprocessing and transportation.

The site evaluation report related to the Jaitapur nuclear power complex was also criticized on the basis that it did not take into account the geological structure of the ground on which the project is sited, and for an alleged error of method in managing spent fuel (Dome 2012). According to a statement issued by the National Committee in Solidarity with the Jaitapur Struggle, on August 26, 2012, the Nuclear Power Corporation of India Ltd refused to disclose the costs and resultant tariff for the proposed power project, which is also of significant public concern.

Similarly, the Kudankulam nuclear operator refused to provide the plant's safety assessment report to S.P. Udaykumar (the convener of the People's Movement Against Nuclear Energy in Tamil Nadu), even after the Fukushima disaster (Chauhan 2012). The refusal was based on the argument that the required information would prejudicially affect the sovereignty and integrity of India under Article 8(1) (a) and commercial confidence clauses under Article 8 (1) (d) of the Right to Information Act.

In response to the appeal filed by Udaykumar in February 2012, the Central Information Commission (CIC) ruled that the purpose of a safety assessment report of a proposed nuclear plant site was to protect the public and the environment from the radiological consequences of radioactive releases due to nuclear power plant accidents. Disclosure of the site evaluation and safety assessment reports was meant to enable citizens to attain an integrated understanding of the project, including environmental and safety risks and hazards (Udaykumar 2012). Nevertheless, the NPCIL refused to make the safety assessment reports of the Kudankulam plant publicly available, arguing that the CIC did not hear their side of the matter. On these grounds, the NPCIL filed a writ petition (a legal filing that a party makes to an appeals court to argue that an issue is reviewed quickly) to the Delhi High Court. It is still pending before the Court, but the CIC's order to make public the safety analysis report and site evaluation report of the Kudankulam nuclear plant has been temporarily suspended (NPCIL 2013).

## Discussion

The effect of the Fukushima Daiichi nuclear disaster was felt in India exactly at the time the Government was planning to commission the Kudankulam nuclear plants and set up the new plants by acquiring land from citizens in populated areas. This nuclear power expansion and its flawed planning process was set also against the background of the notorious 1984 Bhopal gas leak, earlier major nuclear accidents, and many Indian nuclear incidents. Public activism emerged to focus on the seemingly widening gap between controversial nuclear technoscience and the public. Prominent issues included fair compensation, liability of plant owners and suppliers, adequacy of fair process of public hearings, environmental impact assessment and safety, and transparency of disaster management plans and safety assessment reports, as well as weak and poorly implemented legislation.

These issues were democratically informed ones and affected the lives of many local communities. Did Indian science and technology governance in the nuclear energy industry evolve to address these issues after the Fukushima protests? When Bhadra (2013: 241) argued that "Kudankulam became a lightning rod for debates about nuclear energy and democracy . . ., " Bhadra highlighted the gap between governance of nuclear science and democracy; which widened due to the non-transparency policy and decision-making of the Department of Atomic Energy. This lack of transparency is a key reason why public activism spread from nuclear plant sites to across the country (Choudhury 2012), bolstered by the Fukushima effect of what can happen when the technology development and democracy gap becomes too wide.

Land acquisition post-Fukushima also remains an important policy issue fueling public activism, as the State can acquire anybody's private land for nuclear power plants in the so-called public interest. Hindmarsh (2013b) advances his concept of "place-change planning" for nuclear plant siting; it appears to augur well for India, as it poses "a facilitative conduit to better and more legitimately engage local place-based communities for more plural, social-situated understandings around where, when and what to site in targeted situations, all of which aims for enhanced social and environmental and energy outcomes" (Hindmarsh 2013b: 15; for more detail see Hindmarsh 2013c). Engaging local communities as such is a *sine qua non* (an indispensable and essential action, condition, or ingredient) for technoscientific decision-making affecting local communities.

Such engagement with local communities and discussing their concerns on siting, fair compensation, safety, and disaster management plans in a transparent manner would forge a closer connection between nuclear science and democracy and forward the cause of Vishwanathan's concept of "cognitive justice" (the right of different forms of knowledge to co-exist without being marginalized by official, state-sponsored forms of knowledge), to improve technoscientific decision-making (Vishwanathan 2005: 92).

The lessons from this investigation of the Fukushima effect are that government agencies in India responsible for nuclear energy management, regulation, and development—like the Department of Atomic Energy, the Atomic Energy Regulatory Board, and the Nuclear Power Corporation of India Limited—need to engage with the public in a constructive, inclusive, way, not portray them, negatively and inaccurately, as "unreasonable, unscientific, politically motivated and funded by foreign agencies" (Mukharji 2012: 268). This language instead appears to reflect more the activities of government and is inappropriate to apply, with the Fukushima disaster an apt example of ill-conceived planning (see also Wolsink 2007).

The language of recommendations made by the Nuclear Power Corporation of India Ltd and the Atomic Energy Regulatory Board in the aftermath of Fukushima (see Table 6.1 above) is another good demonstration of an inappropriate top-down or "closed" policy style (see Sovacool 2010): it is full of epistemic technicality, and does not envisage pro-active public engagement in the event of actual disaster. Essentially, a key Fukushima effect for India is that the vocabulary of nuclear science and technology needs to now go hand-in-hand with public knowledge, perspectives, and concerns, as a knowledge partnering approach (e.g. see Eversole 2015).

## Conclusion

My research aim has been to investigate the "Fukushima effect" on India's nuclear energy and environmental governance. Accordingly, this chapter focused on nationwide public concerns about the existing norms and practices on nuclear issues affecting lives post-3/11. Fukushima effects on technopolitical decision making practices can be observed very clearly. The disaster at Fukushima informed and broadened civic contestations around controversial nuclear power plant siting at the very time when government was pushing forward with a number of new nuclear power plant projects. Many policy responses also illustrate this effect, including advances on the questions of liability in the event of nuclear disaster, and overall safety issues of nuclear power plants. In addition, issues related to nuclear power have been raised in national parliament debates and at numerous public gatherings—in New Delhi, Chutka, Jaitapur, Kudankulam, Kovvada, and Gorakhpur—and continue to be raised in active movements contesting NPPs. Although decision-making is still dominated by the "ivory towers" of big science and governmental bodies, there has been subsequent increased institutional acceptance of social concerns related to land acquisition, and the independence of regulatory bodies on nuclear safety, after the Fukushima disaster.

At the same time, such advances are not markedly seen in other important areas of civic concern, including public hearings for environmental impact assessment, emergency preparedness, and transparency issues in relation to government policy and decision-making, especially in relation to nuclear safety. The officialdom in the nuclear establishment still tends to actively shield relevant information from the public under the protective clauses of the Atomic Energy Act of 1962. Any semblance of meaningful participatory public hearings has not been in evidence, as exemplified at Mithi Virdi, Kovvada, and Chutka. Many of these hearings subsequently became challenged in the judiciary and criticized by informed citizens. Emergency preparedness exercises, apparently only tokenistic, are still carried out inside nuclear power plants to the exclusion of those living in close proximity to them.

Transparency, safety, and participation in relation to nuclear power development and management thus need to be addressed in a more meaningful and engaging manner in order to address urgent livelihood concerns of the public. Only then will civil society develop any measure of faith in technoscientific procedures related to nuclear power. In conclusion, my analysis of the "Fukushima effect" in India shows that the proposed governmental momentum on nuclear energy development has been and is still being actively pressured and influenced by social concerns about its risks and vulnerability, which were all too easily brushed aside before the Fukushima disaster.

## References

AERB. *AERB Committee to Review Safety of Indian Nuclear Power Plants against External Events of Natural Origin.* New Delhi: Government of India, 2011.

AERB. *National Report to the Convention on Nuclear Safety: 6th Review Meeting of Contracting Parties,* New Delhi: Government of India, 2013.

Agarwal, P. *India's Nuclear Development Plans and Policies: A Critical Analysis.* New Delhi: Northern Book Centre, 1996.

Anderson, R. S. *Nucleus and Nation.* Chicago and London: University of Chicago Press, 2010.

Bajaj, S.S. "Regulatory Practices for Nuclear Power Plants in India", *Sadhana* 38 (2013): 1027-1050, accessed September 25, 2015, https://goo.gl/Af3xCw.

Baxi, U., and A. Dhanda. *Valiant Victims and Lethal Litigation.* New Delhi: Indian Law Institute, 1986.

Bedi, P., and S. Gangwani. "The Land Acquisition, Rehabilitation and Resettlement Bill, 2011," PRS Legislative Research, 2012, accessed August 13, 2014, www.prsindia.org.

Bhadra, M. "India's Nuclear Power Problem," *The Cairo Review of Global Affairs* 5 (2012): 70–81.

Bhadra, M. "Fighting Nuclear Energy, Fighting for India's Democracy," *Science as Culture* 22 (2013): 238–246.

Bidwai, P. "People vs Nuclear Power in Jaitapur, Maharashtra," *Economic and Political Weekly* 46 (2011): 10–14.

Boora, S. "365 Days of Hunger Strike Against Unsafe American Reactor Project in Andhra Pradesh," *Dialogues and Resources on Nuclear, Nature and Society,* December 17, 2013, accessed September 25, 2015, http://goo.gl/BZFjXj.

Chauhan, C. "Disclose Kudankulam Nuclear Reports, CIC Tells Government," *The Hindustan Times,* May 1, 2012, accessed February 25, 2015, http://goo.gl/Pgdum1.

Choudhury, U. "The Impact of the Fukushima Daiichi Nuclear Crisis on Anti-Nuclear Movements in India," *Social Alternatives* 31 (2012): 39–44.

Counterview. "New Land Acquisition Act Fails to Deal with Historic Injustices Committed in the Name of Development, Is Aimed at Facilitating Corporates," November 21, 2013, accessed February 15, 2015, http://goo.gl/b11fdu.

CAG. *Report No. 9 (Scientific Departments) Performance Audit on Activities of Atomic Energy Regulatory Board (Department of Atomic Energy).* New Delhi: Government of India, 2012.

CNN. "Tsunami Death Toll," February 22, 2005, accessed November 11, 2014, http://tinyurl.com/3qben.

Deccan Herald News. "PM Scans Safety Measures at N-Plants," *Deccan Herald,* June 1, 2011, accessed February 25, 2015, http://goo.gl/iBzQ8Y.

Devraj, R. "India: Fukushima Revives Debate over Nuclear Liability," *Global Information Network,* March 30, 2011, accessed February 25, 2015, http://goo.gl/BsP8Te.

Doherty, B. "Harsh Criticism for India's Nuclear Safety Regime," *Sydney Morning Herald*, December 20, 2013, accessed February 26, 2015, http://goo.gl/QJ83yS.

Dome, R.C. "Location of Reactors," *Unstarred Question No. 1691*. New Delhi: Lok Sabha [Lower House of Parliament], 2012.

Dutta, M. "Desirability of Nuclear Power is the Real Question," *The Hindu*, September 27, 2012, accessed February 26, 2015, http://goo.gl/zqExPB.

ET Bureau. "Jaitapur Protests Intensify, One Killed in Police Firing," *The Economic Times*, April 19, 2011, accessed February 26, 2015, http://goo.gl/cvNwkW.

Eversole, R. *Knowledge Partnering for Community Development*. New York: Routledge, 2015.

First Post. "Medha Patkar Raises Pitch against Nuclear Plant at Srikaulam," *F. India*, November 27, 2013, accessed November 10, 2014, http://tinyurl.com/p2kfqex.

Hindmarsh, R. (ed.). *Nuclear Disaster at Fukushima Daiichi: Social, Political and Environmental Issues*. New York: Routledge Studies in Science, Technology and Society, 2013a.

Hindmarsh, R. "Nuclear Disaster at Fukushima Daiichi: Introducing the Terrain". In *Nuclear Disaster at Fukushima Daiichi: Social, Political and Environmental Issues*, edited by R. Hindmarsh, 1–22. New York: Routledge, 2013b.

Hindmarsh, R. "Megatechnology, Siting, Place and Participation." In *Nuclear Disaster at Fukushima Daiichi: Social, Political and Environmental Issues*, edited by R. Hindmarsh, 57–77. New York: Routledge, 2013c.

The Hindu. "Residents Oppose Nuclear Plant at Kovvada," *The Hindu*, June 9, 2013, accessed November 11, 2014, http://tinyurl.com/qdgenax.

The Hindu. "Kovvada Nuclear Power Plant May Take off Soon," *The Hindu*, June 13, 2014, accessed November 12, 2014, http://tinyurl.com/owtrsga.

The Hindu Special Correspondent. "MoEF Urged to Reject Gujarat CRZ Clearance for Mithi Virdi Nuclear Plant," *The Hindu*, August 10, 2013, accessed February 26, 2015, http://goo.gl/9K9gVt.

Human Rights Watch. "India: End Intimidation of Peaceful Protesters at Nuclear Site," May 11, 2012, accessed November 11, 2014, http://tinyurl.com/lfvdm6q.

The Indian Express. "Anti-nuke Activists Step Up Jaitapur Protest." *The Indian Express*, April 26, 2011, accessed February 26, 2015, http://goo.gl/TRBNRP.

Indian Express News Service. "Activists Send Notice to Union Forest Minister over Public Hearing," *Indian Express*, March 5, 2013, accessed February 26, 2015, http://goo.gl/XkGCV1.

International Bank for Reconstruction and Development (World Bank). *The Great East Japan Earthquake: Learning from Megadisaster*s. Washington DC: World Bank, 2012.

Jacob, J. "N-Liability Law Still a Hurdle," *Hindustan Times*, July 20, 2011, accessed February 26, 2015, http://goo.gl/JyM6ck.

Jayaraman, K.S. "India's Nuclear Future Put on Hold," *Nature* October, 2011. DOI: 10.1038/news: 581.

Jha, A. "Indo-U.S. Treaty on Peaceful Nuclear Co-operation, 2007: How Safe Is the New-Born Baby of Indo-U.S. Love Affairs?" *Indian Journal of International Law* 48 (2008): 435–448.

Kamble, M. *People's Report on Social Impact of Jaitapur Madban Nuclear Power Plant*. Mumbai: Tata Institute of Social Sciences, 2010.

Kavula, S. "Kovvada Update: 50 Days of Hunger Strike against Proposed FE Reactors," *Dialogues and Resources on Nuclear, Nature and Society*, February 3, 2013, accessed November 11, 2014, http://goo.gl/Xn64RN.

Kobayashi, T. *FY2020 Nuclear Generating Cost Treble Pre-Accident Level-Huge Price Tag on Fukushima Accident Cleanup*. Tokyo: Japan Centre for Economic Research, 2011.

Krishnakant, R., Prajapati, A., and S. Desai. "Expert Appraisal Committee Meet Summary Record Fails to Reflect Anomalies at Public Hearing of Mithi Virdi Nuclear Power Project," *Counterview.org*, accessed November 11, 2014, http://goo.gl/63aiOW.

Kumar, P. "Radiation Detection," *Un-starred Question No 1674*. New Delhi: Lok Sabha [Lower House of Parliament], 2012.

MEF. *Environment Impact Assessment Guidance Manual for Nuclear Power Plants, Nuclear Fuel Reprocessing Plants and Nuclear Waste Management Plants*. New Delhi: Government of India, 2010.

Morehouse, W. *Science in India*, Mumbai: Popular, 1971.

Mukharji, P.B. "Cultures of Fear: Techno-nationalism and the Postcolonial Responsibilities of STS," *East Asian Science, Technology and Society: An International Journal* 6 (2012): 267–274.

Mukherji, B. "India Official: Won't Relax Nuclear-Liability Rules," *The Wall Street Journal*, October 3, 2013, accessed February 26, 2015, http://goo.gl/6gP6RE.

NPCIL. *Interim Report on Safety Evaluation of Indian Nuclear Power Plants Post Fukushima Incident*. New Delhi: Government of India, 2011.

NPCIL. *Plants under Operation*. New Delhi: Government of India, 2014.

NPCIL v. Udaykumar, S.P. *Delhi High Court*. Writ Petitions (Civil) 3353, 5886/2013; CM Appl. 6380/2013.

Nuclear Accident Independent Investigation Commission. *The Official Report of the Fukushima Nuclear Accident Independent Investigation Commission*. Tokyo: National Diet of Japan, 2012.

Pallavi, A. "Deaf to Nuclear Plant Protests," *Down to Earth*, November 30, 2012, accessed November 11, 2014, http://tinyurl.com/ogvw4qy.

Parthasarathy, K.S. "Historical, Regulatory and Safety Aspects of Nuclear Energy in India," *Atoms for Peace* 2 (2008): 102–119.

Pathak, K.K. *Nuclear Policy of India: A Third World Perspective*. New Delhi: Gitanjali Prakashan, 1980.

Planning Commission. *First Five Year Plan (1951–56)*. New Delhi: Government of India, 1951.

Rahman, M. "How Homi Bhabha's Vision Turned India into a Nuclear R & D Leader?" *The Guardian*, November 1, 2011, accessed February 26, 2015, http://goo.gl/L0lS2Y.

Rajalakshmi, T.K. "A Law and Its Losers." *Frontline*, Dec–Jan 29–11, 2013.

Raju, S., and M. Ramana. "It's Better to Be Safe Than Sorry," *Hindustan Times*, February 5, 2014, accessed February 26, 2015, http://goo.gl/5vDPpQ.

Ram Mohan, M.P. "Nuclear Liability Law of India: An Appraisal of Extent of Liability, Right of Recourse and Trans-boundary Applicability," *Journal of Risk Research* 17 (2014): 115–131.

Ramana, M.V. "Flunking Atomic Audits," *Economic and Political Weekly*, 47 (2012): 10–13.

Rees, J. *Hostages of Each Other: The Transformation of Nuclear Safety since Three Mile Island*. Chicago and London: Chicago University Press, 1994.

Rosen, M. "The Critical Issue of Nuclear Plant Safety in Developing Countries," *IAEA Bulletin*, April, 1977, 12–21.

Roy, A., and P. Singh. "Land Acquisition and EIA in India: Time for Structural Reforms." 3rd NLA Meeting on Nuclear Energy and Indian Society: Public Engagement, Risk Assessment and Legal Frameworks. New Delhi: Nuclear Law Association, 2014, 17–18.

Samarth Trust & Another v. Union of India. *Delhi High Court*, May 28, 2010.

Sarma, E.A. "How Serious Is NPCIL about the Safety of Its Power Plants? The Latest Accident at Kudankulam," *Dialogues and Resources on Nuclear, Nature and Society*, March 17, 2014, accessed September 24, 2014, http://goo.gl/pTC5at.

Sharma, D. "India's Lopsided Science," *Bulletin of the Atomic Scientists*, 47 (1991): 32–36.

Sharma, D. "Confronting the Nuclear Power Structure in India." In *Confronting the Experts*, edited by B. Martin, 155–174. New York: State University of New York Press, 1996.

Sisodia, S.S. *Foreign Policy of India: Indira Gandhi Era*. New Delhi: Inter India Publications, 1985.

Sovacool, B.K. "The Importance of Open and Closed Styles of Energy Research," *Social Studies of Science* 40, no. 6 (2010): 903–930.

Sundarrajan, G. v. Union of India. 2013. *Madras High Court*.

Statesman News Service. "Guj Villagers Intensify Protest against N-Plant," *The Statesman*, September 12, 2013, accessed February 26, 2015, http://goo.gl/9FBTZw.

Tharoor, S. "Question Raised by Dr. Shashi Tharoor (Budget session 2014–2015)," *Lok Sabha (Lower House of Parliament) Unstarred Question No. 4819*, August 12, 2014.

Times of India. "Public Hearing on Chutka Nuke Plant Held," *The Times of India*, February 18, 2014, accessed February 26, 2015, http://goo.gl/s3dXYH.

Times News Network. "Govt Approves Ordinance to Ease Land Acquisition Act to Push Reforms," *The Times of India*, December 29, 2014a, accessed February 26, 2015, http://goo.gl/66W500.

Trivedi, S. "Public Hearing on MP's Chutka Nuclear Power Project Cancelled," *Business Standard*, May 24, 2013, accessed February 26, 2015, athttp://goo.gl/ExOh7N.

Udaykumar, S.P. (Dr.) v. Nuclear Power Corporation of India Limited. *Central Information Commission*, CIC/SG/A/2012/000544/18674, April 30, 2012, http://goo.gl/gjskuo.

Vishwanathan, S. "Knowledge, Justice and Democracy." In *Science and Citizens: Globalization & the Challenge of Engagement*, edited by M. Leach, I. Scoones and B. Wynne, 83–94. London: Zed Books, 2005.

Wisconsin Project on Nuclear Arms Control. "India Nuclear Milestones: 1945–2009," *The Risk Report* 16 (2010), accessed November 12, 2014, http://tinyurl.com/k24syno.

Wolsink, M. "Planning of Renewable Schemes: Deliberative and Fair Decision-Making on Landscape Issues Instead of Reproachful Accusations of Non-Cooperation," *Energy Policy* 35 (2007): 2692–2704.

Yadav, Y. "Disaster Management Is a Disaster in Itself." *The New Indian Express*, August 3, 2014, accessed February 27, 2015, http://goo.gl/aA4Edt.

# 7 Nuclear Exceptionalism in the Former Soviet Union after Chernobyl and Fukushima

*Andrei Stsiapanau*

Since its first development, civil nuclear power has been promoted as a safe mode of energy technology, promising an abundance of green and cheap electric power. However, the environmental, social, and political effects of nuclear accidents and disasters have changed the safety images and frameworks of civil nuclear use and its development. It is now accepted that two major nuclear disasters—at the Chernobyl Nuclear Power Plant (NPP) on April 26, 1986, and at the Fukushima Daiichi NPP on March 11, 2011, have impacted adversely on the safety of nuclear energy production. Whereas Chernobyl's effect on nuclear safety was limited to the specific Soviet designed RBMK reactors,[1] the Fukushima effect went worldwide. For example, after Fukushima, European Union countries launched stress tests and peer-reviews of European NPPs in 2011 and 2012.[2] Among post-Soviet countries, Russia and Ukraine conducted stress tests based on the Western European Nuclear Regulators Association stress model.

The Fukushima disaster reemphasized nuclear safety on international energy political agendas, and appears to have produced an uncertain or patchy effect on the expansion of nuclear power development internationally. Here, in the post-Soviet region, while the Chernobyl disaster significantly changed the development of nuclear programs in Soviet and post-Soviet states, the Fukushima effect appears less pronounced. It is the scope and extent of the effect that Fukushima has had on nuclear power development in former Soviet countries that is the focus of this chapter.

The overall patchy effect of Fukushima was evident where, for example, international concerns—particularly in Asian countries close to Japan, and in western countries—were raised almost immediately after the three-reactor Fukushima meltdown. In contrast, Belarus and Russia signed an intergovernmental agreement only three days after Fukushima, on March 15, 2011, to construct a nuclear power plant in Belarus.[3] Later in 2012, the Russian authority on nuclear safety control, Rostechnadzor, declared that stress tests confirmed the safety of the Russian nuclear power plants against external influences, and that a Fukushima-type disaster would not be possible there. These perhaps controversial reactions to Fukushima indicate that its impact was neither chronic nor linear. For different "nuclear territories" a Fukushima effect would mark either a break or continuity with the Chernobyl effect and be deeply or not so deeply embedded in nuclear "technopolitics."

Potter (1990), in his report about the Chernobyl impact on Soviet decision-making, emphasizes that Soviet responses to Chernobyl involved subsequent technical and political changes. The technical responses to nuclear safety involved a set of engineering solutions that aimed to make existing Soviet-type reactors safer. In turn, the political responses saw organizational changes, including the creation of the Soviet Ministry of Atomic Energy, which affected the future of existing Soviet-type reactors. On April 20, 1989, the Council of Ministers of the Union of Soviet Socialist Republics (USSR) had already decided to stop construction of four planned Chernobyl-type reactors. Even so, Soviet nuclear physicists and engineers continued to pressure for "inherently safe reactors," while advocates of the RBMK concept (Chernobyl-type) continued to argue for its retention (Potter 1990: 78). In addition, escalating anti-nuclear activism in Russia, Ukraine, Lithuania, and Armenia saw more emphasis on political responses to the Chernobyl disaster (Dawson 1996).

These immediate post-Chernobyl contradictions between political and technical responses can be explained by the concept of Soviet nuclear exceptionalism, a "difference manifested in political claims, technological systems, cultural forms, institutional infrastructures, and scientific knowledge" (Hecht 2012: 6). When the Chernobyl disaster occurred, the Soviet nuclear network—comprised of industry, the military, science, workers, and the broader population—immediately faced the threat of a new, uncertain future which challenged the Soviet leadership. According to Wellock (2013), the ensuing Chernobyl reactor-closure debates generated a technopolitical controversy. Western engineers disagreed that RBMK reactors could be technically upgraded to the necessary safety level, and raised the issue that the Soviet regulatory agency "had the contradictory mission of promoting nuclear power and ensuring reactor safety" (Wellock 2013: 7). The inconsistent logic involved in these debates contributed to the technopolitical context within which Soviet nuclear engineers and physicists tried to reverse the Chernobyl effects on nuclear safety and "remake" nuclear energy safe again (see also Molyneux-Hodgson and Hietala in this volume for the Finnish case of nuclear exceptionalism).

Thirty years after the Chernobyl disaster and 25 years after the collapse of the USSR, the Fukushima disaster thus invited the question of what impact Fukushima had on the technopolitical logic of the post-Soviet nuclear programs. Was there a continuing bias towards developing civil nuclear power to the detriment of enhancing public safety, or have both development and safety aspects of nuclear power been addressed in a balanced way?

In addressing this question, within the broader investigation of a Fukushima effect, it is obvious that the Fukushima disaster has reenergized the safety controversies that emerged after the Chernobyl disaster. As such, I argue that Fukushima has produced a very strong technopolitical effect, rather than a more pronounced political effect or technical effect, on the development of post-Soviet nuclear programs. The Fukushima disaster fine-tuned conditions of nuclear exceptionalism. How, in post-disaster conditions and amongst public controversy, would the nuclear industry be re-constructed, the nuclear territories be re-mapped, and the nuclear cities and communities be re-built? In short, the political authorities turned to technical arguments of safety to ensure public opinion to support post-Soviet nuclear build programs. This Fukushima effect clearly positions nuclear safety as a technopolitical response.

In more depth, what are the direct linkages between nuclear power plant projects, political rhetoric and decisions, and the Fukushima effect? To answer this question, I analyze the trajectories of the NPP construction programs in post-Soviet states, from their decline after Chernobyl to their renaissance after Fukushima. My analysis addresses the relationship between nuclear technology and politics through the framing and bounding of Fukushima issues in relation to both technical and political responses, where I describe the narratives that framed the development of civil nuclear policy in the post-Soviet region in terms of technopolitical effect.

Some context for my investigation is provided by science, technology and society (STS) studies, where technopolitical effect aligns with themes of technology and democracy, and science and technology governance, particularly good governance. In the associated field of history of technology, there has been some coverage of USSR nuclear programs and technological choices (e.g. Josephson 2005; Schmid 2005), and of nuclear infrastructures, including plutonium sites in the US and USSR (Brown 2013). In political science, studies have focused on social mobilization during various periods of industrial development and political transformations in the USSR and in post-Soviet states (e.g. Babcock 1997; Dawson 1996; Mandrillon 1991; Marples 1991; Raviot 1995).

In such context to detect the Fukushima effect on post-Soviet nuclear programs I focus on three case studies. The first case is the 2300 MW Baltic Nuclear Power Plant,[4] which started construction on February 25, 2010, in the Kaliningrad region. The second is the 2400 MW Ostrovets NPP in Belarus, which followed Belarus President Alexander Lukashenko signing a decree to build the first NPP in the country, and resulted in a high dependency on Russian energy resources. The third is the planned Visaginas NPP in Lithuania, close to the decommissioned Ignalina NPP. In July 2011, the Lithuanian government announced that the Japanese company Hitachi GE Nuclear Energy would be the strategic partner and investor for the future Visaginas NPP. Despite the results of a 2012 referendum rejecting (65%) the NPP construction, Lithuanian political authorities still plan its construction to reduce the country's current dependence (50%) on imported electricity.

On one hand, these three cases mark the transformation of the energy system in the region, and accompany various modes of energy politics and policies, following the collapse of the USSR. On the other hand, they emphasize the persistence of nuclear programs and a nuclear renaissance between and after the two major nuclear disasters of Chernobyl and Fukushima. Overall, they demonstrate different levels of political lock-in,[5] public participation, and scientific controversies. Civil nuclear technologies are important in regard to their technical, technological, and innovative parameters, but my focus is on their political decision-making and technical thinking trajectories. These trajectories are conditioned by Chernobyl and Fukushima, and qualified by the embodiment, entanglement, or interplay of the technology and politics in these nuclear disaster aftermaths.

In following this focus, which informs the broader investigation of the Fukushima effect, I investigate two major periods: the pre-Fukushima period—before and in between each major nuclear disaster—and the post-Fukushima period. As part of the first period, I background nuclear power developments in the Soviet Union. I then outline nuclear safety issues highlighted by Chernobyl and Fukushima, and, finally, I outline the effect of Fukushima on nuclear technopolitics in the post-Soviet region.

## Nuclear Power Development and Safety Issues in the Soviet Union

The Russian nuclear industry has a long history rooted in military programs developed within the context of Soviet-era (or the Cold War era) ideology, national politics, and geopolitics. As Paul Josephson (1996, 2005) noted in his remarkable research on Soviet nuclear programs, civil nuclear power informed the strong political and ideological commitment to link political and social development with technological and innovative progress. As Josephson (1996: 297) commented: "For citizens, scientists, and officials alike, successes in atomic energy provided undeniable confirmation that at long last society had embarked on the final leg of the long journey to communism."

In the USSR, the development of the civil nuclear industry was informed by three key stages: (i) 1950–1960, the period of competitive research and innovation within Cold War contexts; (ii) 1960–1980, the period of civil nuclear policy elaboration and implementation; and (iii) 1986–1999, the immediate post-Chernobyl policy era. These periods illustrate how the Soviet nuclear industry was shaped first by competitive research and innovation and then by specific nuclear safety issues and problems. These trajectories reveal the technopolitical logic behind

Soviet nuclear projects: to build nuclear reactors inherently safe and compatible with the Soviet political system.

In the 1950s, the Soviet nuclear industry focused on designing a reactor with high power output. Before the world's first nuclear power plant opened in 1954 in Obninsk, a range of possible power reactors were investigated.[6] On May 16, 1950, the Council of Ministers of the USSR agreed to the construction of three experimental reactors based on different technologies: a graphite moderated and water-cooled reactor; a graphite moderated and gas-cooled reactor; and a beryllium moderated reactor cooled by gas or liquid-metal.

These technologies were based on prototype submarine nuclear reactors developed in the Cold War confrontation between the USSR and the US, alternatively within Soviet-era technopolitics (Schmid 2011), or Soviet nuclear culture (Josephson 1996). Josephson studied Soviet nuclear culture in three dimensions: geopolitical, public, and engineering. The geopolitical

> centered on the politics of the Cold War world: it was vital to national security strategies; it facilitated the Soviet Union's rise to one of two great powers; and it served as a symbol of the superiority of national economic and ideological systems, in the Soviet case by demonstrating that its scholars were indeed capable, as a slogan proclaimed, of 'reaching and surpassing' the west in prestigious areas of science.
>
> (Josephson 1996: 298)

Overall, the Cold War era permeated Soviet technological development, especially the civil nuclear program, and entangled post-Soviet nuclear developments in its politics.

In 1954, dual purpose reactors that could combine civil power generation with the generation of weapons-grade plutonium were trialed. One high-power channel-type reactor was a water-graphite reactor with zirconia and steel pipes and a prototype of the later RMBK reactor. The other was a water-furrow reactor, a prototype of the water-water power reactor.[7] Further development of the civil nuclear industry was carried out through state scientific and technical programs and saw a large-scale technopolitical system eventuate. The initial goals for civil nuclear development were to increase energy capacity for mass production; nuclear safety concerns were to come later. These goals reflected the nuclear program as "a facet of the deeper political and administrative structures and research processes that were paradigmatic for science and technology" (Josephson 1996: 323).

The fuller formation of civil nuclear policy in the USSR started with a shift from the science and innovation programs to nuclear policy programs and the creation of pilot reactors, the first ones being at the Novovoronej and Beloyarsk NPPs, which followed the Obninsk plant, the world's first NPP (operational 1954–2002). Within nuclear policy frameworks, various options for policy development as well as different types of reactor designs were considered, as framed first by the economic and industrial needs of energy production and consumption.

The soviet nuclear physicist Igor Kurchatov, a leading atom bomb developer, declared the goal of the first program of nuclear development for the period of 1950–1960 to be the generation of 2000–2500 MW of energy (Sidorenko 2010). This ambitious plan, however, became reduced to the construction of the Novovoronej and Beloyarsk NPPs. In 1962, the civil nuclear program restarted and an extensive program of NPPs saw the serial production of the first generation of VVER (water-water power prototype) and RMBK (graphite-moderated boiling water) reactors (Chernavsky 1988; World Nuclear Association 2014; Zheludev and Constantinov 1980).

In 1973, a fast breeder reactor was launched in the Kazakh Soviet Republic at the Shevchenko NPP. Another was commissioned in the Russian Republic in 1980 at the Beloyarsk NPP. The Soviet civil nuclear program was implemented in the Russian, Armenian, Ukrainian, and Kazakh Soviet Republics. This nuclear program geography was designed to meet existing energy needs and the needs of the industrial development of the Soviet Union. While the RBMK-type reactors were sited mostly in the European (western) part of the USSR, the VVER reactors were located in the eastern part. From and into the 1980s, construction of NPPs with a variety of different reactors followed.

Overall, the Soviet technopolitical system, within which nuclear power was embedded, was built as a monumental iconography of the victory of science and technology over the power of nature. Josephson (1996) noted that the Soviet technological "style" differed from European and US styles in reflecting the mass production of monumental industrial objects, so-called gigantomania. Combined with these two aspects, standardization and centralized bureaucracy created the conditions for "functional designs in which safety and comfort played a secondary role, and environmental issues were rarely raised" (Josephson 1996: 300). Nevertheless, Soviet engineers considered safety design an inherent factor of reactor design:

> In reactor design and construction, Soviet engineer-physicists demonstrated their conviction that reactor technology could be made inherently safe. They designed the Chernobyl-type channel graphite (RBMK) reactor with a positive void coefficient . . . ignoring the chance of 'highly improbable' accidents (like the one at Chernobyl that occurred during an ill-advised experiment). The RBMK was also designed to reload fuel without a complete shutdown; there are reinforced concrete plugs on top of the reactor for ease of refueling that could blow apart during an explosion.
>
> (Josephson 1996: 308)

It was only in 1983, after a set of nuclear incidents and hazards of different degrees that a government body to regulate nuclear safety—the State Committee for Supervision of Safety in Industry and Nuclear Power—was formed.[8] The initial misbalance between reactor design and safety control in the Soviet nuclear industry was seen in the rapid growth of civil nuclear energy production as well as in the implementation of multiple technical, organizational, and administrative structures to ensure its operation and development, accompanied by the lack of an adequate "safety culture."

The Chernobyl disaster in 1986 made this misbalance visible. Some scholars argued the lack of safety design stemmed from underdevelopment of Soviet understandings of design uncertainty and risk, accompanied by a strong commitment to technical and authoritative opinions: "Before Chernobyl, Soviet notions of nuclear risk had been technical, involving numbers and probabilities, while uncertainty was equated with incompetence or lack of experience" (Schmid 2004: 354).

After Chernobyl, reactor design considerations changed from a primary focus on scientific and technological preoccupations informing the choice between different types of reactors, to a focus more on safety considerations and the future of the RBMK reactors. In other words, safe reactor design became a sort of transnational technopolitical concern, with much pressure from other countries. International experts and politicians referred to this lack of safety culture as a condition to stop the export of Soviet-designed reactors into Europe (Wellock 2013).

In Russia, however, three nuclear power plants with the RMBK Chernobyl-type reactors remain in operation today: the Kursk, Smolensk, and Leningrad NPPs. The Smolensk NPP, the last RMBK reactor, was launched in 1990, four years after the Chernobyl disaster. Are these reactors still "inherently safe?" What happened to the issue of nuclear safety? In this case, within Soviet politics, the nuclear reactor's design was an inherently political technology "strongly compatible with particular kinds of political relationships" (Winner 1986: 22). That is, the "inherently safe" reactor was seen as an adequate response to the technopolitical system in the contexts of Cold War militarization and industrialization, technological mass production, and public perceptions of an entwined communist future and advanced scientific and technological development. Moreover, the role of science, technology, and innovation in the communist political process was framed and understood as a tool of infallible expertise and unerring knowledge.

Post-Chernobyl, such expertise was seen to inform a change from the "Soviet era" technopolitical logic of the inherently safe reactor, to the new notion that nuclear safety should not just be focused on the *inside* of the reactor but also on the *outside* of it. Such logic was also technopolitical because it linked Soviet engineering thinking to protect the reactor from the exterior to the discursive politics of not shutting down all NPPs with RBMK or Chernobyl-type reactors. Thereby, the Chernobyl disaster for a long time became an integral part of the technical safety logic of designers, engineers, and builders of nuclear plants and redefined the contours of almost all new types of reactors.

## The Chernobyl and Fukushima Effects on Safety Controversies

In the post-Chernobyl period, Soviet and Russian nuclear engineers keenly searched for technical solutions to protect against radioactive leaks in case of a nuclear reactor accident. For example, in 1995 the Technological Research Institute of St Petersburg (NITI) developed a core-catcher prototype for the nuclear water-water reactor VVER 640 within the Russian state program called "Ecologically pure energy."[9] Although this reactor type was not put into Russian service, the technological process of the 1990s was to do the maximum possible to avoid situations like the Chernobyl disaster. The core-catcher technology, developed in response to the Chernobyl accident, can thus be considered a *Chernobyl effect*.

The core-catcher was a large metal frame with a weight of 739 tons. Placed at the bottom of the reactor's pit and filled with a special sacrificial material, in case of an accident, it stops the leakage of molten fuel onto the foundation of the protective layer of the reactor. This would prevent nuclear fuel from leaving the reactor and thus protect the exterior environment and population against radioactive leakage. Invented by Russian engineers, two core-catcher reactors were first installed in the Chinese Tianwan nuclear power plant in 2011 as a project of Atomstroyexport—the Russian Federation's nuclear power equipment and service export monopoly and part of the state corporation Rosatom. In March 2014, three years after Fukushima, it was announced that China and Russia had struck a deal to supply two more of Russia's revolutionary reactors for the Tianwan NPP (World Bulletin 2014), which can now be described as both a Chernobyl and a Fukushima effect, with the latter a reflection of the shifting technopolitics of nuclear power development in China post-Fukushima (see also Xiang in this volume).

Launched in 1971, the Fukushima Daiichi NPP was positioned in a seismically active site, and was not equipped with a passive safety system that worked without electric power, or with any reliable external protective system against radioactive leakage. The Russian president Dmitry Medvedev, in his video blog of March 24, 2011, spoke about the Fukushima disaster's consequences for nuclear power development. He emphasized that civil nuclear development depends on new and innovative systems of reactor protection and cited the example of the passive protection system at Kudankulam NPP. This plant, in the southern Indian state of Tamil Nadu, was constructed by Russian engineers, with the first reactor coming online in October 2013.[10] In regard to its safety, Alexander Bychkov, the Deputy Director General of the International Atomic Energy Agency (IAEA) and Head of the Department of Nuclear Energy in Russia, highlighted the technopolitical safety effect of Fukushima: "Nuclear reactors have become safer than they were before the accident [Fukushima], like in many other industries" (Bychkov 2013: 7).

This revised nuclear safety narrative of engineers, experts, and political authorities underlines the linkage between the safety system technology that protects a reactor from possible radioactive leakage and current nuclear program development. In other words, the Fukushima effect is linked to the post-Chernobyl logic of the core-catcher, which is becoming a key element of the safety security system of the "AES 2006" nuclear reactor design, the core reactor design for the Belarusian NPP, the Baltic NPP, and the NPP in Saint Petersburg.

What these innovative installations are trying to prevent is not the potential cause of a nuclear meltdown or radioactive leakage but its possible consequences. This is a kind of logic embedded in technological objects invented to protect the environment and the populations from possible nuclear accidents or incidents. What, we may ask, does it imply more exactly? First, that nuclear technology is completely and inherently safe and accidental situations are impossible. Secondly, if an accident happens, it is not due to the technology but, for example, to human or unforeseen technical elements of its operation and perhaps to external factors such as the earthquake and tsunami in the case of the Fukushima disaster (see Hindmarsh 2013). Enhanced technical safety thus becomes a part of the engineering logic—how to prevent the environment and nearby populations from possible nuclear accident but not how to stop that accident from happening, except for the add-on of enhanced safety systems.

In 2009, a branch director of the company Atomenergoproekt (a designer and contractor of a number of Russian and foreign nuclear power plants), described the core-catcher as a "guarantee for society that in the case of an accident the fission products do not come out in the environment."[11] Post-Chernobyl, the core-catcher was used to illustrate and promote the revised engineering logic on safety. Post-Fukushima, the core-catcher system became a *technopolitical object*, as it reflected innovation and technological efforts to reduce political and social uncertainties and controversies in regard to the safety of AES-2006 NPPs.

On one hand, this claim offers a very strong narrative about how technical solutions are trying to resolve social uncertainties and inquiries. On the other hand, this is an example of the technopolitical logic of technology protecting society from possible radioactive leakage, and the civil nuclear use from possible fears and risks. In other words, it illustrates the technical approach of *containability*: innovative protection components, when integrated into nuclear reactors, make the consequences of a nuclear accident containable. Enhanced containability is now forecasted and embedded into the new security system of the reactor. Containability is claimed in the post-Fukushima nuclear safety narrative to produce nuclear damage "reversibility" to ensure that a similar accident will never happen in the future.

What are, in sum, the political qualities of the core-catcher? I argue that the technical logic of the core-catcher—to catch potential radioactive leakage—meets the political logic of lock-in, to block public discussions and controversies within the contested terrain of nuclear technopolitics. The core-catcher is thus more than the technical part of the safety security system; it is an inherent part of the technopolitical nuclear safety discourse.

According to some experts,[12] almost 50% of Russia's reactors operate today without environmental impact assessment or public discussion. In this mode of functioning, nuclear reactors operate beyond limits stated in initial project documentation. This situation reflects both the state of the nuclear industry in post-Soviet Russia and the accompanying political situation. The political authorities appear to have increasingly turned to the strategy of decision-making lock-in to political and economic interests when they face technical, environmental, or energy controversies; notably, the same strategy also occurred in pre-Fukushima Japan (Hindmarsh 2013). So what effect has Fukushima had on such decision-making?

## The Fukushima Effect on Nuclear Decision-Making

The Fukushima effect on nuclear decision-making in the former USSR as investigated in this chapter concerns three countries, although in Russia an enclosed region: the Baltic NPP (near Kaliningrad in Russia), the Ostrovets NPP (Belarus), and the Visaginas NPP (Lithuania) (see Figure 7.1). The distance between the Baltic NPP and the two others is almost equal—about 300 kilometers—and forms a quasi-geometric "space" for energy, technology and politics interplays. As shown in Figure 7.1, Belarus and the Baltic countries are in the center of a region circled by a number of RMBK reactors still working, under construction, planned, or decommissioned. In the northern part of this region, in Russia, there are the old-fashioned reactors at the Leningrad NPP, with two newly designed reactors under construction. In Lithuania, there is the decommissioned Ignalina NPP, in the Eastern part of Russia is the Smoleskaya NPP, and in the southern part of Ukraine the disastrous Chernobyl-style reactors. The new NPPs in the region are planned to be constructed using Russian and Japanese nuclear technologies, and are in different stages of progress—the Baltic NPP is suspended, the Ostrovets NPP is under construction, and the Visaginas NPP is under discussion. In the following, I discuss each of these case studies in turn, and then together.

*Figure 7.1* Nuclear Power Plants in the Former USSR

### The Baltic NPP (Kaliningrad Region)

The decision to construct the Baltic NPP in Kaliningrad—an enclave of Russia on the Baltic Sea, sandwiched between Poland to the south and Lithuania to the north and east—in 2010 ignored the existing state of the energy system as well as all discussions about and existing projects of energy development in the region. In the beginning of the 1990s, after the USSR collapsed in 1991, the political and energy situations changed. The westernmost Kaliningrad region, near the border of Lithuania (now an EU member), found itself in an enclave situation. The Kaliningrad region was receiving energy from the common power grid and was dependent on transmission lines going through Lithuania. The regional political authorities were afraid of possible energy isolation, which technical experts and scientists refer to as "isolated regime."

To respond to this challenge, various projects were developed to create an autonomous energy system for the Kaliningrad region. The nuclear option had never been considered before, due to the technical "incompatibility" of nuclear power with the energy system. A NPP could only work in basic mode, that is, in only providing the necessary level of consumption. To go beyond this level would mean additional stations—shunting thermal power plants that act as transformers in energy transmission networks—which did not exist in the region, and were very costly to build.[13]

Nevertheless, on February 25, 2010, construction of an AES-2006 2300 MW Baltic NPP began, but was suspended in June 2013 due to social mobilization and NGO activity as well as a lack of potential foreign investors. Social activists and organizations developed different types of public campaigns, from local protests and referendum initiatives[14] to petitions to potential investors.[15] A poll conducted between May 15 and 21, 2013, by the independent Kaliningrad Monitoring Group, showed 48% of local residents opposed the plant's construction, while only 30% supported it (Bellona 2013). In addition, "pressure against the project has been provided from Kaliningrad's immediate neighbor Lithuania, which has repeatedly voiced its disapproval of the Baltic NPP and expressed concerns about its safety" (Bellona 2013).

The Baltic NPP project was also increasingly framed by social activists as a nuclear propaganda campaign by the Russian State Atomic Energy Corporation (Rosatom) to rebrand *nuclear exceptionalism* in Fukushima's post-disaster context. For example, head of the fund ECOSOCIS and member of the Alliance of European Green Parties Alexei Kozlov argued that the Baltic plant could be identified as a "great showcase" built by Rosatom in areas most visible to Europe—the westernmost region of the Russian Federation.[16] According to that view, the Baltic nuclear power plant was politicized to illustrate to western publics and beyond the safety (and superiority) of post-Soviet AES-2006 nuclear technology.

The key post-Fukushima effect in this case can be seen as an effect on the political system to be open (though slightly) to social mobilization on safety uncertainties about civil nuclear use. However, it also implied that the Baltic NPP project could potentially be restarted if the right political conditions prevail, which appears increasingly uncertain in the Baltic region (Bellona 2013).

*Ostrovets NPP (Belarus)*

The Ostrovets NPP case in Belarus illustrates another impact of the Fukushima disaster on nuclear decision-making embedded within a highly authoritarian political context.[17] The political decision to construct the nuclear power plant in Belarus appears on the political agenda as inevitable from an energy point of view (given a lack of national energy resources, just like Japan), and from a technological point of view (given the government's refusal to develop renewable sources of energy), as the following statement by Mikhail Mikhadzyuk, Belarus' Deputy Minister of Energy (March 23, 2011), indicated:

> We must understand correctly, even in light of the events that occurred in Japan [the Fukushima disaster], Belarus needs a nuclear power plant. This is a new qualitative leap in the development of the country; brand new technologies are coming to Belarus . . . Belarus does not possess its own energy resources, so we have no other alternative. There is criticism, but no concrete suggestions: how to do without nuclear power.

However, in Article 10 of the Independence Declaration (June 23, 1990, No. 193-XII) and in Article 18 of the Constitution of the Republic of Belarus (November 24, 1996), the Republic of Belarus was declared a neutral and nuclear free territory. The contrary proponent position then stated by Mikhadzyuk for nuclear power development in Belarus then raised the question of the legitimacy of the proposal and of the political decision about its construction.

The formulation of the nuclear free territory, though, referred to nuclear weapons, not civil nuclear use. In 1983, the first project on building a nuclear cogeneration plant near Minsk began, but was suspended following the Chernobyl disaster.[18] Just over a decade later, in 1998, Belarus abandoned development of a civil nuclear program with a 10-year moratorium. This decision was made possible then as the authoritarian regime in Belarus under President Alexander Lukashenka had not yet consolidated its power. By 2008, when the decision about the construction of a nuclear plant in Belarus was made on January 15 at a session of the Security Council of Belarus headed by Lukashenka, that power had been consolidated. No discussions on alternative projects within the civil or scientific community were publicly enabled. The legitimacy of this decision, however, remains the subject of civil disagreement and social mobilization.

The first acts of social mobilization were accompanied by declarations in the media, organization for the proposed referendum and for the collection of signatures of citizens. For example, the environmental NGO Ecohome and the Belarussian Party of Greens made a joint declaration on November 19, 2009, against the NPP and the lack of public discussion.[19] Then, as the government announced the state of progress of the project in 2010, the anti-nuclear campaigns multiplied and acquired a more grassroots character that complemented expert and NGO mobilization, especially in areas close to the planned NPP. A key strategy was the collection of citizen signatures against the construction of the NPP. For example, by autumn 2009, the group in the region Goretsky had collected over 4000 signatures: 350 in Ostrovetsky, and 4000 in Vitebsk (Novikova 2010).

But the main mobilization activities concentrated on public hearings and discussions as a condition required by the Aarhus and Espoo Conventions that Belarus had signed in 2009, and which had followed the 2008 decision to construct the NPP. The conventions stipulated that in any decision related to the development of a nuclear power plant, the public had the right to get comprehensive information and to participate in decision-making processes. Belarus was found to consistently be in a state of non-compliance with the provisions of these conventions from 2010 onwards, as well as in regard to cross-boundary environmental impact assessment procedures about the proposed NPP with Lithuania.[20]

The double process of expertise and public deliberation has now been launched: from one side the official institutions are trying to organize a process of public hearings and from the other side the NGOs are trying to create a platform for an alternative discussion. This tentative arrangement has not been a successful one; instead of cooperation, it has generated additional tensions between governmental bodies and civic institutions. With the authoritarian regime in power, quite simply, civil nuclear decision-making in Belarus reflected a more closed policy model limiting and restraining public participation (on the latter, see Sovacool 2010).

### Visaginas NPP (Lithuania)

In the Lithuanian case, the shutdown of the two-reactor RBMK Ignalina nuclear power plant after Chernobyl was a determining decision for future energy and nuclear policy developments. Following this nuclear disaster, because of similarities in reactor design, robust anti-nuclear national mobilization stopped the construction of a third reactor at Ignalina and influenced the process of decommission (Dawson 1996). Lithuania agreed to close the plant as part of its accession agreement to the EU. Reactor 1 was closed in December 2004, and the second unit, which supplied about 70% of Lithuania's electricity, was closed on December 31, 2009.

Before the closure of the second unit, on October 12, 2008, a referendum was held in Lithuania. The results—88.7% were against the shutdown—however, had no legal ramifications because of insufficient voter turnout (47.6%). But its political meaning was unclear. Why did the Lithuanian authorities take such a step in the context of the irreversibility of the political decision to close the Ignalina nuclear power plant? The reasons and explanations may be found within the internal political games between contesting political forces, the conservatives and social democrats, or in the area of democratic practices and procedures, where the Lithuanian authorities wanted either to demonstrate the vitality of democratic decision-making or to show the EU the value of the people's choice. Therefore, holding this referendum posed as an attempt to turn back the process of political irreversibility within democratic procedures. Such a strategic democratic game in the context of technology could have been a winning one in 2008, except for the insufficient voter turnout, and may have even been, with more voters, successful in moderating the accession deal to the EU.

The political conditions, though, had changed by 2012, when another referendum was held on the construction of a newly proposed NPP. Several months after the Fukushima disaster, in July 2011, the Lithuanian government announced that Japanese Hitachi GE Nuclear Energy would be the strategic partner and investor in a new Visaginas NPP that would use the infrastructure of the Ignalina NPP as its foundation. While this decision marked less continuity with western technologies (of Europe and the US), it also marked a break with Russian technologies previously central to Lithuanian development since the 1950s, not only in nuclear power. Nevertheless, this announcement fell in the robust period of post-Fukushima debates, with intense social mobilization and sharply divided public opinions pressuring the Lithuanian government to hold a referendum.

In the October 2012 referendum, the Lithuanian people rejected the planned construction of a nuclear power plant at Visaginas. The referendum was held in conjunction with legislative elections; the results marked a change in political attitudes from conservative to social-democratic orientation. At the same time, these results marked the Fukushima effect on the political process as an effect of the openness of nuclear decision-making toward the social mobilization.

After the Ignalina NPP shutdown, anti-nuclear activism declined, but after Fukushima it renewed, and developed new forms of a transnational character. For example, on April 26, 2011, the Lithuanian Green Movement made broader technopolitical claims: the nuclear safety issue was not directed at only the national NPP project but to the wider region, specifically to the Baltic NPP, the Visaginas NPP, and the Ostrovets NPP. The Chernobyl and Fukushima disasters were grouped together in the contestational narrative of the Lithuanian Green Party, Russian Ecodefense and Belarussian Ecohome, which organized a set of common protest actions. This transnational form of protest was effective and made the Lithuanian referendum possible.

To recap: these projects all started from a political lock-in. As such, the decisions to build these NPPs all seemed irreversible. They were represented as irrevocable and inevitable, which reflects the rhetoric of technological determinism (Dickson 1974). However, these three nuclear programs were also pursued in different political regimes, ranging from authoritarian to more democratic. Eventually, they were all confronted by Fukushima effects. In Belarus, the decision to build reactors more or less corresponded to the consolidation of the Lukashenka regime. Its nuclear program grew out of debates and discussions in the 1990s and was cemented within authoritarian political process. In Russia, where the nuclear industry is one of the pillars of state corporatism, the Baltic NPP was represented as a project of openness to foreign investment but not public participation. This civic lack of openness reflected the inherent political nature of the Russian nuclear industry as a closed system or, according to Mumford (1964: 2), as "system-centered, immensely powerful, but inherently unstable."

Finally, the Visaginas case was an example of how, in a more democratic regime, political lock-in faces new democratic procedures that do not—or perhaps cannot—ensure that a political decision will be revised according to issues of democratic legitimacy. The Lithuanian authorities—after the Ignalina NPP shutdown and the negative responses to the construction of the new NPP—have (as of the time of writing) postponed the final decision about the nuclear development program in this country and appear to not dare end the nuclear story in Lithuania.

## Conclusion

The aim of this chapter was to follow the trajectories of nuclear exceptionalism in the post-Soviet States in the post-Chernobyl and post-Fukushima disaster periods, particularly in investigating the scope and extent of the effect that Fukushima had on the so-called nuclear renaissance (or rebuild) in former Soviet countries. Three case studies of the nuclear renaissance were analyzed in this chapter: in Russia the Baltic NPP, in Belarus the Ostrovets NPP, and in Lithuania the Visaginas NPP.

Starting from my first question about the Fukushima impact on reactor safety issues, I proceeded to the nuclear trajectories analysis within national properties of decision-making to outline the technopolitical effect of the Fukushima disaster as a "new type of major nuclear disaster making the close interactivity of the social, technological and natural disaster" (Hindmarsh 2013: 217). I showed that nuclear disaster does not necessarily introduce a long-term policy change into nuclear programs. On the contrary, the Fukushima disaster as well as the Chernobyl disaster illustrated both short-term structural changes within national post-Soviet nuclear policies and long-term technopolitical effects.

As to the latter, the Fukushima effect transferred nuclear safety issues into the field of nuclear technopolitics and helped identify linkages between the technical and political logics within the nuclear exceptionalism discursive terrain. I also found that nuclear exceptionalism took various forms. It played a part in political strategies to pursue nuclear program development despite technological, environmental, and public participation controversies. In this way, post-Soviet nuclear technopolitics can be described as slightly open decision-making in Lithuania, and as political lock-in, closely associated with the characteristics of the political regimes, in Belarus and Russia.

Nuclear exceptionalism was also a part of nuclear engineering logic to redesign the reactor to be inherently safe after the Chernobyl and Fukushima disasters. The new technical component of the core-catcher, for example, was technology designed to protect the reactor from social uncertainties and to make the consequences of a nuclear disaster technically containable and politically convertible to mitigate post-accident nuclear technopolitics.

In sum, all these forms of nuclear exceptionalism illustrate the exceptional technopolitical qualities of civil nuclear production and use based on the logic of exclusion of nuclear safety issues from public debates and participation, scientific discussions, and alternative energy projects. That is, nuclear exceptionalism represents the strategic re-branding of civil nuclear power challenged by the consequences of nuclear disasters and the (re)introduction of the nuclear safety issues onto the political agenda to protect its development. The Fukushima effect played a key role in this technopolitical rebranding, while at the same time improving the safety of new reactors based on the addition of the core-catcher to the reactor design.

## Notes

1 The Soviet type of the LWGR (Light Water Graphite Reactor) and BWR (Boiling Water Reactor) according to official IAEA classification, accessed October 10, 2014, http://goo.gl/bchJnr.
2 Communication from the Commission to the European Council and the European Parliament on the comprehensive risk and safety assessments (stress tests) of nuclear power plants in the European Union and related activities, accessed October 10, 2014, http://tinyurl.com/pfcc9yc.
3 *Россия и Белоруссия подписали межправительственное соглашение о строительстве Белорусской АЭС* [*Russia and Belarus signed the Intergovernmental Agreement about the Construction of the NPP in Belarus*], accessed February 10, 2012, http://tinyurl.com/o7zou87.
4 To get some idea of how much electricity can be generated by a 2300 MW plant, Comanche Peak NPP in the US can meet the annual energy needs of almost 1.2 million homes (see http://goo.gl/tH8LjP, accessed April 3, 2015).
5 Barthe (2006) in his research on nuclear waste discussions in France expands the theoretical and methodological frameworks for the analysis of political

decision-making and scientific debates. He developed the idea of a "lock-in process" as a process of extraction of certain decisions from discussions, and attributes them as a-temporal necessity.

6  For example, a light water reactor with a natural circulation boiling coolant tank (VC-50), and a prototype of the reactor with an organic carrier, were designed but never implemented. A power plant with a reactor cooled by dissociating gas was under development. A power reactor moderated by heavy water and gas cooled engine was established and operated in Czechoslovakia.

7  LWGR (light water graphite reactor) and PWR (pressurized water reactor) according to the IAEA official classification, accessed October 10, 2014, http://goo.gl/L7CH4f.

8  From 1964–1979, several incidents occurred at the Beloyarsk NPP; from 1974–1975 at the Leningrad NPP; and in 1982, a generator explosion occurred at the Armenian NPP.

9  Interview with Sergey Bodrov of NGO "Green World" (Sosnovy Bor, Saint Petersburg, June 21, 2013).

10  See http://blog.da-medvedev.ru/post/153, accessed May 12, 2012.

11  Interview with Sergej Egorov, branch director *Atomenergoproekt* Sosnovy Bor, Saint Petersburg, accessed April 24, 2013, http://tinyurl.com/k6tzomf.

12  Interview with Sergey Bodrov (Sosnovy Bor, Saint Petersburg, June, 21, 2013).

13  Interview with Gnatyuk Victor (July 12, 2013, Kaliningrad).

14  A regional initiative of the referendum on the construction of the Baltic nuclear power plant, guided by the environmental and civil activist Kostayev. This initiative was blocked and Kostayev was later jailed for business fraud.

15  A series of protests and public campaigns was organized by the NGO "ECODEFENSE."

16  See "Эксперт: проект Балтийской АЭС обречен на гибель ['The Expert view: The Baltic NPP is doomed']", accessed October 12, 2014, http://goo.gl/iXlQK1.

17  See http://goo.gl/gGJjQP, accessed April 4, 2015.

18  The Institute of Nuclear Energy of the Academy of Sciences of the Belarusian Soviet Republic developed a proposal on the construction of a heating nuclear power plant (AST) for Belarus in Minsk, Vitebsk, and Mogilev.

19  See http://goo.gl/uoQEOJ, accessed April 4, 2015.

20  See http://goo.gl/jKwPt1, accessed April 5, 2015.

# References

Babcock, G. "The Role of Public Interest Groups in Democratization: Soviet Environmental Groups and Energy Policy-Making, 1985–1991." (PhD diss., University of Chicago, 1997).

Barthe, Y. *Le pouvoir d'indécision. La mise en politique des déchets Nucléaires* [*The Indecision Power. Political Development of Nuclear Waste*]. Paris: Economica, 2006.

Barthe, Y. "Nuclear Waste: The Meaning of Decision-Making." In *Making Nuclear Waste Governable: Deep Underground Disposal and the Challenge of Reversibility*, edited by L. Aparicio, 9–27. Paris: ANDRA, 2010.

Bellona. "Baltic NPP Debacle: Construction Reported Halted, Possibly Mothballed," *Bellona*, May 30, 2013, accessed March 4, 2015, http://goo.gl/b3TKBi.

Brown, K. Plutopia. *Nuclear Families, Atomic Cities, and the Great Soviet and American Plutonium Disasters*. New York: Oxford University Press, 2013.

Bychkov A. "Nuclear Power Today and Tomorrow," IAEA Bulletin 54, no. 1 (2013): 7.

Callon, M., Lascoumes, P., and Y. Barthe. *Acting in an Uncertain World: An Essay on Technical Democracy*. Cambridge, MA: MIT Press, 2011.

Chernavsky, S.Y. "Nuclear Energetics: Development Prospects and Forecast Problems". In *Nuclear Energetics: Development Prospects and Forecast Problems*. International Center of Scientific and Technical Information, Working Consultative Group Attached to the President of the USSR Academy of Sciences for studying new issues of long-term energetics development, 3–25. Moscow: MCNTI, 1988.

Dawson, J.I. *Eco-Nationalism: Anti-Nuclear Activism and National Identity in Russia, Lithuania, and Ukraine*. Durham, NC: Duke University Press, 1996.

Dickson, D. *Alternative Technology and the Politics of Technical Change*. London: Fontana Press, 1974.

Hecht, G. *The Radiance of France: Nuclear Power and National Identity after World War II*. Cambridge, MA: MIT Press, 1998.

Hecht, G. (ed.). *Entangled Geographies: Empire and Technopolitics in the Global Cold War*. Cambridge, MA: MIT Press, 2011.

Hecht, G. *Being Nuclear. Africans and Global Uranium Trade*. Cambridge, MA: MIT Press, 2012.

Hindmarsh, R. (ed.). *Nuclear Disaster at Fukushima Daiichi: Social, Political and Environmental Issues*. New York: Routledge Studies in Science, Technology and Society, 2013.

Jasanoff, S. *Designs on Nature: Science and Democracy in Europe and the United States*. Princeton, NJ: Princeton University Press, 2005.

Josephson, P. "Atomic-powered Communism: Nuclear Culture in Postwar USSR," *Slavic Review* 55 (1996): 297–324.

Josephson, P. *Red Atom: Russia's Nuclear Power Program from Stalin to Today*. Pittsburgh: University of Pittsburgh Press, 2005.

Mandrillon, M.-H. "Les voies du politique en URSS. L'exemple de l'écologie ['Political paths in the USSR. The example of ecology']," *Annales. Économies, Sociétés, Civilisations* 46, no. 6 (1991): 1375–1388.

Marples, D. *Ukraine under Perestroika: Ecology, Economics, and the Workers' Revolt*. London: Macmillan Press, 1991.

Marples, D., and J. Marilyn. *Nuclear Energy and Security in the Former Soviet Union*. Boulder: Westview Press, 1997.

Mumford, L. "Authoritarian and Democratic Technics," *Technology and Culture 5*, no 1 (1964): 1–8.

Novikova, T. *Участие общественности в процессе принятия решений по созданию и размещению АЭС в Беларуси* [*The public participation in the decision making process regarding the construction and placement of the NPP in Belarus*], 2010, accessed October 10, 2014, http://goo.gl/hTUPU4.

Potter, W.C. *Soviet Decision-Making for Chernobyl: An Analysis of the System Performance and Policy*, 1990, accessed January 10, 2015, http://goo.gl/e96pXi.

Raviot, J.-R. "Ecologie et pouvoir en USSR: le rapport à la nature et à l'espace, une source de légitimité politique dans le processus de désoviétisation [Ecology and power in the USSR: the relationship with nature and space, a source of political legitimacy in the de-Sovietization process]." (Ph.D. diss., Institute of Political Studies, Paris, 1995).

Raviot, J.-R. "L'écologie et les forces profondes de la perestroika ['The ecology and the driving forces of Perestroika']," *Diogène*, 194 (2001): 152–159.

Schmid, S. "Transformation Discourse: Nuclear Risk as a Strategic Tool in Late Soviet Politics of Expertise," *Science, Technology and Human Values* 29, no. 3 (2004): 353–376.

Schmid, S. "Envisioning a Technological State. Reactor Design Choices and Political Legitimacy in the Soviet Union and Russia." (Ph.D. diss., Cornell University, 2005).

Schmid, S. "Nuclear Colonization?: Soviet Technopolitics in the Second World." In *Entangled Geographies: Empire and Technopolitics in the Global Cold War* edited by G. Hecht, 125–155. Cambridge, MA: MIT Press, 2011.

Sidorenko, V.A. *Об атомной энергетике, атомных станциях, учителях, коллегах и о себе* [*About nuclear energy, nuclear power plants, about my teachers, colleagues and myself*], Moscow: Izdat, 2010.

Sovacool, B.K. "The Importance of Open and Closed Styles of Energy Research", *Social Studies of Science* 40, no. 6 (2010): 903–930.

Wellock, T.R. "The Children of Chernobyl: Engineers and the Campaign for Safety in Soviet-designed Reactors in Central and Eastern Europe," *History and Technology: An International Journal* 29, no. 1 (2013): 3–32.

Winner, L. *The Whale and the Reactor: A Search for Limits in an Age of High Technology*. Chicago: University of Chicago Press, 1986.

World Bulletin. "China to Use Russian Nuclear Reactor as Prototype," *World Bulletin*, March 24, 2014, accessed April 3, 2015, http://goo.gl/PLxnYB.

World Nuclear Association. "Nuclear Power in Russia 2014," 2014, accessed October 10, 2014, http://tinyurl.com/npol4jm.

Zheludev, I.S., and L.V. Constantinov. "Nuclear Power in the USSR," *IAEA Bulletin* 21, no. 2 (1980): 34–45. Accessed October 10, 2014, http://goo.gl/CYVFal.

# 8    Socio-technical Imaginations of Nuclear Waste Disposal in UK and Finland

*Susan Molyneux-Hodgson and Marika Hietala*

The European Union (EU) states are in the process of developing long-term management plans for high-level nuclear waste, deciding how and where to dispose of their long-lived nuclear inventory. The EU's preferred solution is for each nation to build and use a deep underground waste repository, as built into European policy directives (EC 2004, 2011). The topic of waste is important to reflect upon critically; many countries pin hopes on the (re)emergence of nuclear power as part of the energy mix in responding to global climate change. Yet, as Findlay points out, "[a] major constraint on a global expansion of nuclear energy is the abiding controversy over high-level nuclear waste disposal" (2010: 18); a controversy heightened by the Fukushima disaster, which occurred on March 3, 2011.

The "triple disaster" rendered the entire six-reactor plant at Fukushima Daiichi as "nuclear waste" and, like Chernobyl, the everyday job of managing a power plant was superseded by the complex task of organizing and administering contaminated material. Waste disposal programs for *planned* wastes are thus progressing in policy-making and sociopolitical contexts that now include the *unexpected* wastes of Fukushima as a potent reminder of the risks and hazards of nuclear power. In particular, decisions on the siting of nuclear waste repositories, some potentially close to power plants, are salient. In this chapter, then, we consider the impacts of Fukushima 3/11 on the ongoing waste management projects in two European states, the UK and Finland. We contribute to the socio-technical debates through posing two questions: what can we learn about the policy and practice of global waste disposal programs in light of Fukushima? And to what extent did this major nuclear event impact localized debates of repository siting?

Many nuclear power plants across Europe are reaching the end of their operating lives, and some countries brought forward shutdown dates in response to Fukushima (Spiegel Online 2011). Longstanding issues of spent fuel disposal and the handling of waste resulting from decommissioning have thus been creeping up the political agenda for some time. In line with EU regulation, both the UK and Finland need to dispose of existing stockpiles of waste produced over many years of civil power generation programs. The need to find final disposal solutions for civil nuclear waste is seen by some actors as pressing (BBC 2006).

The world has only one geological repository for the final disposal of nuclear waste.[1] In all other countries we are aware of, including the UK and Finland, high-level waste is currently stored above ground. Proposals for new power generation stations are predicated on the existence of suitable options for the management of the wastes produced. Some commentators view the building of repositories as problematic, saying they enable new build programs to proliferate and serve to merely "hide" the waste out of view rather than more actively manage the problem.

In this chapter we examine debates around waste management and final disposal of waste and explore the extent to which Fukushima has shaped the views expressed by publics and officials. Specifically, we want to examine the role(s) that Fukushima Daiichi played in shifting, bolstering, or negating different perspectives on the ongoing waste disposal projects in the UK and Finland, including in the locale of potential repository sites. We draw on earlier analyses of socio-technical imaginaries (Jasanoff 2004) to inform our analysis of the empirical data. An imaginary can be understood as a collective way of thinking how a future may look and fostering the ways in which that future might be brought into being. An imaginary thus acts as an instrument that can foster change, and enact a particular vision and realize collective ideals. By analyzing public accounts reported in news media, and the official positions recorded in policy and industry documents, we seek to understand positions on the management of nuclear waste post-3/11.

In the following section we review existing literature on disposal debates and expand on the idea of socio-technical imaginaries as a means to explore how disposal is framed in two national contexts. We then present a summary of the disposal contexts, histories, and imaginaries that dominate in the UK and Finland. Results of the analysis of local media reporting and public documents relating to the disposal projects are then presented, first for Finland and then for the UK. We then discuss how this analysis of empirical data relates to the aforementioned imaginaries and offer insights on the impact of Fukushima on the progress of the national disposal projects.

## Approach and Methods

Teräväinen et al. (2011) compared pre-Fukushima public debates on nuclear new build in Finland, the UK, and France. Through analysis of national media and public statements, the authors observed how the use of different discourses was guided by state orientations to nuclear power. They noted how the use of specific discourses in the media, such as "technology-and-industry-knows-best" in Finland, included normative claims about the constitution of the "common good" and about the organization of state-market-civil society relations.

Lehtonen (2013) also presented an analysis focusing on the national media in Finland, the UK, and France, which aimed to understand, in broad terms, any shifts in national nuclear policy following on from Fukushima. This author noted that whereas Fukushima triggered changes in France—at least in the public discourse around nuclear power (see also Szarka in this volume)—little change was visible in the UK and Finland. Lehtonen (2013) further described how pro-nuclear environmentalists were evident in public discourse in the two countries and that demonstrations of "othering," of presenting the Japanese case as different from their own contexts, occurred in all three countries. These findings from analyses of national debates provide a useful background to our concern with the effects of Fukushima on more localized debates and on the discussion of disposal projects.

In our analysis, we were inspired by Jasanoff and Kim (2009), who invoked "sociotechnical imaginaries" as a way to relate politics and technologies in relation to nuclear. National socio-technical imaginaries are described as "collectively imagined forms of social life and social order reflected in the design and fulfilment of nation-specific scientific and/or technological projects" (2009: 120), such as nuclear power. These visions of science and technology in society, governance regimes, and science-society relations perform as "tools to anticipate futures and understand presents and pasts" (Burri 2009). Imaginaries can shape the kind of engagements that the public may or may not be expected to enact and "sociotechnical imaginaries can be identified, illuminated, and critiqued through cross-national comparison" (Jasanoff and Kim 2009: 120).

The comparison of Finland and the UK presented here is selected as part of a larger research program.[2] Our overarching aim is to understand the cultures of waste management in these two nations, which are united by EU regulation in issues of disposal. At the same time, they have both long (UK) and short (Finland) histories as nuclear nations and differing histories of engagement with waste matters. Jasanoff and Kim suggest that innovation policies and regulation arrangements are one place in which imaginaries can be examined: "S&T policies . . . provide unique sites for exploring the role of political culture and practices in stabilizing particular imaginaries, as well as the resources that must be mobilized to represent technological trajectories" (2009: 121).

We thus propose to analyze representations of waste and disposal programs—including repository siting decisions, policy processes, and community reactions—found in official accounts. Given the way that public engagement is framed by national imaginaries, we also consider media accounts from potential repository locations as a space in which nuclear nationhood can play out.

The news media in the regions concerned—Cumbria, UK and Satakunta, Finland—formed a significant part of the empirical material to be analyzed here, complemented by key policy documents. The corpus of Finnish data included coverage on Fukushima in the *Satakunnan Kansa* newspaper and in the online coverage of *YLE Satakunta* (Finnish Broadcasting Company, Satakunta region). In 2011, 76 news pieces carried Fukushima as a reference point. The number fell to 29 in 2012 and 19 in 2013. We also analyzed "The Finnish EU Stress Test" country report (STUK 2011), as well as the newsletters of the Finnish body responsible for implementing the disposal policy, Posiva (Posiva 2014). These newsletters are circulated in communities closest to a repository site to keep them up to date on the progress of the disposal project. In the UK, a single portal provided the online archives covering local newspapers (News & Star, Whitehaven News). It was used to access accounts from all local papers between 2011 and 2014, the period of our investigation. Minutes of local council meetings and written responses to national consultations were available from local government websites, and publications from the national government and its agencies (e.g. the Nuclear Inspectorate) also provided reports which were analyzed.

As a broad sample of public representations, these texts allowed us to probe issues of waste management at multiple levels involving a wide range of actors. This enabled us to tease out different aspects of our exploration, such as how actors occupying different positions on waste utilized local and global "matters of concern" (Latour 2004) to negotiate and justify their views. In sum, we wanted to explore what narratives were evident in the local media during key periods, such as around and immediately following the Fukushima disaster, as well as more recently and at key decision points in waste management plans, and then to position these narratives in relation to the longer-term imaginaries at play in the two nations.

## Disposal Contexts and Imaginaries in Finland and the UK

While deep geological disposal was identified in the 1950s as the most suitable permanent option for managing waste (NAS 1957), the final disposal of nuclear waste has remained an afterthought at best: "Almost six decades after commercial nuclear energy was first generated, not a single government has succeeded in opening such a repository for civilian high-level nuclear waste" (Findlay 2010: 18).

Because of its chemical characteristics, the long cooling periods required, and the relatively small quantities of waste produced annually, there has been little urgency historically to implement final disposal programs (Chapman and Hooper 2012). However, with aging nuclear facilities, increasing decommissioning activities, existing stockpiles of spent fuel, and plans for new power plants emerging, disposal is becoming a more important policy issue (OECD 2012).

## Policy Contexts

The first operational civil power reactor in the world began generating in 1957 (at Calder Hall, Cumbria, UK), and a further four reactors were operational at that site by the mid-1970s. Yet the management of nuclear waste was only first seriously addressed in the UK in the 1976 Flowers Report:

> There should be no commitment to a large programme of nuclear fission power until it has been demonstrated beyond reasonable doubt that a method exists to ensure the safe containment of long lived, highly radioactive waste for the indefinite future.
>
> (RCEP 1976: 131)

Since then, a number of policy and practical attempts to deal with high-level waste have begun but none, so far, have reached fruition in the UK. In contrast, Finland began generating civil nuclear power much later than the UK, not until the mid-1970s, and Finnish plans for an operational geological repository have progressed much further.

Currently, all countries in the EU are undergoing the policy process associated with the fulfillment of EU directives on radioactive waste disposal, including spent fuel (EC 2004, 2011). The directives require final disposal of all high- and medium-level wastes into deep geological repositories, although the timeframes for each country in the EU are different and the processes and procedures towards the implementation of final disposal are decided at a national level, by national implementation bodies. The character and volume of waste requiring final disposal varies greatly between countries, depending on factors such as the number of reactors and histories of engagement in weapons programs. Different countries also categorize waste in differing ways and quantities are reported using a variety of measures. A brief comparison of the current situations in the UK and Finland is provided in Table 8.1.

*Table 8.1* A Comparison of Key Features of the Nuclear Industries in UK and Finland

|  | UK | Finland |
|---|---|---|
| **First reactor** | 1956 (Calder Hall) | 1977 (Loviisa 1) |
| **Number reactors** | | |
| Operational | 16 | 4 |
| Shutdown/decommissioned | 26 | – |
| **New Build** | | |
| Planned | 11 | 2 |
| Under construction | – | 1 |
| **Volume of high level waste** | 3,300 tons | 1,930 tons |

Sources: OECD 2014; NDA 2014; WNA n.d. a, b

By 2014, only four of the 28 countries of the EU—Finland, France, Switzerland, and Sweden—had made notable progress with the policy directive, mainly by identifying an underground repository site for the final disposal of waste or by undertaking preliminary research in underground laboratories. Under the single EU directive, each nation-state is taking a different approach towards identification of their repository site. Some nations are driven primarily by geological considerations and others by existing political and social conditions. Public consultation and "host community involvement" is an integral element of the decision making process on the siting of any proposed waste repository (EC 2011: 7). However, each country enacts stakeholder consultation in different ways.

The predominant design concept for deep geological disposal, displayed in Figure 8.1, involves a series of barriers, including the geological environment where the disposal facility is built and an engineered barrier system (EBS). The EBS, which typically consists of a variety of components including the waste form, disposal canister, buffer, backfill, plugs, and seals, plays "a central role" in ensuring safety for thousands of years (CORDIS 2014).

The events of 3/11 significantly raised concerns over nuclear safety across Europe as well as elsewhere in the world. Two weeks after Fukushima, the heads of EU Member States agreed that the safety of all European power plants should be reviewed. The European Commission and the European Nuclear Safety Regulators Group (ENSREG) were invited to develop a "comprehensive and transparent risk assessment" with the full involvement of member states. Known as the EU Stress Tests (EC 2013), these were made to assess both natural and man-made risks to nuclear safety in power plant operations. While assessing the ability to cope with incidents such as earthquakes, floods, and human error, the tests omitted some risks, including security breaches. The tests were conducted by both plant operators and by independent regulators and underwent a peer review process during the remainder of 2011 and through 2012. While

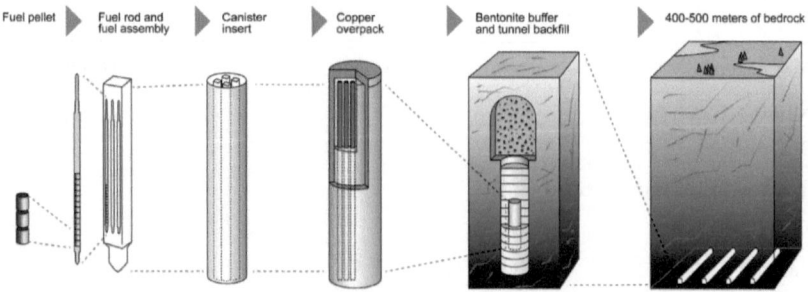

*Figure 8.1* Multi-barrier Disposal Concept Adopted by Posiva, Finland

Source: Image courtesy of Posiva Oy

demonstrating a certain level of responsiveness to the nuclear disaster in Japan, these assessments merely reinforced the existing technocratic approach to the handling of nuclear matters, bringing in more technical processes and checks and bolstering the idea that safety can be ensured through managerial procedures.

## The Disposal Situation in March 2011

In Finland, the "nuclear waste problem" was considered to be solved by the time of 3/11. Consultations on finding a disposal site were officially completed. The municipality of Eurajoki in Satakunta, Western Finland, had been selected as the site for the Finnish repository in 2001, and excavations of an underground characterization facility, Onkalo, had begun in 2004. In the UK at the time of 3/11, the process to identify a repository site was at a crucial stage. It was mid-way through an evidence gathering exercise with multiple stakeholders, including a public consultation process required by EU Directive.

The UK process had begun in 2008 with the publication, by the national government, of the White Paper "Managing Radioactive Waste Safely." The collation of evidence from 2008 onwards would result in a major report—finally published in August 2012—intended to form the basis of a decision on where to site the proposed geological repository for final disposal of the UK stock of waste. However, the whole process, begun in 2008, was halted, unresolved, in mid-2013 when one of the several layers of local decision-making decided to withdraw from the siting process discussions (Cumbria County Council 2015).

## Public Perspectives

According to regular surveys conducted by the UK Energy Research Centre, "UK public attitudes to nuclear power have become more favourable" since Fukushima 3/11 (e.g. ESRC 2013). Yet this deceptively simple statement hides a more conflicting set of views and nuanced engagements between the public and the nuclear sphere. These same surveys also show that nuclear power remains at the bottom of the UK public's preferred set of energy generating options and that proposed nuclear new build projects, such as in the southwest of the country, face strong opposition. In contrast, in Finland, while renewable energy technologies are favored over nuclear power, trust in the nuclear industry and nuclear safety is high. Generally, Finnish attitudes toward nuclear power have remained positive, although marginally less so than before Fukushima 3/11 (Energiateollisuus 2013).

Opposition to new build projects in the southwest UK, surveys find, is as much to do with the erection of electricity pylons as it is about the nuclear power plant and the failure on the "other side of the planet" that Fukushima represented (e.g. ESRC 2013). At several public debates, the specter of long-term disruption from construction traffic has dominated local concerns. In contrast, opposition to new nuclear plants in the north of Finland has stemmed from the problem of Russian involvement in the project, with both longstanding and more recent geopolitical issues between the two countries brought to the fore. These points serve to highlight the often complex meaning-making that people undertake when considering questions of "nuclearity" (Hecht 2012), with numerous "non-nuclear" matters brought into the debates. We suggest that the repercussions of events worldwide, including those of Fukushima, interact with, rather than dominate simplistically, more localized concerns (see Hindmarsh 2014 for similar findings on wind power).

While the public bring in arguments that contextualize and situate nuclear schemes—e.g. the politics and economics of foreign ownership, and the everyday disorder created by large building projects—official bodies have been criticized for ignoring the "radioactive character" of nuclear projects. For example, in approaches to the public consultation elements of repository siting, Gregson (2012) has argued that in the UK, high-level waste has been presented by the implementing authorities as "ordered, disciplined, contained and tamed," and as though the waste is "devoid of radioactivity." Official bodies are thus accused of *cleansing* the waste before it can be discussed publically. This stands in contrast to the situation at Fukushima, where disorder and a lack of containment were immediate concerns (Hindmarsh 2013).

### Toward National Imaginaries

The UK and Finnish contexts for final disposal present an interesting contrast in several ways, such as in terms of the volume and character of the waste requiring attention, the roles played by geology, and the differing methods used in public deliberation for the siting of waste disposal facilities. Historically, the UK has seen itself as at the forefront of "nuclear nations." It has reactor designs not seen elsewhere in the world; it has traditionally exported expertise and know-how; and its reprocessing facilities have serviced the world. This sense of technical reputation has been bolstered more recently by the pursuit of an ambitious new build program, thereby reinforcing a national vision of technology-led solutions to energy needs. Given this, it is perhaps in an awkward position currently, in that implementation of geological disposal is not further forward. Indeed, in some respects it has barely started.

Over the disparate lengths of time that the UK and Finland have been nuclear nations, both countries have prioritized technical arguments over other forms of rationality. This has been consistent over time, regardless of changes to political government and shifts on nuclear policy. The dominant imaginaries are of *engineering approaches to safety* and of *complex political/regulatory oversight* by multiple organizations comprised of technical experts. The public is invited into discussions and consultation only at times and places allowed by the policy process, which is itself shaped by expert voices.

While both the UK and Finland must respond to the same EU directive, we have found that the implementation of the geological disposal policy has experienced a smoother ride, so far, in Finland. In the next two sections we analyze in greater depth some of the aspects of the national approaches to disposal by considering responses to Fukushima and draw into this empirical data gleaned from local media and official reports. In this way we can "flesh out" the imaginaries further and identify and better understand any Fukushima effects.

## Finnish Narratives on Fukushima

In this section, we present our analysis of the local media, and official and industry accounts from Finland. At the time of the Fukushima accident, site characterization of the underground repository was well underway and the implementing authority, Posiva, was preparing a construction license application. The emergence of a new nuclear plant operator—Fennovoima—on the scene in 2010 had re-ignited the debate around deep geological disposal, in terms of whose waste could be disposed. Posiva, invoking arguments about decreased safety, had categorically refused to accommodate Fennovoima's waste in the final repository, against Fennovoima's expectations. These debates, however, are not seen to have influenced the general course of the main disposal project (YLE 2012). The final disposal site in the Satakunta region, of which the existing underground facility of Onkalo would form a part, continues to be developed. The application for final repository construction was subsequently submitted in 2012 and was approved in early 2015 (WNA n.d. a).

In our analysis of available archives and public documents, we observed no change in the national position on nuclear waste disposal in relation to 3/11. Indeed, in the local news media we analyzed, the Finnish disposal project was hardly discussed in conjunction with 3/11. The events at Fukushima were primarily used to bolster a pre-existing narrative of the exceptionalism of Finnish nuclear safety culture more generally. Indeed, following the IAEA's General Director's visit to Onkalo in 2012, a member of the Green Party expressed frustration over the official nuclear talk during the visit. Discussions had focused on the progress made by the Finnish disposal project, while a critical evaluation of Fukushima's impact on the project and the future of the Finnish nuclear industry were entirely absent (Meri 2012). Interestingly, the mobilization of a "safety culture" exceptionalism is remarkably similar to the arguments employed by Japanese authorities and regulators in their response to events at Three Mile Island in the USA and Chernobyl (Hara 2013: 27).

The absence of discussion and debate in relation to nuclear waste disposal was observable across all the local media. Only Posiva made a direct connection between Fukushima and the ongoing disposal project. In its newsletter published in March 2011, Posiva asserted that in the wake of Fukushima "the justification for final disposal has grown increasingly strong. Storing spent [nuclear] fuel cannot go on forever" (Seppälä 2011). Apart from this intervention from industry, and its utilization of the events at Fukushima to *support* the Finnish disposal project, discussion in the public realm focused more on Fukushima's potential impact on nuclear new build projects, but even here the message can be described as one of "business as usual."

A few days after the disaster, Finnish Prime Minister Mari Kiviniemi asserted that Fukushima would not have any influence on Finnish nuclear policy, and maintained that the government would back up new build projects even after 3/11. This stance was also evident among local political actors. For example, one local politician stated that in terms of economic gain and regional development, nuclear power is "akin to winning the lottery" (Ståhle 2011). Another asserted: "[H]ere in the cold North, nuclear power is needed to fill the needs of our industry and wellbeing" (Lehtonen 2011).

The local population also appeared unshaken by the Fukushima event. As *Satakunnan Kansa* reported, five out of six people in the region did not change their attitudes towards nuclear following 3/11 but continued to support nuclear power (Pesonen 2011). Meanwhile, the CEO of the Finnish Forest Industries Federation expressed relief over the government's continuing support for nuclear power to fuel the nationally important forestry industries, and went on to underline that while the "lessons from Japan should not be ignored, the plants and their locations [in Finland and Japan] are entirely different" (Suni 2011c).

Indeed, these allusions to forms of engineering ("the plants") and to the geography ("their locations") were referenced in much public discussion. Beginning with the early news reports, Fukushima and its repercussions were effectively and efficiently kept contained *within* Japan, with arguments emphasizing differences in geology, engineering and regulatory practices between Japan and Finland. The assertion by the prime minister (Ståhle 2011) that Fukushima was a sign of "how vulnerable nuclear power is in those kinds of regions," was echoed by a retired nuclear CEO who also pointed to geology, maintaining that Fukushima was a "good plant in the wrong place" (Peltonen 2012). Writing a year after the accident one commentator argued that because "the power plant did not just blow up on its own" and the accident "was caused by a massive tsunami and an earthquake," there was no need for critical discussion over the future of nuclear power in geologically stable Finland (Helariutta 2012; cf. Hindmarsh 2013).

The concept of an inherent superior engineering expertise and "safety culture" in the Finish nuclear industry were offered as reasons why Fukushima should not impact Finnish nuclear policy and practice. For example, a representative of the Finnish Radiation and Safety Authority (STUK) asserted that "Fukushima would not have been built in Finland even in the 1970s," when the regulations were more lenient (Suni 2011b). In a similar vein, a professor of nuclear technology pointed to safety culture as a "fundamental difference" between the two countries; while a further commentator saw Fukushima as a "manifestation of a conscious technological risk" (Suni 2011a).

In the local media, much was written about the failure of Fukushima's emergency diesel generators. The number of emergency generators was seen to justify the superiority of Finnish safety culture. At Fukushima there had been two emergency generators per reactor to power the reactor cooling system in case of accidents, in contrast to the four back-up generators required per nuclear facility in Finland. A STUK representative maintained that because of this more comprehensive emergency backup system, a "corresponding cluster of accidents in Finland [was] extremely improbable" (Suni 2011a).

The accident at Fukushima, while being attributed to Japanese indifference to safety, was thus also used to validate Finnish safety culture. Some commentators observed how STUK's standing had shifted internationally in the nuclear industry's eye since 3/11. The Authority had earlier gained an international reputation of being squeamish about details and regulations and had, for instance, been accused of being overly meticulous for purposefully slowing down and unnecessarily complicating the construction of the third reactor at Olkiluoto in Eurajoki (Ahvenjärvi 2011). By the end of 2011, however, another commentator had noted that following 3/11 these kinds of accusations had quieted down, and that STUK had much wider support for its activities. Other countries were reported to have quietly adopted STUK practices, while STUK commented that this shift in view "confirmed to us that we have been on the right track all along" (Saarinen 2011).

The mobilization of geography, geology, safety culture, and institutional oversight can thus be understood as promoting a sense of exceptionalism. This narrative of "Finnish exceptionalism" was also visible in the reporting on the application of the EU Stress Tests. While the local media described the EU as having been "woken up" to nuclear safety issues by 3/11, risk assessments at Olkiluoto were discussed in the media as *routine*, with the Stress Tests described as being little more than "homework" (Suni 2011d). The same view was discernible in Finland's country report to the EC. In the report, STUK held that

> the issues raised within the 'Stress Tests' have been a part of Olkiluoto 1 and 2 PSA [Performance and Safety Assessment] studies since the 1990's, and risk-based improvements have been carried out by plant modifications and through revising and updating the emergency operating procedures.
>
> (STUK 2011: 175)

While the tests revealed some deficiencies in the emergency water supply system at Olkiluoto, they were not reported as exposing any significant weaknesses in nuclear safety. Rather, 3/11 and the subsequent stress tests were described as enhancing the Olkiluoto plant by calling for an update of the emergency backup system to ensure that Olkiluoto would be better equipped to survive long power outages (Ahonen 2011).

It is apparent that Fukushima did little to change pre-existing narratives in Satakunta; rather, the accident was used as an opportunity to reinforce them. Lehtonen (2013) has suggested that the efficiency of Finnish reactors and the absence of nuclear incidents in Finland might in part explain Finnish reactions to Fukushima. Indeed, in the local news media, significant space was given to the comparison between Finnish and Japanese, and also other European, safety cultures. In these comparisons, the preparedness to act in emergencies and the prevention of accidents were described as better at the Olkiluoto plant than at Fukushima or across Europe. Trust in Finnish engineering and regulatory practice was observable in local media discussion. Alongside this, the underlined stability of the Finnish bedrock, and trust in engineering expertise and the notion of safety as routine were central to the construction of a narrative of "it would never happen here" in the local news media.

Reactions to Fukushima can also be understood through the imaginary of nuclear power as a pillar of the Finnish welfare state, tied in to a national sense of technical prowess that can deliver energy self-sufficiency. The "growth, welfare and employment connotations of nuclear power" (Lammi 2004: 14) were evident in the local media discussions after 3/11. In Satakunta, and Eurajoki in particular, dependency on the fortunes of the nuclear industry is magnified as the Olkiluoto plant and Posiva are the major employers in the locale and are major taxpayers in Eurajoki, thus directly contributing to the economic wellbeing of the region. The media discussion around Fukushima was equally, if not more, about the future shape of Finnish society and industries, and more locally about the future of Satakunta rather than about Fukushima itself. Giving up nuclear power was thus considered a greater risk to the future welfare of the region, and the nation, than those risks presented by ongoing commitment to nuclear power.

The importance of engineering and nuclear power in the imaginations of the Finnish welfare state may also open up the quiescence around nuclear waste disposal in "local nuclear talk" after Fukushima. Lammi (2004) has observed that, even when the government Decision-in-Principle on nuclear waste disposal was made in 2001, the risks of nuclear waste disposal and nuclear power more generally took a backseat to the development narrative of nuclear power in public discussion. After the Decision-in-Principle and political consensus reached on the national level, opponents to nuclear waste disposal lost the opportunity to highlight the risks of nuclear disposal in future discussions. The Decision-in-Principle presented to the nation a solution to the "waste problem," and, in a Latourian sense, the idea of deep geological disposal of nuclear waste in Eurajoki had become a deeply entrenched matter of fact by 3/11. This acceptance of geological disposal is confirmed by the negligible impact of the more recent debates about Fennovoima's right to dispose of its waste in Onkalo on the logic of geological disposal and the practical disposal project itself. With the politics of nuclear waste disposal already "cold" by March 2011, Fukushima presented Posiva with an opportunity to further argue for the increased safety that deep geological disposal would provide as opposed to the risks—demonstrated at Fukushima—of keeping nuclear waste above ground.

## UK Narratives on Fukushima

The Fukushima event occurred during a critically important period for decision-making in the UK, relating to the siting of the national deep geological disposal facility. Earlier siting processes had failed, and resulted in major changes to the governance structures and public bodies responsible for nuclear waste management. Subsequently, a new policy was published by the UK government in 2008, which began a revised consultation process to identify a site for the national repository. The policy involved the setting up of new bodies to administer policy implementation (e.g. the Managing Radioactive Waste Safely partnerships). It would require several levels of stakeholders to agree to move forward, including potential host communities, represented by district councils; and regional local governments, represented by elected councilors sitting on county councils. The majority of high level waste (~70%) is currently stored on a site in north-west England, and only communities in that area—West Cumbria—formally expressed an interest in taking a direct part in the policy process to explore siting options. Although any community in the country could take part, few local authorities even consulted locally on whether to participate, leaving West Cumbria as the only option.

The West Cumbria Managing Radioactive Waste Safely Partnership (WCMRWS), which involved a wide range of organizations and community groups, began work on its major siting report in 2009. In late summer 2012, the partnership published its 280-page "definitive report" of evidence to be considered by local councils and politicians. The report—the outcome of a three-year process of collating information and gathering evidence from "about 2,300 people and organisations" (WCMRWS 2012a)—had the aim of enabling the various councils in the vicinity of the potential site location to decide on whether to engage further in discussions with national governmental bodies on identifying an underground disposal site within the broader community.

Following a series of surveys, the WCMRWS concluded that, "across Cumbria there are more people in favour of taking part in the search for a suitable site than people who oppose talking part" (WCMRWS 2012b: 10). Even so, the final report stated "a lack of trust appears to us to be at the root of many of the key concerns raised by the public and stakeholders" (WCMRWS 2012b: 8). In January 2013, the two district councils voted to continue to engage with the government process, but a "no" vote at the county council level halted the activity of looking for a deep geological disposal site in West Cumbria, once again leaving the UK without a way forward. The WCMRWS site selection process launched by the 2008 policy was subsequently closed.

The timeframe of the WCMRWS process meant that events at Fukushima took place in the middle of the data-gathering and deliberation exercises. The rejection of the policy process in 2013 prompted governmental reviews of the entire process and a consultation on a revised policy and implementation process was begun. At no point, however, was Fukushima (or other related nuclear events) mentioned in either news reports on the disposal project or in official reports, consultations, and responses.

Analysis of local media following the "no" decision in Cumbria in January 2013 demonstrates a diversity of perspectives within the community, and that those responses were immediate. The local Member of Parliament (Jamie Reed, who sits in the national government) vowed that "the fight is not over" to build a repository (News & Star 2013a); workers at the current storage site held emergency meetings to find ways forward (Duncan 2013); and the joint Trade Unions "branded the [no] decision . . . short-sighted and insulting" (News and Star 2013b). The senior county council member with the nuclear brief even resigned. People associated with the industry and local government in the immediate locale of the potential sites were reported to be astounded by the "no" vote at the regional level. They quickly organized to meet with the national government in efforts to "revive the county's search," describing their view of the no vote as "bitterly disappointing" (Coleman 2013). The future of local jobs and the wider economy of West Cumbria were seen to be at risk. However, campaigners against the site being in Cumbria were dismayed that local politicians were acting to promote the repository site in spite of the WCMRWS process stopping: "They told us we could trust the Government's Managing Radioactive Waste Safely process but now that it's ended, they want to continue and overturn the county council's decision" (anti-nuclear campaigner, quoted in Coleman 2013).

It seems striking that Fukushima did not figure in either the final report, decisions on whether to continue or not to engage in the process, or the review of the whole process after closure. To explore why this may be the case we need to backtrack to how the events of 3/11 were reported at the time. We find that local media reports on the Fukushima events centered on specific human story links between the two regions of Cumbria and Tōhoku. For example, one report discussed a UK teacher "caught up in the horrors" of the quake (although the teacher was not from

Cumbria). Nuclear-related stories included topics such as the nuclear reprocessing plant in Cumbria-Sellafield preparing to send personnel and equipment to Fukushima to assist. Also reported was the potential threat to jobs at Sellafield as the subsequent shutdown of the entire Japanese nuclear program meant that imports for reprocessing, from one of Sellafield's largest customers, were cancelled. Jamie Reed relayed a warning to local folk "against a knee-jerk reaction to the county's nuclear energy plans following the disaster in Japan" (Bourley 2011); one of the first reactions to 3/11 was this message of "don't panic" from the national representative for the region. This highlighted the region's sensitivity to its identity and position as a center of nuclear expertise.

National reviews prompted by 3/11 and responses from regulators and industrial bodies included the set of European Stress Tests (EC 2013), as mentioned earlier in relation to Finland. In the UK, the nuclear inspectorate led a review of all civil nuclear sites to ensure compliance with key requirements (e.g. access to seven days of cooling water). A separate review was also commissioned and, in October 2011, the UK government released a final report on Fukushima to Parliament, produced by the Chief Nuclear Inspector (Weightman 2011). The government announced:

> there are no fundamental weaknesses in the UK nuclear licensing regime or the safety assessment principles . . . the UK practice of periodic safety reviews of licensed sites provides a robust means of ensuring continuous improvement in line with advances in technology and standards . . . the events at Fukushima reinforce the need to continue to pursue decommissioning of former nuclear sites with utmost vigour and determination.
>
> (DECC 2011)

In response to the report, the Secretary of State for Energy and Climate Change stated: "The report makes clear that the UK has one of the best nuclear safety regimes in the world and that nuclear power can go on powering homes and businesses across the UK, as well as supporting jobs" (DECC 2011).Industry responses were briefly reported in the local media (e.g. Irving 2012a, 2012b), and a national paper (the *Daily Mail*) reported on one UK reactor being shut down due to inadequate sea defenses (Steere and Duell 2014). Overall, the outcomes of the stress tests were low-key and the main commissioned report served to confirm a robust safety regime according to the regulatory definitions.

Over the year or so following 3/11, there were several reports of networks, involving a sense of support and camaraderie, being built between the professional nuclear communities at the Sellafield and Fukushima sites. Relationships built between the two sites were primarily based on a mutual exchange of expertise—particularly around the decommissioning processes—and business opportunities for the Cumbrian region.

Finally, we note that Fukushima *did* appear as part of public debates in relation to recent announcements on new nuclear build projects in West Cumbria. In May 2014, agreements were signed to move forwards the building of three new UK reactors with the potential to "result in 21,000 jobs over the construction period" and "sustain around 1000 permanent jobs over the course of the reactors' lifetimes" (North-West Evening Mail 2014a). The announcement was "slammed" by the Green Party and branded "dangerous": "As we saw at Fukushima, nuclear is a dangerous technology" (North-West Evening Mail 2014b). Yet again, local concerns of employment and the region's reputation as a center of nuclear expertise had continued to dominate the debate. Thus, despite global catastrophes and voices such as the Green Party, local affairs won out.

## Conclusion

In this chapter we set out to investigate the policy and practice of waste disposal programs in light of Fukushima 3/11. We considered the cases of the UK and Finland, paying particular attention to national plans for long-term waste disposal, through explorations of publicly available accounts relating to nuclear waste disposal. Analysis of reports from media, governmental, and industry outlets aimed to determine whether Fukushima shaped ongoing disposal debates. Overall, we found that Fukushima did little to change the trajectories or scope of disposal debates. This relative absence of impacts found in both countries was explored using the lens of socio-technical imaginaries. Events of 3/11 did not challenge existing imaginaries in either country. Rather, the effect of Fukushima was more to inform arguments constructed by key actors in the debates that *bolstered specific aspects of existing imaginations*, such as the superiority of western European safety cultures, which, in turn, reinforced the discourse of nuclear power as being central to the economic wellbeing of remote regions.

That nuclear disasters around the world can have variable impacts in different countries is not new (see Felt 2013 for a description of how Chernobyl cemented the Austrian imagination as a nuclear-free country while other European countries continued with their programs). However, it is intriguing that the impact on disposal debates in two European countries that were in the midst of the planning stages of major waste disposal programs was negligible. The impact of the disaster on disposal debates was practically invisible in the local news media around the anticipated repository sites.

In the UK, this absence is perhaps more intriguing, given that 3/11 occurred during a critical period in the decision-making process to site the national waste disposal facility. It appears that Fukushima was not deemed relevant to the debate—for whatever reasons—on the location of a final disposal site. This supports Gregson's (2012) assessment that the politics of nuclear debates beyond an immediate locality tend to go "cold." In other words, a disaster on the other side of the world proved insufficient to override existing narratives that had dominated the siting process for some years, such as local narratives of job creation (in support of the site) and of landscape protection (against the site).

Fukushima has been described as a "new type" of nuclear disaster, resulting from the conjunction of natural and chronic technological failure (Hindmarsh 2013: 3). European countries such as Finland and the UK, successfully mobilized a narrative of "never here" to minimize potential impacts from Fukushima. The possibility of the new type of techno-disaster was kept at bay by a series of tropes, including: "we don't have earthquakes," "we don't have tsunamis," "we have infallible engineers" (Lammi 2004), "we have better safety cultures," and, "we have more pressing local concerns" (e.g. jobs).

Deep geological disposal projects are described by their implementers as "engineered facilities" and current proposals to dispose of waste involve multiple, engineered barriers (CORDIS 2014). Thus, the notion of *engineering* is central to official positions on the management of waste across Europe. The apparent alignment between official discourses on the primacy of the engineering approach to waste disposal and the valuing of the engineering identity in Finnish society goes some way to explaining why Fukushima had little noticeable influence on nuclear talk in Finland. The engineering community in this country is held in sufficiently high regard that an exceptionalism argument could be successfully maintained. In contrast, in the UK, a legacy of distrust in organizational and political expertise relating to nuclear affairs has led to a position where many will never believe what is said regardless of what is being said.

The impacts of a sudden event such as the Tōhoku earthquake, a direct cause of the Fukushima disaster, on long-term processes such as solutions to the final disposal of waste, therefore, appear to be minimal for the two countries we have considered. Neither public views nor policy processes were found to shift in any meaningful way. Rather, the most noticeable effect of the Fukushima disaster was its enrollment to reinforce existing policies, entrenched imaginaries, and associated trajectories of action. We are therefore left with a reflective question: is there any form of nuclear event that would prompt greater national reflections on the management of waste disposal programs?

## Notes

1 The Waste Isolation Pilot Plant (WiPP) in New Mexico, USA, nominally began operations in 1999. This underground repository handles only defense-related nuclear wastes and is not involved in civil nuclear disposal projects. The repository has since suspended usual operations due to two incidents in February 2014 (see www.wipp.energy.gov/wipprecovery/recovery.html, accessed September 14, 2014).

2  Marika Hietala's doctoral research on nuclear waste disposal cultures in the UK and Finland is part of the "Nuclear Societies" PhD program at the University of Sheffield. The program is funded by the Economic and Social Research Council as part of an initiative to create a multi-disciplinary community of researchers capable of critically engaging with future research agendas relating to nuclear energy in society.

# References

Ahonen, H. "Olkiluotoon vaaditaan järjestelmä pitkän sähkökatkoksen varalle ['A system required at Olkiluoto in case of long power outage']." *Satakunnan Kansa*, December 31, 2011, accessed August 15, 2014, http://tinyurl.com/k4ga34d.

Ahvenjärvi, S. "Ihminen on heikoin lenkki ['Man is the weakest link']." *Satakunnan Kansa*, March 16, 2011, accessed August 15, 2014, http://tinyurl.com/ohsbkke.

BBC, " 'Urgency needed' on nuclear waste," July 31, 2006, accessed January 6, 2015, http://tinyurl.com/h4g53.

Bourley, A. "Cumbrian MP Warns Against Knee-Jerk Reaction to Japan Nuclear Crisis," *News & Star*, March 15, 2011, accessed November 4, 2014, http://tinyurl.com/qyuxeuy.

Burri, R.V. "Sociotechnical Imaginaries and the Politics of Science and Technology," presented at Society for the Social Studies of Science (4S), Washington, DC, USA, October 2009, accessed November 20 2014, http://tinyurl.com/nbnzac5.

Chapman, N., and A. Hooper. "The Disposal of Radioactive Wastes Underground," *Proceedings of the Geologists' Association*, 123, 2012: 46–63.

Coleman, P. "Cumbrian Politicians to Meet Minister in Bid to Revive Nuclear Store Site Search," *News & Star*, February 4, 2013, accessed November 4, 2014, http://tinyurl.com/m9donnk.

Community Research and Development Information Service (CORDIS). "Long-term Disposal of Radioactive Waste," 2014, accessed November 4, 2014, http://tinyurl.com/lv7r457.

Cumbria County Council. "Managing Radioactive Waste Safely," accessed January 6, 2015, http://tinyurl.com/pyfjwph.

Department for Business, Enterprise & Regulatory Reform (BERR), Department of the Environment, Department for Environmental Food and Rural Affairs (DEFRA) and Welsh Assembly Government. *Managing Radioactive Waste Safely: A Framework for Implementing Geological Disposal*, 2008, accessed November 4, 2014, http://tinyurl.com/lxty8et.

Department of Energy and Climate Change (DECC). "Announcement: Final Fukushima Report Published," October 11, 2011, accessed September 12, 2015, www.gov.uk/government/news/final-fukushima-report-published.

Department of Energy and Climate Change (DECC). *Implementing Geological Disposal: A Framework for the Long-term Management of Higher Activity Radioactive Waste*, 2014, accessed November 4, 2014, http://tinyurl.com/qgx4ztd.

Duncan, I. "Sellafield Workers Hold Emergency Meeting after Council 'No' Vote," *News & Star*, January 31, 2013, accessed November 4, 2014, http://tinyurl.com/nf6rdqo.

Economic and Social Research Council (ESRC). "Less Opposition to Nuclear Power," 2013, accessed November 4, 2014, http://tinyurl.com/kzfmxqy.

Energiateollisuus *Suomalaisten Energia-asenteet 2011* [*Finnish energy attitudes 2011*], 2012, accessed, August 15, 2014, http://tinyurl.com/nxnukas.

Energiateollisuus. *Suomalaisten Energia-asenteet 2013* [*Finnish energy attitudes 2013*], 2014, accessed August 15, 2014. http://tinyurl.com/kbcp5uv.

European Commission (EC). *Council Directive 2011/70/EURATOM of 19th July 2011 Establishing a Community Framework for the Responsible and Safe Management of Spent Fuel and Radioactive Waste*. Brussels: EC, 2011.

European Commission (EC). *Geological Disposal of Radioactive Wastes Produced by Nuclear Power: From Concept to Implementation. Report of the European Commission*, EUR 21224. Luxembourg: EC, 2004.

European Commission (EC). *Implementing Geological Disposal of Radioactive Waste Technology Platform*, 2013, accessed November 4, 2014, http://tinyurl.com/kegarwf.

European Commission (EC). "Nuclear Energy: Stress Tests," 2013, accessed November 4, 2014, http://tinyurl.com/6vp6v8f.

Felt, U. *Keeping Technologies Out: Sociotechnical Imaginaries and the Formation of a National Technopolitical Identity*, 2013, accessed August 15, 2014, http://tinyurl.com/mrlz3nk.

Findlay, T. *The Future of Nuclear Energy to 2030 and Its Implications for Safety, Security and Nonproliferation*, 2010, accessed November 21, 2014, http://tinyurl.com/nbym535.

Gregson, N. "Projected Futures: The Political Matter of UK Higher Activity Radioactive Waste," *Environment and Planning* 44, no. 8 (2012): 2006–2022.

Hara, T. "Social Shaping of Nuclear safety: Before and After the Disaster." In *Nuclear Disaster at Fukushima Daiichi: Social, Political and Environmental Issues*, edited by R. Hindmarsh, 22–40. New York: Routledge Studies in Science, Technology and Society, 2013.

Hecht, G. *Being Nuclear: Africans and the Global Uranium Trade*. Cambridge, MA: MIT Press, 2012.

Helariutta, T. "Fobiaa syystä ja syyttä ['Phobia with and without reason']." *Satakunnan Kansa*, April 16, 2012, accessed August 15, 2014, http://tinyurl.com/ndzptqx.

Hindmarsh, R. "Nuclear Disaster at Fukushima Daiichi: Introducing the Terrain." In *Nuclear Disaster at Fukushima Daiichi: Social, Political and Environmental Issues*, edited by R. Hindmarsh, 1–21. New York: Routledge Studies in Science, Technology and Society, 2013.

Hindmarsh, R. " 'Hot Air Ablowin': 'Media-speak', Social Conflict, and the Australian 'Decoupled' Wind Farm Controversy," *Social Studies of Science* 44, no. 2 (2014): 194–218.

Irving, A. "Calder Hall Passes European Stress Tests," *News & Star*, January 19, 2012a, accessed November 4, 2014, ttp://tinyurl.com/o2dd3on.

Irving, A. "Lessons from Fukushima," *News & Star*, May 31, 2012b, accessed November 4, 2014, http://tinyurl.com/ozccayp.

Jasanoff, S. (ed.). *States of Knowledge: The Co-production of Science and Social Order*. New York: Routledge, 2004.

Jasanoff, S., and S.-H. Kim. "Containing the Atom: Sociotechnical Imaginaries and Nuclear Power in the United States and South Korea," *Minerva* 47 (2009): 119–146.

Kuvaja, S. "Kun mahdoton tapahtui ['When the impossible happened']." *Satakunnan Kansa*, March 16, 2011, accessed August 15, 2014. http://tinyurl. com/nuad8jv.

Lammi, H. "Tarinat kovasta ytimestä ['Stories from the hard core']." In *Ydinvoima, valta ja vastarinta*, edited by M.Kojo, 11–50. Helsinki: Like, 2004.

Latour, B. "Why Has Critique Run out of Steam? From Matters of Fact to Matters of Concern," *Critical Inquiry*, 30, no. 2 (2004): 225–248.

Lehtonen, H. "Harri Lehtonen, SDP," *Satakunnan Kansa*, April 6, 2011, accessed August 15, 2014, http://tinyurl.com/pa9mx5v.

Lehtonen, M. "Reactions to Fukushima in Finland, France and the UK—Rupture or Continuity in the Nuclear Techno-Politics?" 2013, accessed, August 15, 2014, http://tinyurl.com/q3ny34a.

Meri, T. "Pääjohtaja tietää" ['Director general knows']." *Satakunnan Kansa*, August 28, 2012, accessed August 15, 2014, http://tinyurl.com/plc95uw.

National Academy of Science (NAS). *The Disposal of Radioactive Waste on Land*. Publication 519, Washington: National Academy Press, 1957.

News & Star. "Nuclear Store the Fight Is Not over Vows Cumbrian MP," *News & Star*, January 31, 2013a, accessed November 4, 2014, http://tinyurl. com/pjjudsw.

News & Star. "Sellafield Unions Condemn Council's Veto of Nuclear Dump," *News & Star*, February 3, 2013b, accessed November 4, 2014, http://tinyurl. com/nnlmfwu.

North-West Evening Mail. "New Cumbria Nuclear Power Station in Cumbria Could Create 21,000 Jobs," *North-West Evening Mail*, May 2, 2014a, accessed November 4, 2014, http://tinyurl.com/nhdvda6.

North-West Evening Mail. "Green Party Slams Cumbria Nuclear Power Station Plans," *North-West Evening Mail*, May 2, 2014b, accessed November 4, 2014, http://tinyurl.com/n9sznby.

Nuclear Decommissioning Authority (NDA). *Radioactive Wastes in the UK: A Summary of the 2013 Inventory*, 2014, accessed November 4, 2014, http:// tinyurl.com/l5gerro.

OECD. *Geological Disposal of Radioactive Waste: National Commitment, Local and Regional Involvement—A Collective Statement of the OECD Nuclear Energy Agency Radioactive Waste Management Committee, Adopted 2012*, 2012, accessed January 6, 2015, http://Tinyurl.com/94febu2.

OECD. *Radioactive Waste Management Programmes in OECD/NEA Member Countries: Finland*, 2014, accessed November 4, 2014, http://tinyurl.com/nst37hr.

Pelkonen, J. "Amano: Suomi johtava maa ydinjätteen loppusijoituksessa ['Amano: Finland the leading nation in final disposal']." *YLE Satakunta*, August 28, 2012, accessed August 15, 2014, http://tinyurl.com/ob8kgg7.

Pesonen, A. "Mitä mieltä? ['What do you think?']." *Satakunnan Kansa*, March 14, 2011: 5.

Posiva Oy, "Posiva tutkii ['Posiva investigates']," 2014, accessed January 6, 2015, http://tinyurl.com/kbkr9fw.

RCEP (Royal Commission on Environmental Pollution). *Nuclear Power and Environment, Sixth Report*, 1976, accessed August 15, 2014, http://tinyurk. com/o8z2eb4.

Rantanen, J. "Posivan Onkalo voi jäädä välivarastoksi ['Posiva's Onkalo might remain as interim storage']." *Raumalainen*, December 19, 2012, accessed August 15, 2014, http://tinyurl.com/pgwylv8.

Saarinen, E. "Japanin järistyksen jälkeen STUK sai ymmärtäjiä ['After the Japan earthquake STUK gained sympathisers']." *Satakunnan Kansa*, October 5, 2011, accessed August 15, 2014, http://tinyurl.com/pqkbql5.

Seppälä, T. "Siruja ['Splinters']." *Posiva tutkii 2/2011*, 2011, accessed August 15, 2014, http://tinyurl.com/qzcbeg5.

Spiegel Online, "Saying Goodbye to Nuclear: Merkel Takes First Steps towards a Future of Renewables," *Spiegel Online*, April 15, 2011, accessed January 6, 2015, http://tinyurl.com/ouoyj3y.

Ståhle, J. "Keskustan päättäjät eivät kadu päätöksiään ['Centre party do not regret their decisions']." *Satakunnan Kansa*, March 13, 2011, accessed August 15, 2014. http://tinyurl.com/q7xetga.

Steere, T., and M. Duell. "'Britain's Fukushima': EDF Shut Nuclear Reactor for Five Months over Fears of Similar Crisis to Meltdown at Tsunami-Hit Plant," *Daily Mail*, March 19, 2014, accessed November 4, 2014, http://tinyurl.com/opcc3ce.

STUK (Radiation and Nuclear Safety Authority). *European Stress Tests for Nuclear Power Plants: National Report FINLAND. 3/0600/2011*, 2011, accessed August 15, 2014, http://tinyurl.com/neqxgqt.

Suni, K. "Suomessa turvalaitteet on varmistettu nelinkertaisesti ['In Finland safety systems secured four fold']." *Satakunnan Kansa*, March 13, 2011a, accessed August 15, 2014, http://tinyurl.com/pjboh7v.

Suni, K. "Riskistä tuli totta ['Risk became reality']." *Satakunnan Kansa*, March 16, 2011b, accessed August 15, 2014, http://tinyurl.com/q8sf66d.

Suni, K. "Ehdokkailta turvatakuuehtoinen kyllä ydinvoimalle ['Safety dependent yes to nuclear power from candidates']." *Satakunnan Kansa*, March 18, 2011c, accessed August 15, 2014, http://tinyurl.com/phzgj9u.

Suni, K. "EU:n testiläksyt voimayhtiöille tehtäväksi jo huhtikuuksi ['EU test homework for power companies to be done by April']." *Satakunnan Kansa*, March 23, 2011d, accessed August 15, 2014, http://tinyurl.com/p2vkfsd.

Teräväinen, T., Lehtonen, M., and M. Martiskainen. "Climate Change, Energy Security, and Risk: Debating Nuclear New Build in Finland, France and the UK," *Energy Policy*, 39 (2011): 3434–3442.

Weightman, M. "Japanese Earthquake and Tsunami: Implications for the UK Nuclear Industry," September 2011, accessed November 4, 2014. www.onr. org.uk/fukushima/final-report.pdf.West Cumbria Managing Radioactive Waste Safely (WCMRWS). *E-bulletin No. 19*, 2012a, accessed November 4, 2014, http://tinyurl.com/qbaqj46.

West Cumbria Managing Radioactive Waste Safely Partnership (WCMRWS). *The Final Report of the West Cumbria Managing Radioactive Waste Safely Partnership*, 2012b, accessed November 4, 2014, http://tinyurl.com/32zca4s.

World Nuclear Association (WNA). "Nuclear Power in Finland," n.d.(a), accessed November 4, 2014, http://tinyurl.com/nepq734.

World Nuclear Association (WNA). "Nuclear Power in the United Kingdom," n.d.(b), accessed November 4, 2014, http://tinyurl.com/pdozmom.

YLE. "Pekkarinen: No Fennovoima Nuke without Waste Solution," *YLE*, March 29, 2012, accessed August, 15, 2014, http://tinyurl.com/qdkndxz.

# 9 Germany's *Energiewende* after Fukushima

## Nuclear Politics at the Forefront of Change

*Detlef Jahn and Sebastian Stephan*

The meltdown of the nuclear reactors at Fukushima had an important impact on the energy policies of countries far from the accident. For some countries in Europe, such as Belgium, Switzerland, and, above all, Germany, the future of nuclear power reached a turning point. In other countries, such as Sweden, neither energy policy nor public opinion was dramatically affected by Fukushima. This chapter describes the German decision to phase out nuclear power and explores why Germany was willing to change its energy system, and contrasts this against nuclear energy policy in Sweden, which was unaffected (see also Fjaestad et al. 2013).

The German decision was embedded in broader conceptual changes to the country's energy policy called *Energiewende* in German. *Energiewende* means a turning point in energy policy or an energy transition. It implies an exit from the nuclear and fossil-fuel energy that has dominated Germany for decades. Because of the recent change in German energy production and consumption, the term *Energiewende* is already well-established in scientific and journalistic circles.

Bolstering such change was the Fukushima nuclear disaster on March 11, 2011. The disaster had a significant impact on German energy policy within a short time. In less than four months, the government and the Second Chamber, the Federal Council (*Bundesrat*),[1] decided to end nuclear power in Germany (Mez 2012). Eight nuclear reactors lost their operating licenses immediately. For the remaining nine reactors, a clear phasing-out plan was determined such that the last German nuclear power station is due to be shut down in 2022. This government decision was supported by nearly 80% of the population. Only 8% disagreed with the decision and 12% were undecided (Hoening 2011).

However, even if the immediate political situation after the catastrophe in Fukushima was turbulent, the phasing out of nuclear energy after the Fukushima reactor meltdowns did not reflect a new trend in Germany. In fact, it denoted the end of a long-lasting debate about nuclear energy development. The phase-out decision, rather, indicates that the nuclear disaster in Japan tipped the political balance in Germany to accelerate the energy transition and strengthen support for renewable energy as the alternative energy source for the future. The scope of the German decision is unique in Europe, and more broadly, worldwide.

Why did Germany take this position, which went so much further than other countries? To address this question we compare Germany with Sweden. Sweden is an interesting case because, although it was the first country with a substantial nuclear power program to decide to phase out nuclear power (in the early 1980s it decided to phase out nuclear power by 2010), this decision was revoked in the first decade of the new millennium and was unaffected by the Fukushima disaster. A May 2011 Swedish parliamentary bill from the oppositional Green Party to immediately start nuclear decommissioning was rejected because the Parliament wanted to avoid a panicky post-Fukushima decision.[2]

To understand Germany's position and its difference from Sweden we use the approaches of the political opportunity structure and the window of opportunity. As in other fields of energy policy, it is advantageous to combine structural and process theory in order to explain policy change (Breukers 2006; Karapin 2010, 2014; Laird and Stefes 2009). The political opportunity structure approach has been helpful to explain the mobilization of anti-nuclear power protest and its outcomes (Kitschelt 1986). The major elements of the approach are constituted by the openness or closedness of the institutional political system; the stability or instability of a broad elite consensus on a policy or issue; the presence or absence of elite allies for the anti-nuclear protest groups; and the state's capacity and propensity for the marginalization of contesting views (McAdam 1999; see also Eisinger 1973). The political opportunity structure thus comprises specific configurations of resources, institutional settings, and historical precedents—in this case, for anti-nuclear power protest—and therefore provides the stage for political actors.

In contrast, the window of opportunity approach offers a tool for analyzing the political process in a crucial event (Kingdon 1984; also Jenkins-Smith et al. 2014; Zahariadis 2007). In Kingdon's approach, major change in the policy-agenda-setting process is likened to the convergence of various streams flowing through the political landscape at any one time. First, the policy problem must be perceived by a broad base of political actors as a problem (*the problem stream*). In turn, policy elites select solutions, often involving alternatives, and pressure for change (*the policy stream*). Finally, the political stream finds politicians willing to make significant policy change (*the political stream*). When these three streams come together forcefully, a window of opportunity is opened and agendas for change can be better realized.

In sum, this analytical framework is useful to analyze the reasons why countries decided to modify or not modify their energy systems as an effect of Fukushima and why Germany reacted so distinctively. In comparing the sociopolitical impact of Fukushima on Germany against Sweden, it is necessary to trace or map the way post-Fukushima decisions were made, and look at the driving (energy) technology governance forces behind national policy-making and the consequences for each country.

The chapter is structured in the following way: first, we overview Germany's 2011 phasing-out decision, as a direct reaction to Fukushima. Here we show that Fukushima, in being considered as a technological and political problem in Germany, constituted a problem stream for a window of opportunity to change policy. In order to understand this decision, we go back to the phase-out decision of the former government at the turn of the millennium. This decision offered an alternative to a nuclear path in Germany and constituted the policy stream for a window of opportunity.

In the second section, we compare the situation in Germany and Sweden in order to emphasize the "special" German case. We employ the concept of political opportunity structure to explain the differences between the anti-nuclear power movements and their allies for policy change. We show that the political stream for a window of opportunity was based on structural differences that emerged in the course of the last decades. Section three then compares Germany and Sweden with Belgium and Switzerland, two countries where Fukushima also led to a change in energy policy, although to a lesser extent than in Germany. Here we show that the three policy streams were not as prevalent as in Germany and, therefore, the phasing-out decisions were less ambitious. The conclusion summarizes the results and speculates about the consequences of the German *Energiewende*.

## The German Phase-Out of Nuclear Energy as a Fukushima Effect

The Fukushima nuclear disaster of March 11, 2011 had a rapid and profound effect on Germany's energy policy. On June 30, 2011 the First Chamber, the *Bundestag* (or federal parliament), agreed to a gradual phase out of nuclear power with the support of 513 to 79 votes (Mez 2012). On July 8, 2011, the *Bundesrat* approved this decision. On August 1, 2011, the President signed the bill, and on August 6, 2011 it was enacted.[3] As we discuss below, this decision seemed panicky, but it had a significant pre-history. Nuclear energy had been questioned particularly from the mid-1990s. We start our analysis by tracing the stages of the 2011 phasing-out decision (Bennett et al. 2014). This decision can be seen as ironic because the government coalition between the Christian Democrats (CDU/CSU)[4] and Liberals (FDP), who agreed to phase out nuclear energy after Fukushima, had previously dismissed the 2002 phasing-out plans of the preceding "Red-Green" coalition between the Social Democrats (SPD) and the Alliance90/The Greens (Greens).[5]

## The 2011 Decision to Phase Out Nuclear Energy

In response to Fukushima, the German parliament voted for a phasing out of nuclear power by 2022 and a comprehensive transition of Germany's energy supply towards renewable sources (Jahn et al. 2012). After decades of controversy, the clear majority of political players had decided to change the energy status of Germany. The future of nuclear power is no longer a controversial issue; the debate now focuses on the *Energiewende*.

Before the nuclear phase-out began, Germany, similar to the United Kingdom, Finland and Spain (see Figure 9.1), generated around 25% of its electricity from nuclear energy. In Switzerland, Sweden, Belgium, and France the share was much higher. However, a comparison of the share of nuclear energy of total electricity production between 2010 and 2012 shows that only Germany made substantial reductions in production of nuclear energy over this period. In other countries, nuclear energy generation remained on the same level or even increased, as seen in Finland, France, and the United Kingdom.

The multi-decade energy transition in Germany will involve the expansion of renewable energy production, to take up almost the entire

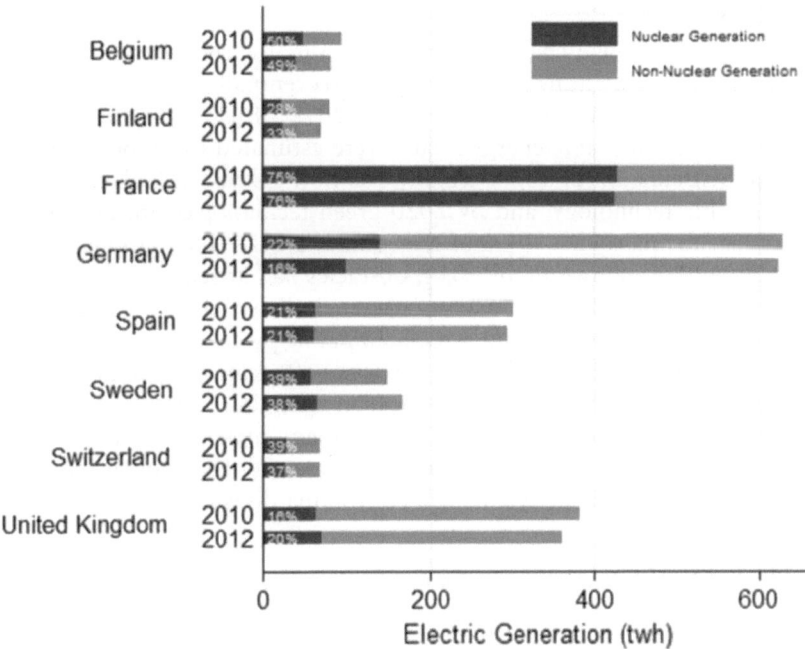

*Figure 9.1* Nuclear and Other Electricity Generation in Selected Countries (in Terrawatt Hours), 2010 and 2012

Source: data extracted from International Energy Agency 2014

energy gap created by the loss of nuclear power,[6] and the provision of the infrastructure needed for expansion (Mez 2012). Another key element of the *Energiewende* is the significant improvement in the country's energy-efficiency efforts. Financing this project, however, is one of the most difficult aspects of the *Energiewende*. In 2013, the average consumer in Germany paid 48% higher energy prices (cents per KWh) than consumers in their European neighbors (Wetzel 2014). One reason for the high prices was the so-called merit order effect, which gives renewable electricity priority access to the grid (Schreurs 2012: 37). Consumers bear the fixed prices through the German Renewable Energy Act reallocation charge.[7] Big industries are protected by the government and exempted from paying the additional costs for renewables. Private households will thus bear the major burden of increasing energy costs.

Energy prices for German industry are now high in comparison to other OECD countries. Up to 2008, German energy prices for industry were not higher than anywhere else, but as a long-term effect of the Red-Green coalition's 2002 decision to phase out nuclear energy, energy prices increased significantly. The same effect occurred after the 2011 phasing-out decision following Fukushima. Energy prices for industry increased considerably more in Germany than in Western Europe and overall for the rest of the OECD, and particularly in comparison to Sweden.

Despite the huge challenges and substantial early economic costs of Germany's energy transition, it promises to eventually result in substantial economic gains (Mecklin 2012; Schreurs 2012: 38). In 2011, future industries in the renewable energy sector were estimated to support 370,000 jobs (Schreurs 2012: 38). Germany has now become a global leader in renewable technology, and by 2020 green technologies are expected to account for more than 14% of the country's GDP (International Energy Agency 2013). However, by 2014, obstacles had arisen to Germany realizing the vision of a new energy mix, including technical and financial problems of conversion, and associated public protest and environmental challenges. Nonetheless, the *Energiewende* offers an opportunity for Germany to benefit from being a pioneer in changing to renewable energy sources. So, in more depth, why did Germany make such a far-reaching decision on its energy policy after Fukushima? We first look at the political situation immediately after the Fukushima disaster.

### The Political Situation in Germany Post-Fukushima

Prior to the nuclear catastrophe in Japan, the coalition government of the Christian Democrats (CDU/CSU) and Liberals (FDP) supported nuclear energy as a necessary transitory technology (*Brückentechnologie*) from fossil fuels to renewables. This policy position, however, provoked strong negative reactions from a large sector of the population even before

Fukushima. This was because, in October 2010, the sitting government abolished the skillfully negotiated Red-Green predecessor government's energy compromise of 2002. Therefore, it was of no surprise that the Greens, many members of the Social Democrats (SPD), environmental organizations, and the anti-nuclear movement responded quickly to the crisis in Japan (e.g. see Figure 9.2). It presented an opportune moment to advocate a faster transition of the national energy system. Pollsters immediately predicted that the CDU and FDP would lose middle-class voters to the Greens if they stuck to their pro-nuclear position. Both government parties subsequently experienced a dramatic decline in electoral support, despite an improving economic situation. Undoubtedly, the disaster at Fukushima translated quickly into a technological *and* political problem, which catalyzed into a problem stream in Germany.

Although the next federal election was not until 2013, German political parties were particularly vulnerable to any decline in electoral support because almost every year there are state elections. These elections determine the majority in the important second chamber, the Federal Council (*Bundesrat*). In 2011, there were federal state elections due in Saxony-Anhalt, Baden-Württemberg, and Rhineland-Palatinate, in addition to local elections in Hesse. The election in Baden-Württemberg, a state with a population of 10.6 million people, was particularly significant because a loss there by government parties would change the majority in the Second Chamber.

*Figure 9.2* Anti-nuclear Protests in Cologne, Germany, March 2011

Source: Bündnis 90/Die Grünen Nordrhein-Westfalen, 2011

Chancellor Angela Merkel quickly responded to the anti-nuclear mood with a change in policy direction (as outlined below in more detail), but without success. The government parties received 10% fewer votes compared to the previous election in Baden-Württemberg and lost government there after 58 years of power. The first Green prime minister in a federal state came to power and formed a coalition government between the Greens and the SPD.[8]

In the campaign leading up to these elections the government, and in particular Chancellor Merkel, tried to encourage its coalition partner, the Liberals, to shift policy direction (Schreurs 2012). The Liberals resisted this attempt. Nevertheless, Chancellor Merkel attempted to retain control of how the nuclear energy issue was being framed in public debate. She announced a three-month moratorium on the nuclear extension plan, a safety check of all nuclear power plants, and the shutdown of the seven oldest plants for this three-month period.

During the moratorium, two commissions were charged with advising the government. The Reactor Safety Commission, in its report on the safety of Germany's nuclear power plants, concluded that all German plants featured sufficient precautions for power supply and extreme scenarios like extensive flooding (Jahn et al. 2012). Nonetheless, nuclear energy carries with it a unique set of risks not comparable with any other form of energy. Although the power plants on German turf are among the safest in the world, the events in high-tech Japan were tragic proof that every technology may contain a residual risk that cannot be foreseen or ruled out.

In turn, the Ethics Commission for a Safe Energy Supply—which included leading figures from politics, science, industry, and religion—surmised that the severity and possible uncontrollability of a nuclear reactor accident had a substantial effect on residual risk; risk that remains despite all other factors including safety being considered adequate at any one time. Consequently, the Ethics Commission recommended the government limit the use of nuclear energy as much as possible. This was because, it argued, there was no solution to the nuclear waste storage problem (see also Molyneux-Hodgson and Hietala in this volume), and the costs of a cleanup after a reactor catastrophe would be enormous. In Fukushima, for example, the multi-billion dollar costs associated with damage to the plant itself were compounded by the high costs of compensating victims, and other associated aspects. Moreover, the Ethics Commission contended that alternative low-carbon energy production options were safer than nuclear energy and supported a shift to a renewable energy-dominated system (Ethikkommission für Sichere Energieversorgung 2011). At the end of the three-month moratorium period, the Commission recommended that nuclear energy be phased out within a decade.

Following this advice, and with increasing pressure from some federal state governments opposing nuclear energy, the plan to shutdown

Germany's nuclear plants was not lifted. The result was the 13th Amendment to the Atomic Energy Act (AEA), which also included support for renewable energy expansion (Rossnagel et al. 2012). This was the solution to the policy problem and thus represented convergence with the policy stream. By 2020, the aim was to double the national share of power drawn from renewable sources to 35%.

In sum, the major driving forces of the policy decision that realized this window of opportunity policy change were the robust anti-nuclear sentiment of many federal states, a consistently strong anti-nuclear movement, and an intense and anxious public discourse response to the Fukushima disaster. These aspects changed the politics stream by motivating the conservative government to change its political commitment to nuclear energy and to integrate the beliefs and perspectives of interest groups and environmental movements into policy.

In addition, the country was already conditioned to support this agenda shift due to a similar phasing-out plan introduced by the Red-Green coalition, which nearly a decade before had received wide popular attention and support. The Red-Green plan remained fresh in the minds of German citizens, as its abolition by the CDU/CSU-FDP coalition three months before Fukushima received wide media coverage. But what was the more detailed history behind the final phase-out of nuclear energy in Germany? This history is important to illustrate why Germany took such a firm and committed position to eventually phase out nuclear energy. This retrospection will show that two of the three streams were already inherent to the political debate and this downstream decision.

### The Red-Green Coalition: The First Nuclear Phase-Out Plans

In October 1998, the SPD and the Green Party formed a governing coalition, changing the national configuration of power after 16 years of CDU/CSU-FDP dominance. When in government, the SPD and Greens crafted a coalition for a nuclear phase-out, which means in theoretical terms that the political stream shifted sides from pro-nuclear to a skeptical view of nuclear. The original purpose of the AEA[9]—the promotion of nuclear power—was replaced by a new policy. The new intention was to phase out nuclear energy in a structured manner until the date of discontinuation. It took two years for the government and the main energy companies to reach agreement, and the new legislation entered into force in 2002.

A significant challenge in amending the original Act had been to clarify the proposed limits on the previously unrestricted operation licenses for nuclear power in accordance with constitutional law (Winter 2013), which gave strong legal guarantees to these licenses. After difficult negotiations, the agreement put a cap of 2623 billion kWh on lifetime production for all 19 operating reactors, equivalent to an average lifetime of 32 years (World Nuclear Association 2014b). In addition, it was agreed that the two least economically efficient reactors, in Stade and Obrigheim, would be shut down in 2003 and 2005, respectively, which occurred. It was also agreed that a high standard of safety had to be maintained for the remaining period of operation for the remaining reactors. In return, the government agreed to respect that nuclear power operation and related waste disposal would be protected from any further political interference.

An accompanying condition for the government on the phase-out of nuclear energy was the growth of a viable renewable energy industry, particularly wind power (Karapin 2014). In this context, it was noted that if energy and investment entrepreneurs were not sufficiently convinced about the environmental, energy, and economic benefits of renewable energy, the move towards this new technology would fail. Therefore, the Renewable Energy Law of 2000 aimed to benefit German industry. The new law replaced the first feed-in law by guaranteeing prices for 15–20 year horizons for renewable energy development.[10] Initially, the energy transition in Germany began as a bottom-up process, with many small industries pursuing alternative energy. After 2000, though, the contribution of renewable energy to the country's electricity production increased from 6% to 16% in less than 10 years. At the same time, the share of nuclear power dropped by around 6% (International Energy Agency 2012a).

Although this transition in the energy mix was supported by a majority of politicians and citizens, intensive debate remained about whether renewables could be the mainstay of Germany's electricity sector. Not all parties supported renewables as be the energy mainstay, and therefore the first phase-out missed a clear problem stream. In addition, a large and important faction of the Christian Democrats continued to view nuclear energy as a necessary transitory technology. In this debate it was argued strenuously by pro-nuclear interests that nuclear energy, as a non-fossil source, would produce electricity without greenhouse gas emissions while renewable energy technologies were developed. Furthermore, it was argued that nuclear energy was cheaper than other electricity forms and that an early shutdown of nuclear power plants would endanger Germany as an industrial state. It was therefore not surprising that a new CDU/CSU-FDP government, elected in 2009, would challenge the energy plans of its predecessor.

### Exiting from Germany's First Nuclear Phase-Out Plans

In 2009, the newly elected CDU/CSU and FDP coalition government was determined to rescind the Red-Green nuclear phase-out policy plans. The counter-policy was driven by two key ambitions: for Germany to remain an industrial powerhouse and for Germany to be a role model for other countries in fighting climate change. If Germany was to proceed with its nuclear phase-out schedule and reduce carbon emissions, it would also likely need to import electricity, which would threaten its autonomy and make it dependent on electricity imports from its neighbors, who mostly used nuclear energy.

The conservative government argued that nuclear and coal had to remain a key part of any secure, cheap climate-friendly energy mix. In December 2010, new legislation was implemented. It gave eight-year extensions for nuclear reactors built before 1980, with newer reactors allowed to operate approximately 14 years longer than envisaged in 2002. This decision would have affected public finance, because of the shorter transitional period during which taxes would be collected, as well as Germany's energy future.[11] Nevertheless, the new legislation did not fundamentally challenge the phase-out policy per se. Instead, it disagreed with the concrete configuration of the agreement negotiated by the previous government. The new legislation intended to dilute the process and gain time.

The decision to water down the nuclear phase-out was unpopular. Tens of thousands of anti-nuclear activists protested against the policy change. However, after the Fukushima disaster, doubts about the future of nuclear energy in Germany reinforced an anti-nuclear political stream. In combination with the two other occurring streams, it opened wide the policy window of opportunity to end the controversy over nuclear power once and for all by a more or less immediate phasing out. Without Fukushima, Germany may have continued using nuclear energy well into the 2030s and beyond. In other words, the most significant effect of Fukushima—as a problem stream—was that policy-makers phased out nuclear energy as soon as it was possible. The political stream against nuclear power was no longer deactivated, and a broader consensus was reached to turn the entire economy to a low-carbon energy structure without nuclear energy, a new policy stream. So how did this far-reaching decision compare with other countries?

## Germany's Phasing Out Decision in Comparative Perspective

Germany's phasing out decision was, in its magnitude and speed, unique. However, other countries, such as Belgium and Switzerland, have also stressed that they will exit nuclear energy but rather in the medium-term future and without broader conceptual changes to the country's energy policy. In contrast, the United Kingdom, Finland, and Sweden intend to increase their nuclear power capacity. Sweden is perhaps most interesting to contrast with Germany. It has a substantial share of nuclear power and decided in 1980, as a consequence of a national referendum, to phase out nuclear energy by 2010. However, Sweden changed its position between

2006 and 2010, just before the Fukushima disaster, and is now a strong promoter of nuclear energy, with plans to build up to 10 new reactors on top of the 10 existing ones when the old ones are worn out. The different political positions on nuclear energy in Germany and Sweden are reflected in public opinion. In Germany, majority public opinion has been against nuclear since 2000. In contrast, public opinion in Sweden has mostly been in favor of nuclear energy. Fukushima also had a greater impact on public opinion in Germany than in Sweden. Between 2010 and 2011, in Germany, disproval of nuclear energy grew by 10%, with more than 70% of Germans favoring an exit from nuclear energy and only 21% remaining in favor of nuclear energy. Even if Fukushima did change public opinion against nuclear energy in Sweden, this change was moderate in comparison to Germany.

So what was the main difference between Germany and Sweden, given that public opinion and the amount of electricity generated by nuclear reactors was similar in both countries? The difference lay in politics and the respective windows of opportunity for policy changes. One major reason for Sweden's radical change in favor of nuclear energy was the influence of the Center Party, which endorses an "eco-humanistic" ideology. In 2006, the Center Party gave up its anti-nuclear position, which it had adopted in the mid-1970s in a coalition government with other right-wing parties. Subsequently, Swedish anti-nuclear forces became decidedly weakened and the anti-nuclear power political stream vanished. To better illustrate why Germany succeeded in change and why Sweden did not, we now analyze the political opportunity structures for nuclear power positions in Sweden and Germany as a prerequisite for the convergence of the three streams of Kingdon's window of opportunity approach.

### The Political Opportunity Structure in Germany and Sweden

Nuclear energy has been controversial in Germany and Sweden since the 1970s, when anti-nuclear movements were catalyzed, alongside the environmental and anti-war movements, as part of the fervent social change of the day (Jahn 1992, 1999; also Holmberg et al. 2013b). Nevertheless, nuclear energy was introduced and used extensively for electricity generation in both countries in the 1960s and 1970s. For example, the share of nuclear power for generating electricity grew from 3% in 1973 to 1989 to 29% in Germany and from 3% to 46% over the same period in Sweden. After Fukushima, however, they went in opposite directions. By 2013, nuclear power contributed 4% more electricity to Sweden's total generation than before the meltdown. In contrast, in Germany, nuclear generation dropped to 15% by 2013.

Until the early 1970s, the nuclear industry had developed virtually unhindered by any political opposition. Powerful nuclear infrastructure was established. It involved state-funded research centers, a nuclear construction industry, and electricity utilities committed to a nuclear future (Jahn et al. 2012). In the beginning, political parties of all persuasions seemed convinced by the claimed technological advantages of nuclear energy. But the anti-nuclear movement, which emerged in the 1970s, challenged the use of nuclear power. However, the politics of nuclear power played out differently in Germany and Sweden, which ultimately illustrated how the two countries deviated in policy post-Fukushima. In the following section we look at each country in turn, before looking at the broader European perspective.

## Germany

Driven by a few particularly questionable nuclear projects, a vibrant and powerful social movement formed in Germany that challenged any expansion of nuclear generation capacity, and indeed aimed to phase out nuclear energy. Protests sprung up over nuclear power plants in Brokdorf, Grohnde, and Wyhl (Glaser 2012; Jahn 1992). After the mid-1970s, robust protest actions expanded from nuclear energy facilities to include the transport of spent fuel and nuclear waste storage sites. The unresolved issue of radioactive waste disposal mobilized many people against nuclear energy. This was in particular the case in Wackersdorf and in Gorleben where giant nuclear reprocessing plants were to be built. These plans, which have never been realized as intended, were—and still are—strongly opposed by the local farming population, with vigorous support for the farmers from anti-nuclear groups around the country (Jahn et al. 2012; Winter 2013).

The powerful Green Party, which was established in Germany at the end of the 1970s, strengthened opposition to nuclear power (Jahn 1992). This political party was narrow in its scope, focusing on environmental issues, including the adverse impacts of nuclear technology. Through Germany's proportional electoral system, the Greens received enough support to access the federal and state political stages, harnessing concern in the country that industrial society was destroying its environmental foundations. Germany is a federal country with 16 states, which each have a parliament. In some states, the Greens have been very successful. In 1983, Green Party candidates were elected into the federal parliament (*Bundestag*) for the first time. Thereby the anti-nuclear movement, through the conduit of the Green Party, was able to robustly challenge the Social Democratic Party (SPD) on environmental issues. In combination with the formation of large environmental NGOs and local and regional green parties, the anti-nuclear movement was thus able to institutionalize itself effectively in the German political system (Frankland et al. 1992; Markovits et al. 1993). These developments changed the political stream of the window of opportunity about nuclear power.

The reactor catastrophe at Chernobyl in 1986 altered the nuclear power problem stream substantially. A strong coalition of anti-nuclear activists strenuously called for major changes to Germany's energy strategies (see Kitschelt 1986). This affected the political positions towards nuclear energy to the degree that traditional parties began to change their stance and the notion of a nuclear phase-out became mainstreamed in the public and political debate. For example, after Chernobyl, the SDP shifted from a pro-nuclear power position to abandon nuclear power within 10 years. From then on, the so-called Red-Green alliance (Offe 1985) advocated a phase-out of nuclear energy. This political alliance affected other party politics, including substantial parts of the trade union movement, and increased pressure against nuclear power (Jahn 1993a). Despite the increased pressure, the coalition government of the CDU/CSU and FDP remained firm in its support for nuclear energy.

Nevertheless, even political proponents of nuclear power were affected strongly by the disaster at Chernobyl. Under Chancellor Helmut Kohl, the government responded to the disaster and a growing concern about air pollution and climate change by establishing a Ministry for the Environment, Nature Conservation, and Nuclear Safety in 1986. Its main responsibility was to develop and regulate renewable energy and reduce greenhouse gas emissions. One of the Ministry's first policy instruments was the electricity feed-in law, which required grid operators to purchase renewable electricity from third-party generators at 65% to 90% of the retail price. This policy response was projected as a key component of the comprehensive policy turn that later aimed for a national energy portfolio dominated by renewable energy (Winter 2013). With the new law in place, non-fossil or nuclear-free electricity sources were provided with another good political window of opportunity for expansion in the policy stream for a shift from nuclear.

Such moves also reinforced that, post-Chernobyl, the nuclear debate became complicated by concerns about climate change and claims that nuclear was a clean energy technology low in greenhouse gas emissions. Proponents of nuclear energy argued that coal electricity-generating plants needed to be replaced by nuclear reactors as part of the fight against global warming. However, this argument received little traction in the environmental and anti-nuclear movements, as it neglected the tremendous environmental impacts of nuclear waste (under the ground), which contested nuclear energy's claimed "cleanness" and ignored the enormous costs and inefficiencies of building and dismantling of nuclear facilities over very long periods of time. Thus, although nuclear energy did not receive much public attention in the post-Chernobyl period, the highly critical attitude towards nuclear power remained (Glaser 2012: 16).

## Sweden

In Sweden, the nuclear power dispute was much less heated. In the 1970s, Swedish political parties and government tended not to resist but rather to incorporate environmental concerns at an early stage of the debate (Jamison et al. 1990). The initial response came from the Center Party and the Communist Party, both of which adopted anti-nuclear stands (Jahn 1993a).

In the 1980s, Swedish environmental organizations strengthened their challenges to the emergent nuclear energy regime. Instead of generating broader ideological alternatives to established society, the movement spread technical and scientific information about environmental risks and hazards and worked on a more pragmatic level. This was because mainstream political parties resisted basic elements of green ideology or thought (Goodin 1992) that emerged in the 1970s and included aspirations of decentralization and no or limited economic growth. These parties also focused more on the technical, scientific aspects of individual issues than on social considerations. In contrast, the counter-cultural milieu of new social movements, which included the environmental movement, was built upon alternative cosmopolitical worldviews that challenged the then-established values of industrial society and politics in Germany (Jahn 2000).

Political ideology and sociopolitical context determined the political strategies of the anti-nuclear power movements in Sweden, as well as in Germany. However, assimilative strategies, including lobbying, petitioning government bodies, and influencing public opinion through referendum campaigns became common practice during the 1980s. This also happened in many other countries, where more conservative environmental groups began following the pragmatic approach that emerged in Sweden. In contrast, the German environmental movement at that time applied more confrontational strategies, including mass demonstrations and direct action featuring acts of civil disobedience, exemplified by trespassing with occupations of nuclear sites and access roads (Joppke 1993; Kitschelt 1986: 67–72).

The different characters of the early environmental movements in Sweden and Germany had fundamental consequences for nuclear politics in both countries (Jahn 1993b). Another factor was that German resistance to nuclear was strengthened through an enduring alliance with the Labor movement; this did not occur in Sweden.

Against the background of these different political opportunity structures, with the Swedish political opportunity structure for change to nuclear power considerably weaker than Germany's, it is then not surprising that German governments responded differently to Swedish governments over the future use of nuclear power. In addition, while Fukushima consolidated the window of opportunity for policy change in Germany, it did not in Sweden, because they were missing a clear and strong political stream questioning nuclear power. Similar but less-pronounced nuclear politics are also found in other European states which have adopted a critical position towards nuclear energy.

## The German *Energiewende* in the Broader European Perspective

The only European country whose response to Fukushima has come close to Germany's so far is Belgium, partly through the parliamentary influence of the Belgium Green Party. Following Fukushima, on October 30, 2011 (BBC News 2011), the new Belgian government announced its decision to phase out nuclear reactors after 40 years of operating (Belgium's first nuclear reactors were installed in the 1970s) and to prohibit the building of new nuclear power plants. In 2011, seven nuclear power plants produced about 56% of Belgium's total electricity (World Nuclear Association 2014a). Belgium aims to close all seven plants between 2015 and 2025 (Kunsch et al. 2014). As with Germany, the political decision in Belgium did not arise out of the blue but reached back to a phase-out law introduced in 2003.

Before Fukushima, three political groups had been arguing intensely about a 2003 government decision to phase out nuclear power in Belgium.[12] In 2007, the Belgium government set up the Commission on Energy 2030, which concluded that nuclear power should be used to meet Belgium's carbon dioxide reduction commitments, to enhance energy security, and to maintain economic stability. Another reason given to reconsider the 2003 phase-out decision was that the phase-out effect was estimated to double the price of electricity. In this context, the government argued that the operating life of the seven nuclear units should be extended (World Nuclear Association 2014a). However, the government was unable to formalize its proposal due to too much political contestation from a range of stakeholders.

When the nuclear disaster hit Japan in 2011, Belgium was without an elected government,[13] which is not that unusual in Belgium's turbulent politics. However, because of the strong problem stream in regard to the challenges of nuclear reactors, and the accompanying policy stream, which Fukushima had the effect of strengthening, a political consensus in parliament was reached to retain the 2003 nuclear phase-out law. This was agreed on the condition that adequate power could be secured from other sources and prices would not rise unduly (Kunsch et al. 2014). Fukushima thus had the notable effect in Belgium of consolidating the finalizing of the controversial discussion about the timing of the nuclear phase-out. Since then, the government has approved energy plans that will subsidize gas-fired generation and offshore wind capacity with taxes from nuclear power.

Switzerland is the third country in Europe in which the Fukushima effect led to significant changes in the level of support for nuclear power. Switzerland rapidly took action, only three months post-Fukushima, in regard to the operation of its five nuclear reactors at four power plants (International Energy Agency 2012b). The Swiss Federal Council recommended that the existing reactors be allowed to continue to operate during their license periods, but that they should close at the end of these licenses, which should not be replaced or extended. The proposal was approved by the upper house in September 2011, pressured by the biggest anti-nuclear protests in Switzerland in 25 years. One protest saw 20,000 people marching near the Beznau commercial nuclear power plant as a reaction to Fukushima.

However, although the Swiss reaction was similar to the German reaction, Switzerland has a different sociopolitical terrain, which demonstrated a less confrontational and enduring nuclear politics. Nuclear energy was never fundamentally questioned in Switzerland or a phasing-out demanded (World Nuclear Association 2014c). On the contrary, the Swiss people rejected two anti-nuclear referendum proposals, which would have led to an earlier phase out of nuclear power. Furthermore, a near majority of the public supported new nuclear power only months before Fukushima occurred. Post-Fukushima, the Swiss cabinet ignored referendum proposals for change, simply observed the protest reactions of the public, and (likely influenced also by neighboring Germany's response and also that of Belgium) subsequently declared that Switzerland was changing its energy system. So far this decision has stuck, despite some publics still supporting nuclear power to meet current and future electricity demand (Schaub et al. 2014). It thus seems the effect of Fukushima in Switzerland, like in Germany, although perhaps less forcefully, has been to prompt more sharply a nuclear phase-out.

In contrast, other European countries, including Finland, the Czech Republic, the United Kingdom, Poland, Russia, and the Netherlands, all plan to use more nuclear energy in the future, while France and Spain want to keep their share of nuclear energy at current levels (Konrad-Adenauer-Stiftung 2012). Perhaps this is because it is too expensive to phase out embedded nuclear energy regimes, and moreover because of the political commitment of these countries to nuclear. Concomitantly, many European states, including Austria, Denmark, and Italy, run their economies without nuclear energy, and have no plans to go nuclear. Perhaps they also have a good choice in this matter, as they do not have expensive embedded nuclear regimes, and now, post-Fukushima, little window of opportunity to shift to nuclear even if they desired.

## Discussion and Conclusion

In summary, the German phase-out policy for nuclear energy resulted from historically robust nuclear politics for change, consolidated by the

Fukushima disaster, which created a political opportunity window highly favorable to anti-nuclear power groups and their political position on nuclear power. In comparison to Sweden, German anti-nuclear groups were able to forge a strong political commitment from the beginning of the nuclear power debate. In Sweden, civic nuclear power concerns were subsumed by the political system to some extent, furthered by the anti-nuclear movement's failure to recruit the strong labor movement as an ally.

The German anti-nuclear position strengthened after Chernobyl in the political stream, with major actors of the political establishment joining forces with anti-nuclear power groups. In Sweden, the Chernobyl effect was not so long-lasting (Jahn 1988). Another reason why German politics were more open to anti-nuclear power groups was because of Germany's strongly risk-averse culture (Pulzer 1989). This feature of Germany's political culture may explain why nuclear disasters in other countries have more long-lasting impacts in Germany than anywhere else.

The study of the German *Energiewende*, again meaning significant improvement of the country's energy-efficiency efforts, shows that the political window of opportunity is a useful way to explain what occurred in the context of changing technology governance from nuclear to renewable energy. The first phase-out decision in Germany in 2003, although well-crafted and highly supported in German society, was abolished by a new right-wing government, as its fast transition seemed likely to put Germany's economy at high risk. Instead, a slower and indeterminate phase-out was instituted. However, the phase-out strategy of the Red-Green coalition government substantiated an alternative energy system without nuclear energy. That meant that the energy policy stream was open for a future without nuclear energy.

Matters then changed with Fukushima. The disaster clearly demonstrated that nuclear energy has very high risks. That activated the problem stream that made nuclear energy appear too risky. With that, Fukushima caused the three streams to came together, a situation which is needed for change: (i) the high and problematic impact of Fukushima made nuclear power redundant in the public's eyes (problem stream); (ii) the Red-Green plan was already highly established as a policy alternative (the policy stream); and (iii) politicians of all persuasions were convinced and willing to make significant policy change (the political stream).

In sharp contrast, in Sweden, Fukushima was mainly seen as a specific Japanese problem. Furthermore, although Sweden decided on a phasing out of nuclear energy in the 1980s, alternative plans for an energy supply without nuclear energy were never established. Thus, post-Fukushima, Sweden lacked the necessary level of strength in problem and policy streams to change the political stream. Major political actors were not willing to phase out nuclear energy.[14] The shift of the Center Party in its nuclear power position even changed the political stream. As a result, the window of opportunity for a shift in Swedish energy policy remained closed after Fukushima.

In sum, Germany is probably the only country where Fukushima has had a long-lasting effect on technology governance. However, it remains to be seen if the *Energiewende* is a model for an ambitious industrial society that has taken a lead in shifting to renewable energy. In other words, if the German economy becomes more environmentally benign through the shift from nuclear (and coal) to renewables, it will be at the forefront of a new energy policy internationally (Jahn 2016). The effect of Fukushima in Germany thus demonstrated a potential to significantly influence science and technology governance in a highly industrialized country, and to contribute to policy change on nuclear energy.

## Acknowledgements

We wish to thank the editors of the book for their very helpful comments and support for this article. We also thank Roger Karapin for his suggestions and encouragement.

## Notes

1 Similar to the Upper House (or Senate) in the Westminster style of government.
2 The bill stated that "the parliament notes that the accident in Japan should not be an excuse for an almost panicky decision on a radical shift in the direction of energy policy." For the full text of the policy statement, see Säkerheten i svenska kärnkraftsreaktorer. In Betänkande 2010/11: NU26, accessed November 8, 2014, http://tinyurl.com/kz3nmq6.
3 The original bill is found at http://tinyurl.com/mbgvn3c, accessed February 11, 2015.
4 The Christian Democrats in Germany are constituted of two parties working closely together. In Bavaria it is Christian Social Union (CSU) and in the rest of Germany it is the Christian Democratic Union (CDU). The former holds almost around 10% of the votes and the latter 30%.
5 See http://tinyurl.com/yza9tku, accessed February 11, 2015.
6 Renewable source electricity went up from 17% to 26% of gross electricity generation over 2010–14 (see: www.ag-energiebilanzen.de/4–0-Arbeitsgemeinschaft.html, accessed February 12, 2015).
7 The German Renewable Energy Act (Erneuerbare-Energien-Gesetz, EEG) was designed to encourage cost reductions based on improved energy efficiency from economies of scale over time.
8 In Saxony-Anhalt, despite a substantial loss of almost 7% fewer votes for the CDU (-3.7) and FDP (-2.9), the CDU-SPD coalition remained in power. In Rhineland-Palatinate, the CDU won 2.4% more votes and the FDP lost 3.6% of votes. The latter fell below the 5% threshold and did not regain any seats. The Fukushima effect has been clear for the Greens, who obtained 10.8% and entered into a SPD-Green coalition, although the SPD lost almost 10% of votes in this election mainly because of corruption. The communal elections in Hesse resulted in a loss of 4.8% votes for the CDU and 1.9% of votes for the FDP while the Greens increased their vote by 9.1% to 18.3%.

9  The Atomic Energy Act (AEA) came into effect in 1960. It is the legal foundation of the peaceful utilization of atomic power and was intended to promote nuclear research and the development and use of nuclear energy.

10  Feed-in tariffs are a policy mechanism designed to accelerate investment in renewable energy technologies by providing them a fee (a "tariff") above the retail rate of electricity. The mechanism provides long-term security to renewable energy producers, typically based on the cost of generation of each technology (see www.nrel.gov/docs/fy10osti/44849.pdf, accessed April 10, 2015).

11  An eight-year extension would have produced between €21 billion and €73 billion of extra profits for utilities (World Nuclear Association 2014b). The government could then raise €2.3 billion a year from a tax on nuclear fuel.

12  The Christian-Democrats and the nationalist Flemish *Vlaams Belang* supported nuclear energy. French-speaking socialists and the Liberals took neutral positions. The Flemish and Walloon Green parties and the Flemish socialists shared a deep rejection of nuclear power and initiated the phase-out.

13  See, e.g. www.theguardian.com/world/2011/dec/01/eurozone-crisis-forces-belgium-government, accessed May 6, 2015.

14  In autumn 2014, the government changed in Sweden. The new Red-Green minority government brought a phase out of nuclear energy to the political agenda. However, the majority in parliament supports a continuation of nuclear energy, so it is doubtful that a radical change in energy policy will take place in Sweden in the near future.

# References

BBC News. "Belgium Plans to Phase Out Nuclear Power," *BBC News*, October 31, 2011, accessed May 6, 2015, www.bbc.com/news/world-europe-15521865.

Bennett, A., and J.T. Checkel. *Process Tracing. From Metaphor to Analytic Tool.* Cambridge: Cambridge University Press, 2014.

Breukers, S., *Changing Institutional Landscapes for Wind Power*. Amsterdam: University of Amsterdam Press, 2006.

Bündnis 90/Die Grünen Nordrhein-Westfalen. "Anti-nuclear-protests in Cologne, Germany, March 2011," March 26, 2011, accessed May 27, 2015, www.flickr.com/photos/gruenenrw/5560595589/in/photostream/

Dryzek, J.S., et al. *Green States and Social Movements: Environmentalism in the United States, United Kingdom, Germany, and Norway*. New York: Oxford University Press, 2003.

Duit, A. *State and Environment: The Comparative Study of Environmental Governance*. Cambridge, MA: MIT Press, 2014.

Eckersley, R. *The Green State*. Cambridge, MA: MIT Press, 2004.

Eisinger, P. "The Condition of Protest Behavior in American Cities," *American Political Science Review* 67, no. 1 (1973): 11–28.

Ethikkommission für Sichere Energieversorgung. *Deutschlands Energiewende—ein Gemeinschaftswerk für die Zukunft. Die Bundesregierung [Germany's energy transition: A collective project for the future]*, 2011, accessed July 28, 2014, http://tinyurl.com/k592nyq.

Fjaestad, M., and P. Hakkarainen. "Sweden, Finland, and German Energiewende." London: Friedrich Ebert Stiftung, 2013, accessed April 21, 2015, http://library.fes.de/pdf-files/id/10163.pdf.

Frankland, E.G., and D. Schoonmaker. *Between Protest and Power*. Boulder: Westview Press, 1992.

Glaser, A. "From Brokdorf to Fukushima: The Long Journey to Nuclear Phase-out," *Bulletin of the Atomic Scientists* 68, no. 6 (2012): 10–21.

Goodin, R.E. *Green Political Theory.* Cambridge: Polity Press, 1992.

Hoening, A. "Umfrage zum Thema Atomausstieg. Deutsche akzeptieren hoeheren Strompreis ['Opinion poll on phase-out: Germans accept higher energy prices']." *RP Online*, September 16, 2011, accessed October 5, 2014, http://tinyurl.com/m8vkres.

Holmberg, S., and P. Hedberg. "Party Influence on Nuclear Power Opinion in Sweden." In *Energy Opinion Compared across Time and Space*, edited by S. Holmberg and P. Hedberg, 55–90. Sweden: University of Gothenburg, 2013a.

Holmberg, S., and P. Hedberg. "The Will of the People? Swedish Nuclear Power Policy." In *Energy Opinion Compared across Time and Space*, edited by S. Holmberg and P. Hedberg. Sweden: University of Gothenburg, 2013b, accessed May 21, 2015, http://goo.gl/xxWyzG.

International Energy Agency. *Energy Balances of OECD Countries 2013.* Paris: OECD Publishing, 2012a.

International Energy Agency. *Energy Policies of IEA Countries—Switzerland 2012 Review.* Paris: OECD Publishing, 2012b.

International Energy Agency. *Energy Policies of IEA Countries—Germany 2013 Review.* Paris: OECD Publishing, 2013.

International Energy Agency. *Energy Balances of OECD Countries 2014.* Paris: OECD Publishing, 2014.

Jahn, D. "Tschernobyl und die schwedische Energiepolitik ['Chernobyl and Swedish energy policy']," *Österreichische Zeitschrift für Politikwissenschaft* 17, no. 1 (1988): 43–51.

Jahn, D. "Nuclear Power, Energy Policy and New Politics in Sweden and Germany," *Environmental Politics* 1, no. 3 (1992): 383–417.

Jahn, D. *New Politics in Trade Unions.* Aldershot: Dartmouth, 1993a.

Jahn, D. "The Rise and Decline of New Politics and the Greens in Sweden and Germany," *European Journal of Political Research* 24, no. 3 (1993b): 177–194.

Jahn, D. "The Mobilization of Ecological World Views in a Post-Corporatist Order." In *New Perspectives on Environmental Policy in Europe*, edited by U. Collier and G. Orhan and M. Wissenburg, 129–155. Aldershot: Ashgate, 1999.

Jahn, D. *Die Lernfähigkeit politischer Systeme [The learning capacity of political systems].* Baden-Baden: Nomos, 2000.

Jahn, D. "The Three Worlds of Environmental Politics." In *State and Environment: The Comparative Study of Environmental Governance*, edited by A. Duit, 81–109. Cambridge, MA: MIT Press, 2014.

Jahn, D. *The Politics of Environmental Performance.* Cambridge: Cambridge University Press, 2016, in press.

Jahn, D., and S. Korolczuk. "German Exceptionalism: The End of Nuclear Energy in Germany!" *Environmental Politics* 21, no. 1 (2012): 159–164.

Jamison, A., Eyerman, R., and J. Cramer. *The Making of the New Environmental Consciousness.* Edinburgh: Edinburgh University Press, 1990.

Jenkins-Smith, H.C. et al. "Advocacy Coalition Framework: Foundations, Evolution, and Ongoing Research." In *Theories of the Policy Process*, 3rd edn., edited by P. Sabatier and C.M. Weible, 183–224. Boulder, CO: Westview Press, 2014.

Joppke, C. *Mobilizing Against Nuclear Energy.* Berkeley: University of California Press, 1993.

Karapin, R. "Explaining Success and Failure in Climate Policies: Developing Theory through German Case Studies," *Comparative Politics* 45 (2010): 46–68.

Karapin, R. "Wind-Power Development in Germany and the U.S.: Structural Factors, Multiple-Stream Convergence, and Turning Points." In *State and Environment: The Comparative Study of Environmental Governance*, edited by A. Duit, 111–146. Cambridge, MA: MIT Press, 2014.

Kingdon, J.W. *Agendas, Alternatives, and Public Policies*. Boston: Little, Brown, 1984.

Kitschelt, H. "Political Opportunity Structures and Political Protest: Anti-Nuclear Movements in Four Democracies," *British Journal of Political Science* 16, no. 1 (1986): 57–85.

Konrad-Adenauer-Stiftung. *Ein Jahr nach Fukushima [One year after Fukushima]*. Sankt Augustin/Berlin: Konrad-Adenauer-Stiftung, 2012.

Kunsch, P.L., and J. Friesewinkel. "Nuclear Energy Policy in Belgium after Fukushima," *Energy Policy* 66 (2014): 462–474.

Laird, F., and C. Stefes. "The Diverging Path of German and United States Policies for Renewable Energy," *Energy Policy* 37 (2009): 2619–2629.

Markovits, A.S., and P.S. Gorski. *The German Left: Red, Green and Beyond*. New York: Oxford University Press, 1993.

McAdam, D. "Concept Origins, Current Problems, Future Directions," In *Comparative Perspectives on Social Movements. Political Opportunities, Mobilizing Structures, and Cultural Framings*, edited by D. McAdam, J.D. MacCarthy and M.Y. Zald, 23–40. Cambridge: Cambridge University Press, 1999.

Mecklin, J. "The German Nuclear Exit: Introduction," *Bulletin of the Atomic Scientists* 68, no. 6 (2012): 6–9.

Mez, L. "Germany's Merger of Energy and Climate Change Policy," *Bulletin of the Atomic Scientists* 68, no. 6 (2012): 22–29.

Offe, C. "New Social Movements," *Social Research* 52, no. 4 (1985): 817–868.

Pulzer, P. "Political Ideology." In *Developments in West German Politics*, edited by G. Smith, W.E. Paterson and P.H. Merkl, 78–98. London: Macmillan Education, 1989.

Rossnagel, A., and A. Hentschel. "The Legalities of a Nuclear Shutdown," *Bulletin of the Atomic Scientists* 68, no. 6 (2012): 55–66.

Schaub, A., and N. Blumenfeld. UNIVOX Umwelt 2013, *gfs-zürich, Markt- & Sozialforschung*, 2014.

Schreurs, M.A. "The Politics of Phase-out," *Bulletin of the Atomic Scientists* 68, no. 6 (2012): 30–41.

Wetzel, D. "Deutsche Strompreise 48 Prozent über EU-Schnitt ['German energy prices 48 percent above EU average']." *Die Welt*, 2014.

Winter, G. "The Rise and fall of Nuclear Energy Use in Germany: Processes, Explanations and the Role of Law," *Journal of Environmental Law* 25, no. 1 (2013): 95–124.

World Nuclear Association. *Nuclear Power in Belgium*, 2014a, accessed July 25, 2014, www.world-nuclear.org/info/Country-Profiles/Countries-A-F/Belgium/.

World Nuclear Association. *Nuclear Power in Germany*, 2014b, accessed July 28, 2014, http://goo.gl/LHTTsn.

World Nuclear Association. *Nuclear Power in Switzerland*, 2014c, accessed August 01, 2014, http://tinyurl.com/pvdlg57.

Zahariadis, N. "The Multiple Streams Framework: Structure, Limitations, Prospects." In *Theories of the Policy Process*, 2nd edn., edited by P. Sabatier, 65–92. Boulder, CO: Westview Press, 2007.

# 10 Swiss Risk Governance of Nuclear Energy after Fukushima, and Citizen Perspectives

*Fabienne Crettaz von Roten*

The Fukushima Daiichi disaster, which unfolded from March 11, 2011 onward, had devastating consequences for Japan and the communities in close proximity. It raised questions about the safety and future of nuclear energy worldwide. In Switzerland, the disaster gained high media coverage (Crettaz von Roten 2013; Kristiansen and Bonfadelli 2014). Many articles focused on what lessons Switzerland might learn from Fukushima. Issues for policy learning included: what are the security risks of Swiss nuclear power plants? Are seismic, social and technical risks of Swiss nuclear power plants adequately evaluated? Is it necessary to build new nuclear power plants in the future? Are the risks involved in nuclear energy socially acceptable?

Nine days after the disaster, two Swiss Sunday newspapers, *Le Matin Dimanche* and *Sontagszeitung*, published the results of a survey conducted March 17–19, 2011, on 506 voters in the French- and German-speaking regions of Switzerland. According to the survey, 87% of the population wanted the country to withdraw from nuclear energy and 74% were opposed to the construction of new nuclear plants. Public opinion appeared to be highly affected by Fukushima. Some commentators, however, argued this would just be a short-term effect that might not lead to any political change.

On May 25, 2011, only six weeks after the disaster, the Swiss Federal Council—the executive government authority—announced its decision to phase out nuclear power by 2035 (see Figure 10.1). Existing power plants would operate until the end of their life expectancy and no new power plants would be constructed. In the summer of 2011, both councils of parliament accepted the ban after intense debate, but allowed research into nuclear energy to continue. In turn, the Federal Council defined its 2050 energy strategy to phase out nuclear energy by developing hydropower and new renewable energy, improve energy efficiency of buildings, appliances, and transport, and reduce energy consumption. In addition, a target of a 13% reduction in energy consumption from 2000 was set by promoting energy conservation.

*Figure 10.1* Cartoon from P. Chapatte Published in the Newspaper *Le Temps* on the First Page of the May 26, 2011 Issue

The two men from the nuclear lobby exclaim "Still another disaster that we did not expect" when seeing D. Leuthard, Swiss Federal Councillor, brandishing her phase-out of nuclear power.

Source: Courtesy of P. Chappatte

For Swiss citizens, the Fukushima effect thus had two immediate components: first, the decision to ban nuclear energy, and second, a decision that demanded a significant reduction in energy consumption. In general, in Switzerland, governance must be adapted to the citizens' points of view; Swiss direct democracy enabled the population to have a say on both decisions.

This chapter illustrates Swiss citizen positions on these two policy decisions during 2011–2013—the period immediately after Fukushima, which informed and defined new nuclear and energy governance in Switzerland. In investigating this topic, I first provide an overview of the politics of nuclear power in Switzerland that led to these topical decisions in the aftermath of Fukushima; I then outline the investigative context of science, technology society (STS) studies; and then I present the investigative methods, the results, and a discussion of the results before concluding.

## Background

Between 1969 and 1984, Switzerland built four nuclear power plants. Nuclear development involved problems. An accident in the Lucens' small pilot prototype power reactor in 1969 destroyed the reactor through a partial meltdown, though the radiation release was within permitted levels

(Marques 2011: 131). Many protests about nuclear power development followed during the 1970s and 1980s. For example, a referendum was held in 1979—the year of the Three Mile Island nuclear accident in the US—with a citizens' initiative for the interrelated issues of the protection of civic rights and nuclear safety. The initiative made the concession-holder, or the company operating a nuclear plant, responsible for any damage during and after construction. The granting of the concession was subject to the approval of voters living within 30 kilometers of the future nuclear power plant. The initiative, however, was rejected. By 1984, developments had deepened against nuclear power, but citizens voted 55% to 45% against an initiative "for a future without nuclear power stations." Anti-nuclear activists had argued that planning should proceed according to the real power needs of the country and not according to the projected needs of an industry that demonstrated uncontrolled expansion (Favez and Mysyrowicz 1987).

Subsequently, anti-nuclear protesters had mobilized against the construction of nuclear power plants in Gösgen and Leibstadt, and in the bordering regions of Creys-Malville and Superphénix in France (Favez and Mysyrowicz 1987). A nuclear reactor construction project in the 1970s at Kaiseraugst, near Basel, was notable in raising intense debate on the role of the state, the competencies of the Confederation and cantons,[1] and the private sector's agenda in regard to energy. Construction began in 1985 but was abandoned in 1988, after years of intense contestation, including large demonstrations and occupation by protesters of the future power station's building for 11 weeks in 1975.

Aligning with activist mobilization, public opinion on nuclear power steadily became more negative. In 1990, a majority of citizens (54.5%) accepted a 10-year moratorium against the construction of new power plants, but an initiative to phase out nuclear power was rejected. In 2000, a vote on a green tax to support solar energy was also rejected. In 2003, in another two referenda, citizens refused to phase out nuclear energy, and chose to prolong the 10-year moratorium.

In contrast to international concerns about nuclear power, particularly after accidents and disasters such as Chernobyl in 1986, and nuclear waste storage, concerns about climate change facilitated a reframing of the then-poor image of nuclear energy by the industry from the early 2000s "as a low carbon alternative to fossil fuels . . ." (Hindmarsh 2013a: 7). By the late 2000s, intense international promotion of nuclear energy as a "clean energy technology" saw a positive trend for nuclear energy acceptance in relation to climate change. This trend coincided with the international financial crisis and the increasingly assertive strategy of the Swiss nuclear lobby (Crettaz von Roten 2013). At the same time, meeting increasing demand for energy remained a problem for Switzerland. Various studies analyzed future energy wants and needs, with and without nuclear (e.g. Kannan and Turton 2012; Ocha and van Ackere 2009).

By early 2011—prior to the Fukushima disaster—Switzerland had five operational nuclear reactors, all located in the German-speaking region of Switzerland (Switzerland is at the crossroads of three cultures defined by language spoken). Nuclear power accounted for 41% of total electricity production in 2011, with hydroelectricity accounting for 54%. However, nuclear faced two more problems, apart from enduring and active citizen opposition: first, the need to replace three old nuclear power plants among the oldest in Europe, with an average age of 35 years in contrast to about 25 years in other countries, and second, the unsolved problem of nuclear waste management and storage. After 2005, nuclear waste was no longer sent to France or Germany, but was temporarily cooled down in large water pools. Due to Switzerland's governance policy of direct democracy and public participation, the population was to be given a say on both issues, with three sites already declared appropriate for nuclear waste storage. Public consultations were planned to occur in the cantons of Berne and Vaud in early 2011. But then the Fukushima disaster struck.

In anticipation of these public consultations, just before 3/11 occurred, the nuclear lobby had run full-page promotional pieces in newspapers, and commercials on television, in which nuclear energy was described as climate friendly, carbon-free, and economically viable. They were funded as part of the Swiss nuclear lobby's annual expenditure of three million Swiss francs on advertising and promoting the benefits of nuclear energy (von Tommer 2010).

However, in the absence of a transparent budget of the industry lobby, some journalists and opposition politicians inquired about the origin of the money for this expenditure, suspecting that the advertisements were financed by citizens' electricity bills (e.g. Angeli 2010). Alongside industry promotions, the media ran stories about nuclear energy issues raised by the green and green liberal parties, which had each put nuclear and renewable energy high on political agendas for the 2011 national elections.

Overall, the Swiss climate for nuclear energy was mostly favorable before 2011, although it was highly contested. This provided an opportunity for the argument of Pidgeon et al. (2008) to be tested. These authors predicted that any major nuclear accident would reverse the internationally positive trend for nuclear energy and mark the end of the portrayed "nuclear renaissance."

The post-Fukushima nuclear energy ban in Switzerland was enacted because of two associated reasons stressed by the Federal Council. First, the decision was rooted in scientific and economic evidence provided by previous studies of energy system scenarios, as stated thus:

> Due to the expected increasing costs of generating nuclear energy (new safety standards, upgrades, revised liability risks, greater financing difficulties due to higher risk premiums for investors), its

competitive advantage with respect to renewable sources of energy is likely to diminish in the longer term.

(Federal Council 2011: 1)

Second: "In view of the earthquake and the tsunami that devastated Fukushima, it feels that the people of Switzerland would like to see a reduction in the residual risk associated with the use of nuclear energy" (Federal Council 2011: 1).

A new kind of political consensus against nuclear energy thus emerged after the Fukushima disaster. The ban on nuclear rebuild was followed by many actions, including immediate safety assessments of existing Swiss nuclear power plants, an increase in the realizable potential of hydropower plants through retrofit, and expansion of existing power plants and planning of new hydropower plants.

## Investigative Context

Nuclear energy has attracted attention in science, technology and society (STS) studies because it is considered as a "megatechnology" that gives rise to technological risk and hazard uncertainties, and subsequent controversies, resistances, and, sometimes, crises, as in the Fukushima disaster (Bauer 1995; Beck 1995; Hindmarsh 2013b). The side effects of technological developments have been addressed under the conceptual frame of "risk" since the 1970s. In Ulrich Beck's (1992) account of the "risk society," Beck described how mega-technologically adverse risks can have novel impacts complex in terms of causation and predictability, latent, often irreversible and "invisible," and can have global effects.

Beck started the German foreword of his book *Risk Society: Towards a New Modernity* (1992) by stressing that the Chernobyl disaster has taught us that one cannot avoid or distance oneself from nuclear danger. At the book's end, Beck called for a new political culture with greater public involvement and accountability to respond to "the growing self-confidence and participatory interest of the citizens" (Beck 1992: 195). In this context, the relationship between publics, policy and a (highly risky and controversial) technology begs scrutiny over time, especially following a major technological disaster such as that of Fukushima, and in a country with a deliberative public participatory governance system and culture such as Switzerland.

The key objective of this chapter is then to investigate the scope and extent of the "Fukushima effect" in terms of its impact on national histories, debates, and policy responses on nuclear power development in Switzerland. One significant effect has already been discussed in regard to Switzerland: the decision made by the Swiss Federal Council to phase out nuclear power. From that "effect," and in regard to nuclear risk governance regulation, what further impact did this phase-out decision have on Swiss society? What exactly was the extent of this dramatic Fukushima effect, particularly in regard to behavioral change on energy consumption? In answering these questions I also contribute to the field of the public understanding of science (e.g. see Wynne 1995), in particular, the subfield of risk perceptions (e.g. Slovic 1987). Accordingly, this chapter explores: (i) the effect of Fukushima on acceptance of nuclear power in the population; and (ii) the effect of Fukushima on attitudes to energy conservation and other energy-saving behaviors, as the Swiss government accompanied the ban on nuclear power with energy conservation obligations.

Two research questions inform this investigation. First, to what extent was Swiss opinion against or for nuclear energy after Fukushima? Second, to what extent was the Swiss population willing to lower energy consumption by adopting energy-saving behaviors? The first question is, of course, also informed by studies on the Chernobyl nuclear disaster, which documented increased opposition to nuclear power and subjective probabilities of occurrence of catastrophic accidents at nuclear power plants in the short- and medium-term after an accident (e.g. Eiser et al. 1989; Rosa et al. 1994). I hypothesized a similar reaction would occur in Switzerland following Fukushima.

Informing my second question, difficulties with achieving pro-energy-saving behavioral change have been illustrated by a number of studies, and appear, in particular, to be due to entrenched habits and various determinants, including cognitive factors (e.g. knowledge and inspiration), attitudinal factors (values, beliefs, attitudes, and norms), personal capabilities (e.g. age, gender, and income), and sociopolitical aspects (e.g. government regulations and community expectations) (Faiers et al. 2007; Sweeney et al. 2013). However, as few studies on energy-saving behaviors were available in Switzerland prior to Fukushima, behavioral change could not be well-investigated before and after Fukushima. Based on the rise in opposition to nuclear energy after Fukushima, though, it would arguably be expected that positive attitudes to energy conservation would emerge with perhaps a medium level of energy-saving behaviors.

## Method

Public perceptions of nuclear energy and energy conservation in Switzerland were measured using a series of quantitative surveys conducted at different times before and after Fukushima by different university research teams. This chapter draws on six surveys, both longitudinal and cross-sectional, conducted from November 2009 to April 2013 (see Table 10.1). The analysis, in mixing together within-participant and between-participant studies, is a useful way to detect insights into the effects of Fukushima.

All surveys included items on acceptance of nuclear power, risks, and benefits; some of them included items on trust towards stakeholders (Surveys 1, 3 and 5), knowledge about energy (Surveys 1, 3 and 6), and behaviors related to energy conservation (Surveys 3 and 6). The survey questionnaires were framed using the psychometric paradigm, which seeks to identify rationality in public perceptions of a risky technology (Slovic 1987).

It is important to note, however, that because of the variable methods and geographical range and targets of the surveys, their results are more illustrative than representative in any comparative or independent sense in relation to investigating the Fukushima effect on Switzerland's nuclear power development.[8] Another complication was that a non-response rate was reported in three studies and another one had a very low response rate. In addition, although the surveys were conducted by academic researchers, some were financed by industry and/or nuclear stakeholders, which invites questions of bias and independence. However, given the unavailability of other data on an otherwise important topic for a

*Table 10.1* The Six Surveys Investigated

**Survey 1.** This longitudinal study was conducted with a mail survey sent to a random sample of households in the German-speaking region of Switzerland; 1232 questionnaires were filled in during September 2010 (response rate 40%), another 790 questionnaires at the end of March 2011 (response rate 77%), and 570 questionnaires in October 2011 (response rate 86%) (Visschers and Siegrist 2012)[2]

**Survey 2.** Based on a CATI telephone survey, the study was conducted on a random sample of Swiss people from 5–14 May, 2011 ($n = 1005$) (response rate non-available) (GFS 2011)[3]

**Survey 3.** A mail survey was conducted at the end of 2011 and the beginning of 2012 on a random sample of the population living in the Italian-speaking region.[4] 1121 filled questionnaires were returned by respondents aged 18 or over (response rate non-available) (Luraschi et al. 2013)

**Survey 4.** A CATI telephone survey was conducted from 6–24 March, 2012 (response rate non-available).[5] The study recorded 806 respondents, aged 18 or over, from the Italian-speaking ($n = 150$), French-speaking ($n = 151$) and German-speaking regions ($n = 505$) (Bonfadelli and Kristiansen 2012)

**Survey 5.** This longitudinal study was based on a CATI telephone survey in the French and German-speaking regions: first wave conducted in October 2009 ($n = 1221$ respondents, response rate 26%), second wave in November 2012 ($n = 564$ persons, response rate 46%) (Siegrist et al. 2014)[6]

**Survey 6.** A mail survey was sent April–June 2013 to a population living in urban Lausanne, in the French-speaking region. The study recorded $n = 665$ respondents, aged 20 and over (response rate 21%) (Crettaz von Roten et al. (submitted for publication))[7]

cross-international study of a country that experienced a strong Fukushima effect, these studies were useful to provide some understanding. Undoubtedly, a follow-up rigorous, independent study that builds a more accurate and valid picture is called for.

In investigating the two research questions, different themes were identified to provide depth of investigation. On the effect of Fukushima on the state of public opinion, against or for nuclear energy, six themes were investigated: acceptance of the nuclear phase out, change of opinion about nuclear power post-Fukushima, perceived risks and benefits of nuclear power before and after Fukushima, social and cultural stratification in regard to these perceptions, and trust toward stakeholders related to nuclear energy. For the second research question related to energy consumption, two themes were investigated: attitudes toward energy conservation and behaviors related to energy saving.

## Survey Results

### Acceptance of Nuclear Phase-Out

#### General State of Opinion after Fukushima

Whether longitudinal or cross-sectional, all surveys analyzed in this chapter indicate a Fukushima effect, as public acceptance of nuclear energy after the disaster was lower than before Fukushima. This decreased level of acceptance was significant. On a 7-point acceptance scale, it dropped from an average acceptance of 4.16 in 2010 to an average of 3.16 in March 2011, the month that Fukushima occurred (Survey 1, Table 10.1). It is also significant that, as details of the scale and impact of the Fukushima disaster continued to be revealed, and international debate about the safety of nuclear energy continued, Swiss acceptance of nuclear energy continued to drop, to a low of 28% in all three regions in March 2012 (Survey 4, Table 10.1) and to only 9% in the French-speaking region in April 2013 (Survey 6, Table 10.1).

In this period, the Federal Council made its decision to phase out nuclear energy and stop all nuclear power plants. Citizens were initially cautious about it, with 84% considering it necessary to avoid knee-jerk reactions related to a nuclear phase-out and with 63% disagreeing (Survey 2, Table 10.1). However, citizens became more supportive over time. In December 2012 (Survey 3, Table 10.1), 62% of Italian-speaking respondents believed nuclear power should be abandoned.

The evolution of public opinion about both nuclear energy and the Federal Council's decisions appeared to be the result of various interrelated and influential phenomena, including the positions of political authorities and parties, and Swiss nuclear power plants approaching their age limit. For example, the Mühleberg reactor of the same generation as the

Fukushima reactors was scheduled to be shut down in December 2012. Survey questions on what should be done about aging plants and possible decommissioning and what should replace them were addressed to a lesser degree by respondents. Regarding existing power plants, 59% of national respondents agreed they should be operated until the end of their life expectancy and 57% were of the opinion that Mühleberg and Beznau (the oldest nuclear power plants) should not be closed immediately (Survey 2, Table 10.1).

Survey 3 (conducted in the Italian-speaking region), conducted at the end of 2011, found opinions remained significantly affected by the disaster. However, while 53% of national respondents considered Switzerland's existing nuclear power plants safe, 89% favored a gradual abandonment of nuclear energy for a shift to renewable sources. This was because 80% of respondents in the survey were frightened by some aspects of nuclear energy. These aspects included radioactive waste management and disposal; technical incidents that threatened the security of the system; nuclear accidents due to external causes (e.g. terrorist attacks, earthquakes); and the consequences of accidents and radioactive contamination of the environment and society. Concomitantly, 33% of respondents were worried about the possibility of an accident in a Swiss nuclear power plant (Survey 4, Table 10.1).

Citizen support for replacement of Switzerland's older nuclear power plants also fell. In 2009, 41% of respondents opposed their replacement—by 2012 this had grown to 55% (Survey 5, Table 10.1)—with 55% supporting the Federal Council's decision to suspend construction of new nuclear power plants (Survey 2, Table 10.1). By September 2012, the Federal Council had announced initial measures that promoted the use of renewables to compensate for the loss of electricity production from nuclear energy. Survey 6, conducted in April 2013 in the French-speaking region, indicated clear support for an energy system built on renewable energies. More than 90% of respondents supported solar energy, more than 80% hydropower and wind energy, and some 60% biomass and natural gas.

## Change of Opinion after Fukushima

The period before 2011, internationally and in Switzerland, was portrayed by the nuclear industry as one of "nuclear build, revival or renaissance" in promoting nuclear energy as a clean (air) energy technology (Crettaz von Roten 2013; Visschers and Siegrist 2012). The Fukushima disaster dramatically challenged perceptions of nuclear energy at the population level, as seen in the results above. This negative effect on nuclear power was also confirmed at the individual level by cross-sectional and longitudinal studies.

The share of Italian-speaking respondents changing their view about nuclear power was evaluated at 33% in the survey conducted at the end of 2011 (Survey 3, Table 10.1); for the German-speaking region, this had increased sharply to 48% by the beginning of 2012. The longitudinal Survey 5, (Table 10.1), conducted in the French- and German-speaking regions, also demonstrated that the share of proponents of nuclear power was smaller in November 2012 than in October 2009. Close to 50% of proponents of nuclear energy before Fukushima (44%) had changed to being undecided about or opposed to nuclear energy post-Fukushima.

### Perceived Risks and Benefits of Nuclear Power

To understand public acceptance of a hazardous technology, it is useful to document its perceived risks and benefits according to the "psychometric paradigm," which "weigh[s] technological risks against benefits in order to answer the fundamental question 'How safe is safe enough?' " (Slovic 1987: 281). Slovic documented that people associate nuclear power with "unknown, dread, uncontrollable, inequitable, catastrophic, and likely to affect future generations" risks (1987: 285). That dread is a common reaction to the possibility of radioactive contamination was also made clear by Hindmarsh (2013a: 6). The surveys reported in this chapter indicate that, after the Fukushima disaster, respondents saw nuclear power as even more risky and of little benefit.

As such, Survey 1 (in the German-speaking region) reported that respondents perceived significantly higher risks and lower benefits associated with nuclear energy in March 2011 than they did before Fukushima, in 2010. Interestingly, the authors of this study reported that the roles of the usual determinants of acceptance remained stable over time: "Both before and after the nuclear accident, perceived benefits had the strongest influence on acceptance of nuclear power stations, compared to perceived risks" (Visschers and Siegrist 2012: 10).

Furthermore, as information about the extent of the problems at Fukushima traveled worldwide throughout 2011, a majority of Survey 3 respondents (56%) in the Italian-speaking region considered nuclear energy so risky that it must be abandoned at any cost and as quickly as possible (Table 10.1). In some alignment, when asked to weight risks and benefits of nuclear energy at the end of March 2012, 41% of respondents from all three Swiss regions emphasized the risks, and 25% emphasized the benefits (the remaining share was ambivalent or undecided) (Survey 4, Table 10.1).

In Spring 2013, some incidents occurred at the almost destroyed Fukushima site, including a power interruption in March, and leaking radioactive water in April. Reporting of these events led to heightened risk perceptions of nuclear energy for a majority of respondents in the French-speaking region of Lausanne. Less than 20% of these respondents approved nuclear energy for energy dependency, safety of operation of nuclear power plants, and global warming; and only 30% for the country's energy needs under the safety guarantee of the Swiss nuclear safety agency (Survey 6, Table 10.1). In addition, an overwhelming majority saw nuclear energy as risky for the environment (82%) or for the population (70%).

## Social and Cultural Stratification

Public perceptions of any risky technology are affected by socio-demographic factors (e.g. gender, age, and education) and cultural characteristics (e.g. egalitarianism and individualism) (Boy 2007). Documenting the effect of these factors is relevant to assessing the consequences of a political decision on technology risk; here, nuclear risk in a direct democracy like Switzerland, where referenda can be launched by citizens.

By the end of 2011, in the Italian-speaking region, male respondents over 65 years of age and with a low level of education were more likely to argue that nuclear power should not be abandoned, adopting the view that it makes an indispensable contribution to Switzerland's energy production (Survey 3). This response was also found in the three Swiss regions at the beginning of 2012: men and those over 65 showed more positive attitudes toward nuclear energy. In contrast, women and citizens with higher education were more likely to consider Swiss nuclear power plants as unsafe (Survey 4). The same gender and age effects were found in Survey 5 in the two main Swiss regions—the German and French language regions—at the end of 2012, and in Survey 6 conducted in Lausanne in 2013. The last survey found, in addition, a significant effect of political orientation: people with a left-leaning political orientation were more opposed to nuclear energy, and held higher risk perceptions than others.

## Trust toward Stakeholders

Because people do not always have sufficient knowledge about the potential risks of a complex technology to enable an informed decision, they appear to rely on their trust in relevant actors to determine the importance of risks and benefits (Siegrist and Cvetkovich 2000). From a sociological perspective, trust, also known as confidence, plays a key role in citizen perceptions in relation to uncertainty and complexity. In complex situations, trust is rooted in the belief of the competence of relevant (e.g. policy and business) actors to carry out their tasks and display appropriate accountability for what is occurring. The influence of trust on regulators, government, and related industry has been frequently highlighted in the high technology risk literature (Boy 2007; Slovic 1999).

Many actors—politicians, policy-makers, electricity utilities, and developers, technicians, and scientists—were involved in what occurred before and after Fukushima (e.g. Hindmarsh 2013a). Surveys in Switzerland probed levels of public trust in nuclear power development and management post-Fukushima. In the surveys reported here, respondents indicated a decline of trust in actors in government, industry, and science related to nuclear energy, but a lesser decline in trust for the Federal Council and the Swiss Federal Office of Energy.

In more detail, Survey 1 (in the German-speaking region) found that over 2010–2011 a decrease in trust occurred toward actors related to nuclear energy. The decrease was higher for nuclear power plant operators (20% decline) than for the Swiss Federal Office of Energy (12% decline) or for scientists working in the field of nuclear energy (12% decline).

A year later, in 2012, Survey 4 (in the three Swiss regions) results for competency and trust related to nuclear energy found similar results to Survey 1: 63% of respondents considered that the Federal Council was most competent; with the following decline for the Swiss Federal Office of Energy (61%), the Federal Nuclear Safety Commission (56%), the Swiss Federal Nuclear Safety Inspectorate (53%), operators of nuclear power plants (51%), and political parties and politicians (43%).

Another year later, in 2013 (in the sample of Lausanne), respondents indicated the highest level of trust in science and technology (79%) and environmental organizations (55%), followed by the local and federal executive authorities—the Lausanne municipality (39%) and the Federal Council (38%)—the electricity (nuclear) industry (27%), and political parties (9%) (Survey 6, Table 10.1).

Typically, nuclear energy opponents trusted environmental organizations more than did nuclear supporters. But, perhaps surprisingly, nuclear energy supporters put more trust in the Federal Council—even though it decided to ban nuclear energy—than did opponents of nuclear energy. This result appears to indirectly reinforce the Federal Council's decision, and suggests a possible referendum to contest the Federal Council's point of view (and support nuclear energy) would likely be rejected.

## Energy Conservation

The second research question of this chapter related to change in energy conservation behaviors, as the Swiss government accompanied the ban on nuclear power with new energy conservation obligations. On the basis of the material available for this analysis, the reasoned-action approach was adopted. This approach follows the view that attitudes guide social behavior; as they are the feelings by which an individual evaluates the positive or negative outcome of a particular behavior (Ajzen 1991). I first investigate attitudes to energy conservation and, secondly, related behaviors.

*Attitudes toward Energy Conservation*

The 2011 study in the Italian-speaking region of Switzerland found that 64% of respondents considered that they could save energy (Survey 3). The majority of respondents were conscious of wasting electricity, but older people (over 65 years) were more likely to believe they did not waste energy compared to other age groups. Attitudes towards specific behaviors varied a lot: some 50% of respondents claimed that they could turn off lights at home or turn off electric devices, but only around 15% said they could reduce the temperature at home by a few degrees or use energy-saving light bulbs.

Two years later, in 2013, when the implementation of energy conservation obligations was being discussed more frequently in the political, media, and public sphere, French-speaking respondents in Lausanne showed mixed feelings about saving energy (Survey 6): 46% found it easy to reduce electricity consumption without it affecting their lifestyle. In contrast, 58% associated higher energy consumption with increased wellbeing. Reduced energy consumption was particularly a worry for women, the lower and upper ends of the age distribution, secondary-educated people and people on the right side of the political spectrum, while negative consequences of reduced energy consumption on personal wellbeing were of particular concern to young people and secondary-educated people.

*Energy-Saving Behaviors*

Survey 3 (the 2011 study in the Italian-speaking region of Switzerland) analyzed energy-saving behaviors: 50% of respondents stated that they regularly turned off television, radio, or computers (25% sometimes and 25% never) but 18% said that they regularly used a clothes dryer, which typically is a high energy-consuming device. The data aligned with the thrust of reasoned-action theory, as those who said they could reduce electricity consumption also indicated higher usage of energy-saving behaviors.

Survey 6 (in Lausanne) found some energy-saving behaviors totally integrated (85% of respondents always or almost always turned off the lights when leaving a room, 71% walked for short trips, 65% practiced water conservation behavior), while others were not yet fully assimilated (41% always or almost always chose local products while shopping; and 41% turned off electrical devices). On the aggregate level, only 14% of respondents did each of these five energy-saving behaviors, while 25% engaged in four out of the five behaviors.

Energy savers were more likely to be female, senior, and politically left-oriented. The results revealed a general consistency with the reasoned-action approach. For example, the more one finds it easy to reduce consumption in one's lifestyle, the more frequently one adopts behaviors such as turning off the lights. These analyses, however, did not adequately explain energy conservation behaviors adopted at the individual level. A better understanding emerged from my analysis of Survey 6, where I identified four groups of energy conservers with various motives.[9] The results are especially useful to provide more targeted information on energy conservation.

The first group, which I nominated as "idealistic energy savers" (28% of the sample), showed the highest energy-saving efforts.[10] They were most worried about energy consumption and found it the least difficult to reduce energy consumption. They were also the most cautious about nuclear energy; they did not fully trust the nuclear safety inspectorate, refused nuclear energy for energy dependency, and associated high environmental risk with nuclear power stations. They had the lowest levels of agreement with the statement that "new inventions will always be found to counteract any harmful effect of scientific and technological developments," which indicates lack of a positivist attitude.[11] This group could be described, based on socio-demographic characteristics, as mostly comprising women, young people, and politically left-oriented people.

The second group, which I nominated as "thrifty energy consumers" (31%), were ranked second place in energy-saving efforts,[12] but followed a different logic. They did not save energy to avoid nuclear power or to limit the harmful effect of a technological invention. This was because they were the most favorable toward nuclear energy and they were the most positivist group. This group could be described as mostly men, senior, and politically right-oriented. Their views may also be due to past periods of deprivation and restrictions (such as the Second World War or the 1970s energy crisis). They also showed the lowest level of knowledge related to energy issues,[13] and wished to have the lowest level of involvement in decision-making about energy (decisions should be made by scientists and engineers, they opined).

The third group I nominated as "inconsequent energy conservers" (26%).[14] Their behaviors can be described as inconsequent because they had the least favorable attitudes toward nuclear energy. They trusted the nuclear safety inspectorate the least, were the least favorable toward nuclear energy for energy dependency, and were the most preoccupied with environmental risks associated with nuclear power stations. Moreover, they had the highest level of knowledge related to energy. While their attitudes were assertive, they were most likely to say that it was difficult to reduce energy consumption; this is why they fit an inconsequent description. They could be described as mostly politically left-oriented.

The last group, nominated as "problem-indifferent energy consumers" (15%), had the lowest level of energy-saving behaviors.[15] They were the least worried by energy consumption at the individual level. They consumed the most, and were typically of the view: "One must guarantee to companies all the electricity they need." They also showed the strongest preference for the highest level of participation in energy decision-making. This group was mostly comprised of middle-aged people (31–45 years old).

## Discussion

This chapter explored how the Fukushima disaster potentially affected the Swiss population in terms of public acceptance of nuclear power and public attitudes towards energy conservation and energy-saving behaviors. This investigation is especially important because the Fukushima effect had two immediate components for Swiss citizens. First was the decision to ban nuclear energy, and second was the decision to encourage energy-saving to reduce energy consumption significantly. This chapter focused on several public surveys on nuclear power and energy conservation conducted in Switzerland between March 2011 and April 2013.

The results illustrate that, post-Fukushima, Swiss citizens were far less supportive of nuclear power and saw nuclear power as more risky and as of less benefit than before Fukushima. They reported a slight to moderate decline of trust in actors related to nuclear energy, with the least decline for the Federal Council after it banned nuclear energy. Survey respondents after 3/11 indicated a general willingness to reduce energy consumption, even if this posed challenges for consumers. Some energy conservation behaviors were thus fully supported, whereas others (the more difficult) were not. Levels of behavior related to energy conservation were found to be informed by different concerns, attitudes, and knowledge related to energy. These two sets of results further confirmed an effect of the Fukushima disaster in Switzerland at both the institutional level (the ban on nuclear energy) and at the individual level, although revealing variations among subgroups of the population.

The results on shifts to lower acceptance of nuclear energy post-Fukushima are consistent with earlier studies on this effect in regard to Three Mile Island and Chernobyl (Eiser et al. 1989; Rosa and Dunlap 1994). Indeed, this outcome aligns to the view of Pidgeon et al. (2008) that the effect of a major accident or disaster in the 21st century nuclear revival period would lower public support for nuclear power, and diverges from those who opined that the effect of Fukushima on Switzerland would be short-term. A large majority of Swiss citizens, it appears, were, and are still, opposed to nuclear energy and therefore agree with the decision of the Swiss Federal Council to ban nuclear energy.

On this point, Switzerland is among the countries that have witnessed a lasting negative effect of the Fukushima disaster on public acceptance of nuclear power (as shown in most studies in this volume). Following Fukushima, electricity consumption did decrease, but the latest available figures indicated that national electricity consumption in 2013 was only 0.7% lower than in 2010. This rate of decrease is obviously not enough to achieve the objectives of the Federal Council (a decrease of 13% compared to 2000). Therefore, important behavioral changes are needed if national policy commitments are to be met. Households represent an important share of national electricity consumption, roughly 31% in 2012. Citizens will therefore need to make a stronger effort to reduce electricity consumption.

Sweeney et al. (2013) stressed the importance of tailored information if one seeks to change behaviors. The analyses in this chapter may help to define better-targeted awareness-raising campaigns, with knowledge tailored to the concerns, beliefs, and attitudes of citizens in order to best realize the Fukushima effect of catalyzing national aims to reduce energy consumption (see also Faiers et al. 2007).

However, to reiterate, while this analysis contributes new and useful comparative insights across Switzerland, the findings can only be illustrative. One should note a general lack of coherent whole-of-Switzerland studies that are also completely independent, as well as longitudinal or cross-sectional surveys related to public perceptions of nuclear technology and shifts to alternative energies in Switzerland.

Another complexity is linked to the difficulty of discriminating between direct and indirect effects of Fukushima. Risk management in the immediate aftermath of the disaster involved various key actors—risk and disaster emergency management and relief agencies, government, media, NGOs, and so on—and their various responses to the disaster, which have all variously influenced public opinions worldwide (e.g. see Hindmarsh 2013a). In regard to Switzerland, Kristiansen and Bonfadelli (2014), for example, reported that although two-thirds of articles in the Swiss German-speaking media did not immediately post-disaster take a position in favor of or against nuclear energy, risk arguments dominated. It is also important to note that the influence of the media on public opinion was mediated by the coverage of reactions of political commentators:

> The Swiss Sunday newspapers reported in the first three months right after the accident a lot about the disaster and a political discourse followed that was then depicted by the media. These various factors combine[d] . . . and in the population, a Fukushima-effect became visible, i.e., a general opinion of nuclear energy as unsafe and an increase of risk awareness.
>
> (Kristiansen and Bonfadelli 2014: 317,
> translated by the author of this chapter)

In further documenting the effect of Fukushima, analyses that mixed together the role of all actors involved in the immediate aftermath of the disaster would thus also be useful to assess Swiss perspectives more accurately.

# Conclusion

The Fukushima Daiichi disaster had political effects worldwide. For Swiss citizens, this took the form of two specific effects: first, the decision to ban nuclear energy, and second, the decision that demanded a significant reduction in energy consumption. This chapter investigated Swiss citizen positions on these two policy decisions during 2011–2013. In summary, what did it find?

This analysis of surveys related to nuclear energy in Switzerland indicates that a majority of citizens oppose nuclear energy and support energy conservation. Therefore, there appears both institutional and societal support of the nuclear power phase-out, with the Swiss citizenry on the whole agreeing with the Federal Council's decision. This point is particularly important in Switzerland, as direct democracy allows citizens to launch citizen's initiatives on any subject at any time. These initiatives play a key role in Switzerland's political and social culture of deliberative and discursive democracy. According to Romerio (2008), no people in Europe have been called to the polls as often as the Swiss population to make decisions in the energy area.

The Federal Council's new general framework for energy, which accompanied the nuclear phase-out, produced many initiatives at the canton, municipal, and industry levels. For example, cantons are now responsible for the implementation of energy efficiency of buildings. In addition, the energy industry has been catalyzed to take a strong role in changing energy consumption behaviors by providing relevant information to consumers; in proposing innovative incentive electricity tariffs or smart metering; and in developing more efficient electricity-generating networks.

Nevertheless, to fully realize the beneficial Fukushima effect, more research is needed on energy transitions with a focus on "science in society," such as information on the efficiency of these measures or scrutiny of public perceptions and behaviors related to energy. Notably, in 2012, the Federal Council commissioned the Swiss National Science Foundation to conduct two large national research programs devoted to energy: "Energy Turnaround" and "Managing Energy Consumption." Both projects, which began at the end of 2014, will also continue to document the effects of the Fukushima disaster in Switzerland, which have both been significant and ground-breaking.

# Notes

1 Switzerland is a confederal federal state. The three pillars of the Swiss political system are federalism, direct democracy, and consensus democracy. The territory is organized into 26 cantons; the cantons themselves are the aggregates of some 2600 municipalities. The cantons have a permanent constitutional status and a high degree of independence; they possess constitutional powers

on education, public security, health, and infrastructure, among other areas. Direct democracy allows popular referenda and initiatives at all political levels.

2  Swissnuclear funded the first stage of the survey.

3  The survey was funded by Economiesuisse, the largest umbrella organization representing the Swiss economy.

4  The survey was funded by the electricity industry.

5  The survey was funded by the Swiss Nuclear Safety Agency.

6  The study was funded by Swissnuclear.

7  The study was funded by the CROSS-research program of the University of Lausanne and Swiss Federal Institute of Technology of Lausanne.

8  The German-speaking region is the largest region with about 65% of the population, and the French-speaking region is the second largest with about 23% of the population.

9  I performed a hierarchical cluster analysis based on level of concern toward electrical energy consumption, attitudes toward energy and energy-saving behaviors, attitudes and risks related to nuclear energy, and frequency of energy-saving behaviors, in using Ward's method based on the squared Euclidian distance.

10  The mean was 23.25 on a 25-point additive scale of energy conservation behaviors.

11  In the field of the public understanding of science, the notion of positivism of Auguste Comte characterizes the belief held by some people that science is the only reliable source of knowledge and the ultimate way to solve problems.

12  The mean level of energy conservation behaviors was 21.99.

13  The 2013 survey included a series of seven questions that covered two topics: nuclear energy and energy conservation. Results indicated that respondents had a fairly good level of knowledge (an average of 46.83% of correct answers was achieved). For example, 64% of respondents gave the correct answer for the origin of radioactivity and 61% for the existence of political decision on nuclear waste in the country, however, only 23% knew where heat mainly escapes in a house (correct answer: the roof). The level of knowledge varied significantly according to gender (men had a higher level of knowledge) and education (higher-educated citizens had a higher level). No differences by age groups or political orientation were found for the questions related to energy conservation, but for nuclear energy older people and those who were left-oriented had a higher level of knowledge.

14  The mean level of energy conservation behaviors was 21.31.

15  The mean level of energy conservation behaviors was 17.17.

# References

Ajzen, I. "The Theory of Planned Behavior," *Organizational Behavior and Human Decision Processes* 50 (1991): 119–211.

Angeli, T. "Die Atomlobby macht Dampft ['The nuclear lobby puts pressure on']." *Beobachter* [Observer] October 1, 2010: 17–18.

Bauer, M. (eds.). *Resistance to New Technology: Nuclear Power, Information Technology, Biotechnology*. Cambridge: Cambridge University Press, 1995.

Beck, U. *Risk Society: Towards a New Modernity*. London: Sage Publications, 1992.

Beck, U. *Ecological Enlightenment: Essays on the Politics of the Risk Society*. Atlantic Highlands, NJ: Humanities Press, 1995.

Bonfadelli, H., and S. Kristiansen. *Meinungsklima und Informationsverhalten im Kontext von Atomenergie und ENSI* [*Climate of opinion and information behavior in the context of nuclear energy and ENSI*]. UniZh report, 2012.

Boy, D. *Pourquoi avons-nous peur de la technologie?* [*Why are we afraid of technology?*]. Paris: SciencesPo les Presses, 2007.

Crettaz von Roten, F. "Society, Politics and Nuclear Energy in Switzerland," *Journal of Scientific Temper* 1 (2013): 101–118.

Crettaz von Roten, F., Clémence, A., and A. Thévenet (submitted for publication). "Understanding Attitudes toward Nuclear Energy: Differences between Asserted and Midpoint Positions."

De Vaus, D. *Research Design in Social Research*. London: Sage Publications, 2012.

Eiser, R., Spears, R., and P. Webley. "Nuclear Attitudes before and after Chernobyl. Change and Judgment," *Journal of Applied Social Psychology* 19 (1989): 689–700.

Faiers, A., Cook, M., and C. Neame. "Towards a Contemporary Approach for Understanding Consumer Behaviour in the Context of Domestic Energy Use," *Energy Policy* 35 (2007): 4381–4390.

Favez. J.C., and L. Mysyrowick. *Le Nucléaire en Suisse* [*The nuclear in Switzerland*]. Lausanne: L'Age d'Homme, 1987.

Federal Council. "Federal Council Decides to Gradually Phase out Nuclear Energy as Part of Its New Energy Strategy," May 25, 2011, accessed April 1, 2015, www.news.admin.ch/message/index.html?lang=en&msg-id=39337.

GFS. *Fukushima als dominantes Element* [*Fukushima as the dominant element*]. Gfs-Bern, 2011.

Hindmarsh, R. "Nuclear Disaster at Fukushima Daiichi: Introducing the Terrain." In *Nuclear Disaster at Fukushima Daiichi: Social, Political and Environmental Issues*, edited by R. Hindmarsh, 1–21. New York: Routledge, 2013a.

Hindmarsh, R. "Megatechnology, Siting, Place and Participation." In *Nuclear Disaster at Fukushima Daiichi: Social, Political and Environmental Issues*, edited by R. Hindmarsh, 57–77. New York: Routledge, 2013b.

Kannan, R., and H. Turton. "Cost of ad-hoc Nuclear Policy Uncertainties in the Evolution of the Swiss Electricity System." *Energy Policy* 50 (2012): 391–406.

Kristiansen, S., and H. Bonfadelli. "Risikoberichterstattung und Risikoperzeption ['Risk reporting and risk perception']." In *Fukushima und die Folgen* [*Fukushima and the consequences*], edited by J. Wolling and D. Arlt, 297–321. Ilmenau: Universitätsverlag, 2014.

Luraschi, M., Galeandro, C., and G. Pellegri. "L'energia elettrica vista dai cittadini. ['Electrical energy seen by citizens']. *L'ideatorio* 5 (2013): 1–40.

Marques, J.G. "Safety of Nuclear Fission Reactors: Learning from Accidents." In *Nuclear Energy Encyclopedia: Science, Technology, and Applications*, edited by S. Krivit, J. Lehr, and T. Kingery, 127–150. New York: John Wiley & Sons, 2011.

Ocha P., and A. van Ackere. "Policy Change and the Dynamics of Capacity Expansion in the Swiss Electricity Market," *Energy Policy* 37 (2009): 1983–1998.

OFEN. "Aperçu de la consommation d'énergie en Suisse au cours de l'année 2013 ['Overview of energy consumption in Switzerland in the year 2013']." OFEN, Berne, June 2014.

Pidgeon, N., Lorenzoni, I., and W. Poortinga. "Climate Change or Nuclear Power—No Thanks! A Quantitative Study of Public Perceptions and Risk Framing in Britain," *Global Environmental Change* 18 (2008): 69–85.

Poortinga, W., Aoyagi, M., and N. Pidgeon. "Public Perceptions of Climate Change and Energy Futures Before and After the Fukushima Accident: A Comparison between Britain and Japan," *Energy Policy* 62 (2013): 1204–1211.

Romeiro, F. "Les controverses de l'énergie: fossile, hydroélectrique, nucléaire, renouvelable ['Energy controversies: Fossil, hydroelectric, nuclear and renewable']." Lausanne: Presses Polytechniques Universitaires Romandes, 2008.

Rosa, E., and R. Dunlap. "Nuclear Power: Three Decades of Public Opinion," *Public Opinion Quarterly* 58 (1994): 295–324.

Siegrist, M., and G. Cvetkovich. "Perception of Hazards: The Role of Social Trust," *Risk Analysis* 20 (2000): 713–719.

Siegrist, M., Sütterlin, B., and C. Keller. "Why Have Some People Changed their Attitudes toward Nuclear Power after the Accident in Fukushima?" *Energy Policy* 69 (2014): 356–363.

Slovic, P. "Perception of Risk," *Science* 236 (April 17, 1987): 280–285.

Slovic, P. "Trust, Emotion, Sex, Politics, and Science: Surveying the Risk-Assessment Battlefield," *Risk Analysis* 19 (1999): 689–701.

Sweeney, J., et al. "Energy Saving Behaviours: Development of a Practice-Based Model," *Energy Policy* 61 (2013): 371–381.

Visschers, V., and M. Siegrist. "How a Nuclear Power Plant Accident Influences Acceptance of Nuclear Power: Results of a Longitudinal Study Before and After the Fukushima Disaster," *Risk Analysis* 33 (2012): 333–347.

Von Tommer B. "AKW-Kampf wird zur bisher teuersten Politschlacht ['Nuclear struggle becomes most expensive political battle']." *NZZ am Sonntag* [New Newspaper of Zürich on Sunday], November 21, 2010: 9.

Wynne, I. "Public Understanding of Science." In *Handbook of Science and Technology Studies*, edited by S. Jasanoff et al., 361–388. Thousand Oaks: Sage, 1995.

# 11 France, the Nuclear Revival and the Post-Fukushima Landscape

## Joseph Szarka

France has consistently promoted nuclear power since the 1970s. The French nuclear industry has expertise over the whole fuel cycle (uranium mining and milling, fuel processing and sale, reprocessing and waste management) and has staked out leadership positions in reactor design, construction, and operation. The key firms in the sector are Areva and Electricité de France (EdF), which are largely state-owned corporations. France has 58 commercial reactors, which in 2013 generated 73.3% of domestic electricity, the highest proportion in the world, and is second only to the US in terms of volume and number of reactors (IAEA 2014). International support for nuclear power has, however, waxed and waned; its rapid expansion in the 1970s was halted in the 1980s. The safety crisis caused by nuclear accidents at Three Mile Island in 1979 and Chernobyl in 1986 left a lasting negative stigma.

The sector also traversed an economic crisis. Regulatory tightening intended to reduce risk delayed construction and increased costs, while liberalization increased market competition. The compounded effect was a slow-down in the building of nuclear power plants (NPPs), with a virtual moratorium in many countries. Yet during the late 1990s, in a context of heightened concern over energy security and climate protection, nuclear proponents staged a comeback. France took on a self-appointed leadership role in promoting a revival, seeking to restore the legitimacy of nuclear power and generalize its use worldwide.

But to what extent did France succeed in leading a nuclear "renaissance"? To investigate this question, this chapter analyzes the ways in which France promoted a nuclear revival by attempting to renew the norms of global nuclear governance. Between the 1990s and 2010s, three main stages can be discerned. In the first stage, a renaissance discourse was constructed, which claimed that nuclear power was cheap and quick to exploit. In a second stage, during the 2000s, new orders for NPPs suggested that the revival was underway. For the French nuclear industry, 2005 was a milestone year, with construction starting on its first NPP since 1991, in Olkiluoto (Finland). This was also the first construction by Areva of its European Pressurized Reactor (EPR), a "generation three+"

reactor. Construction of a second EPR was started in 2007 in France, with two more construction starts in 2009 and 2010 in China, developments which constituted the high-point of the revival. Subsequent to the tsunami that hit Japan on March 11, 2011, the core meltdowns at Fukushima were categorized as a level 7 "major accident" (the highest level on the International Nuclear and Radiological Event Scale). This inaugurated a third phase, in which the revival stalled. Through exploration of the putative nuclear revival, this chapter will argue that the inadequacy of French attempts to renew the norms of global nuclear governance was exposed by a "Fukushima effect."

## Conceptual Framework

To explicate the different stages of the revival, the interplay between ideas, interests, and institutions will be analyzed. Within this interplay, the role of norms is fundamental due to the contested legitimacy of the commercial nuclear sector and to questions over its capacity to assure the safety and security of reactors, the millennial disposal of waste, and the non-proliferation of atomic weapons.

Norms were defined by Katzenstein (1996: 5) as "collective expectations for the proper behaviour of actors with a given identity." A norm is thus a standard of behavior that is expected rather than required. In other words, it is useful to distinguish between *norms* and *rules*. Whereas the former encourage regularities of conduct in line with particular (sometimes implicit) expectations, the latter explicitly prescribe behavior and dictate sanctions for non-conformity (Ostrom and Basurto 2011: 322). Hence, norms are not always formulated as rules (e.g. they do not necessarily stipulate sanctions). Whilst all norms contain the notion of what "ought" to be done, this notion also needs specification. A normative injunction can involve an ethical dimension. However, not all norms have moral content. This is clearly the case for technical norms (such as standard voltage levels), but is also considered to hold for many economic norms. Elster (1989: 99) recalled that the classic difference between *homo economicus* and *homo sociologicus* is that "the former is supposed to be guided by instrumental rationality while the behaviour of the latter is dictated by social norms." This is pertinent for the present discussion since it focuses on economic and political actors who are guided by "instrumental rationality" (e.g. in the pursuit of the maximization of benefits accruing to themselves).

To analyze the instrumental rationality displayed in France's promotion of the nuclear revival, six categories of norms will be examined: (1) cognitive, (2) diplomatic, (3) geopolitical, (4) technological, (5) market, and (6) safety and security. In combination, they provide a framework for the global governance of nuclear power. Norms are in constant evolution due to the influence of vested interests and concerned parties, and the actions of arbitrators at national and international levels.

Advocates and proponents of nuclear power aim to make domestic and global norms more favorable to its expansion, while critics and opponents seek to raise normative barriers to deployment.

To trace developments in nuclear governance, this chapter will fold the six categories of norms into the three stages of the nuclear revival. This chapter reveals how norm modifications contributed to the revival's initiation, rise, and stall, devoting a section to each of those three stages. A final section explores the effects of Fukushima on the nuclear renaissance, allowing conclusions to be drawn from France's attempts to steer global nuclear governance.

## Talking up the Nuclear Revival

This section reviews *cognitive* norms related to nuclear power and explores the ways in which advocates have massaged them to initiate a revival. It probes the nuclear lobby's discourse, pin-pointing French contributions to it. It also analyzes how France sought to modify *diplomatic* norms related to the diffusion of nuclear power in order to facilitate reactor sales internationally.

### Cognitive Norms

Cognitive norms shape the way we think about issues, and discourses are the vehicle for transmission of "right thinking." The Three Mile Island (1979) and Chernobyl (1986) disasters created a negative framing of nuclear power. The nuclear industry understood that a positive framing was required if it were to make a comeback. This started with choice of terminology. Because the concept of "renaissance" gives a favorable gloss, it is the term of choice for nuclear advocates. In contrast, the term "revival" is relatively neutral.

The official French line has systematically talked up the prospects for renaissance. Since the 1960s, French presidents have been the nuclear sector's leading political advocates. Nicolas Sarkozy, as president between 2007 and 2012, followed this tradition. Even before acceding to the presidency, Sarkozy was a champion of nuclear power. As Finance Minister, he oversaw the 2005 Energy Bill, which restarted nuclear build in France. In 2010, at the International Conference on Access to Civil Nuclear Energy, he declared:

> We're in a new nuclear era, one of renaissance. The analogy with this glorious period of European history is no doubt controversial. But there are common elements: the questioning of the old way of thinking, the questioning of irrational fears that goes with faith in science and technology.

(quoted in Renard 2011)

Sarkozy's speech updated the cognitive norms associated with the promotion of nuclear power. Since the 1950s, the discourse of the French political establishment has linked nuclear technologies with economic modernization, the projection of military power, the preservation of independence, and the enhancement of national prestige (Hecht 1998).

Nuclear power has frequently been justified on grounds of economic need. In developing countries, limited access to electricity is a road-block to development. In this framing, utility maximization is the imperative which nuclear power should fulfill. Claimed in the 1950s to be "too cheap to meter,"[1] and presented in the 1970s (notably in France and Japan) as the solution to the oil crisis and a source of national energy independence, nuclear power in the 1990s acquired two further economic arguments. Energy security once again became a policy issue. Increased diversification of supply was seen as an essential hedge, given the geopolitics of oil and gas sourcing, characterized by threats to supply, pinch-points in transportation by sea and land, prospects of "peak oil," and runaway price increases. Nuclear advocates stressed three energy security dimensions: access, availability, and competitiveness. They argued that uranium sources were geographically dispersed and abundant, and that improved reactor design meant lower prices (e.g. Areva n.d.).

Environmental concerns were also exploited. Nuclear power was portrayed as contributing to sustainable development (Rougeau 1998). This was perhaps the most novel component of the renaissance discourse. It sought to overturn the perception of nuclear energy as a threat to future generations due to the unresolved issues of radioactive waste disposal and proliferation risks. Further, advocates sought to "green" nuclear power by claiming that it is free of greenhouse gas emissions. Lepage (2011: 98–103) showed how France attempted at the EU level to assimilate nuclear power to a form of renewable energy and, when that gambit was rejected, rebranded nuclear as a "low carbon" energy source. The objectives were to reposition the nuclear sector as essential for climate protection, legitimize policy support, and attract financial subsidies intended for "clean" technologies and renewable energy. This renewal of cognitive norms provided the nuclear lobby with the discursive means to pressure national governments into abandoning moratoria.

Technological improvements, particularly with the development of generation three+ reactors, allowed the nuclear industry to claim greater safety and security, higher fuel efficiency, and lower waste production. They also enabled claims of increased economic competitiveness, due to purportedly shorter construction times, more attractive investment terms, and cheaper electricity. In this vein, Areva (n.d.) asserted that electricity generated by the EPR would cost 10% less than its previous N4 model due to improved characteristics including: shorter construction time (promising 57 months), larger unit size (1650 MW rather than 1450 MW), longer life (60 years rather than 40), higher availability, more efficient fuel utilization, greater operational flexibility, and less radioactive waste and plutonium production. New usages—water desalination, district heating, electric vehicles, and hydrogen production—were advanced in order to position nuclear technology as essential for an enlarged range of users. The advancement of new usages intersected with emergent social norms related to intergenerational equity and with the technical norms of energy efficiency. This renewed discourse aimed to make the sale and purchase of nuclear technologies into a standard component of energy markets.

In summary, advocates sought to reposition the nuclear sector within policy debates on climate change and sustainability by modifying existing cognitive norms. Their strategies went beyond an exercise in intellectual capacity-building. They supported a number of practical initiatives, with diplomatic norms providing a bridge to nuclear deployment.

### Diplomatic Norms

Diplomatic norms help shape the behavior of actors in international society. In the domain of technology choice, sovereign rights prevail. Every state is entitled to select energy sources and technologies on the basis of its independent preferences, and subscribe to international environmental or climate protection treaties only to the extent that it chooses. The dissemination of nuclear power has followed these generic norms (albeit with reservations regarding the proliferation of nuclear weapons). Consequently, "national interest" figures prominently in energy politics.

The promotion of nuclear energy as crucial to national energy independence forms part of a French political tradition that stretches unbroken from President Charles de Gaulle to the present. In practice, nuclear power has improved French energy independence to a limited degree. Its primary energy source (uranium) is imported, and nuclear power as a proportion of total energy consumed hovers around the 25% mark and may be calculated as even lower, depending on the accounting method (Schneider 2009). Nevertheless, substantial economic advantages are asserted. Electricity prices in France are relatively low, with benefits for households and for the international competitiveness of French firms.

Further, President Sarkozy positioned commercial nuclear power as an economic tool to aid developing countries and hence a "weapon for peace" (quoted in MacLachlan 2008). The discourse of nuclear energy as a weapon for peace can be traced to US President Eisenhower's *Atoms for Peace* speech, given before the UN General Assembly in 1953. The 1970 Non-Proliferation Treaty granted in its article IV "an inalienable right" to all countries to "develop research, production and use of nuclear energy for peaceful purposes without discrimination" (United Nations 1970). The right to buy and develop nuclear power was upheld by emergent economies, even though it sometimes raised international alarm (as the cases of Israel, India, Pakistan, Iraq, and Iran illustrate). However, Sarkozy's translation of the norm as giving France a right to promote nuclear technology was novel, as indicated by his statement that he was

> favourable to the development (of nuclear power in the world) with the obvious rule that we collaborate only with democratic govern- ments and under strictly administered conditions. This kind of part- nership backed by the strength of the French nuclear industry [means we must] maintain the leadership in this domain of our French ven- dor (Areva) and our French nuclear utility (EdF). Construction of the first EPR will help us keep our leadership.
>
> (quoted in MacLachlan 2007)

For Sarkozy, the nuclear renaissance was French-led and guided by instrumental rationality. More specifically, it was about sales of nuclear technology in the short-term, to as many buyers as possible and on a state-to-state basis.

The massaging of diplomatic norms aimed both to legitimize French nuclear sales and change international conduct. France pushed for the mainstreaming of nuclear energy within global climate policy. During negotiations on the 1997 Kyoto Protocol, France wanted EU partners to press for the inclusion of nuclear power within the Clean Develop- ment Mechanism. Only Finland was in favor, with the remaining EU-15 states opposed (Lajoinie 2001: 102–103). Even though the US and China favored the proposal, France was unable to convert a national preference into an international diplomatic norm (Mühlenhöver 2002: 175–176). But this setback did not prevent the French nuclear industry from exploit- ing cognitive and diplomatic norms to promote its export drive.

## Toward a Nuclear Revival

What are the signs by which a nuclear revival can be recognized? Whilst no official list exists, a number of indicators recur in the literature. These indicators include quantitative measures, such as a head count of coun- tries with NPPs, the number of reactors in construction, total capac- ity figures, and terawatt hours of nuclear-sourced electricity generated. Qualitative indicators include the development of a wider range of reac- tor types and end uses, and levels of social acceptance. Since the 1980s, the rate of expansion has been slow on most indicators, and negative on some (Mez 2012). However, the number of NPP construction starts worldwide jumped from around two to four per year to eight in 2007, 10 in 2008, 12 in 2009, and 16 in 2010 (IAEA 2014: 79). But for this incipient revival to become a global phenomenon, evolution in geopoliti- cal norms was crucial.

## Geopolitical Norms

The geopolitics of nuclear trade remains marked by the international relations context in which it emerged, namely the development and use of atomic weapons by the USA during the 1940s, the post-World War Two settlement, and the Cold War and its legacy. For several decades, two categories of nation monopolized nuclear technologies: a "Western" bloc comprising North America, Western Europe, Japan, and Australia (as uranium supplier), and an "Eastern" bloc, comprising the former USSR, its satellite countries, and client states. The primary norm was the absence of East-West nuclear trade. The second was that recruitment to the "nuclear club" was conducted in a North-South direction. Nuclear trade moved down a one-way street, with zones of geopolitical influence acting to ring-fence markets.

As only 31 countries use nuclear power (IAEA 2014), a crucial dimension of a global revival is the recruitment of "newcomer" countries. The International Atomic Energy Agency established a "milestones approach" to the identification of "newcomers" (IAEA 2011). It comprised three phases: (i) pre-project; (ii) project decision-making, particularly inviting bids for a first NPP; and (iii) construction and operation of a commercial NPP. In the decade of the 2000s, some 60 states expressed an interest in nuclear energy to the IAEA—a statistic often cited as proof of the renaissance—yet only a handful of newcomers had proceeded to phase 2 and none to phrase 3 (Findlay 2011: 83–92).

In this uncertain context, France took a proactive role in opening new markets. Over 2007–2008, President Sarkozy became the leading salesman for nuclear technology, with a whistle-stop tour of Middle Eastern and African states. France signed non-binding memoranda of understanding with Libya, Algeria, Morocco, Saudi Arabia, Jordan, and South Africa, but none led to actual contracts. France was more successful in established markets. A turning point came in 2003 with the sale of Areva's first EPR to Finland. The dash for export markets led to construction of the first EPR abroad *before* building a demonstration model at home. The French also looked to key markets to which they had gained entry in the 1990s. In 2007, the highlight of President Sarkozy's state visit to China was the finalization of a contract for two EPRs (WNN 2007). In 2010, an outline agreement was reached for the sale of two EPRs to India (MacLachlan 2010b).

However, Areva's ambitions to sell to the United Arab Emirates (UAE)—a newcomer country—came to nothing. In 2010, the UAE signed a $20 billion contract with the Korea Electric Power Corporation to construct and operate four nuclear reactors, based on a generation-two design. The reasons cited for losing the contract included the higher cost of the EPR (due to its greater safety provisions), its dented reputation after construction problems in Finland and France, and disorganization within the French consortium (MacLachlan 2010a). For the first time, an emergent economy had won a tender from a newcomer country.

This outcome was not in the French script for nuclear revival. It revealed that the geopolitical norms for reactor sales were changing. Ring-fencing based on "East-West" and "North-South" differentiation was a hangover from the Cold War that could not persist in the era of globalization and open markets. However, the change in trade patterns went in a direction that France neither planned nor desired. It raised questions both about French leadership and the nature of the revival itself.

## The French-Led Nuclear "Renaissance" Stalls

Prospects for the renaissance deteriorated sharply during 2010–2011. The main signs of a stall were, first, an inability to close further reactor sales. Second, the new starts announced for Jaitapur in India and Penly in France did not happen. Third, the failure to complete an EPR generated bad publicity. Fourth, the financial position of Areva grew problematic, in part due to its problems in Finland. Fifth, the nuclear disaster at Fukushima led to public acceptance of nuclear power dropping worldwide from 57% to 49% (Hayashi and Hughes 2013: 106), prompting a tightening of safety regulations, a hold on NPP construction, and reluctance to sign new deals. But to what extent did this stall occur because of safety and security issues (arising from Fukushima), or because of technological and economic issues (as revealed by setbacks with the EPR)? To probe this question, we next consider technological, market, and safety norms.

### Technological Norms

All 58 commercial NPPs in France are pressurized water reactors, arising from three production series: 34 x 900 MW, 20 x 1300 MW, and 4 x 1450 MW plants. The construction of two long series of near-identical reactors allowed standardization of design and build, with emphasis on achieving scale economies and greater efficiency in output. This formula is understood to be the key explanation for the French nuclear industry's success in the late 20th century (Finon and Staropoli 2001). It also set a benchmark for future systems.

The EPR, as the "evolutionary" extension of the French fleet, was expected to lay the industrial foundation for the nuclear revival and reproduce the success formula based on standardization and scale economies. The building of a prototype was initially planned for 1998–1999, with a series of eight to 10 reactors to be constructed in France over 2000–2012 (Finon 2009: 213). Due to overcapacity, this domestic program was abandoned. The construction of the first EPR was instead started in Finland in 2005. Although completion was promised for 2009, construction will not be completed until 2018 at best (Rosendahl and

De Clercq 2014). The second EPR was started in 2007 at Flamanville (France), with grid connection announced for 2012. Yet entry into service was put back to 2016, due to the need to learn lessons from Fukushima (EdF 2011). A question mark now hangs over the capacity of Areva and EdF to build a standardized nuclear fleet using the EPR.

One main cause of the delays is the lack of a sufficiently detailed design, entailing problems for regulators and on-site contractors (Sains 2014). The Finnish and French builds have been plagued by expertise shortages, industry bottlenecks, and problems with the supply network (MacLachlan 2009). Project management proved inadequate. Three fatal accidents occurred at Flamanville. Early problems at Olkiluoto were ascribed to Areva's lack of construction expertise. Yet EdF is project manager at Flamanville, where construction problems have dented its reputation as the world's most accomplished builder of nuclear reactors.[2]

*Market Norms*

A major challenge for a nuclear revival is to demonstrate competitiveness in liberalized energy markets. Liberalization involves withdrawal of the subsidies from which the nuclear industry benefited in the past. The revivalist discourse stressed the nuclear sector's capacity to manage economic risk, secure finance from the private sector, and compete on a level playing field (Platts 2012). The nuclear industry thereby advertised its conformity with free market norms, and government went along with its claims (Thomas 2010).

Actual developments have tested these claims and undermined their credibility. The EPRs in Finland and France are not being built on budget. The original estimate of €3 billion for Flamanville has increased to €8.5 billion (EdF 2011). Export credit guarantees and low rate loans were made available for the sale to Finland. The appetite of investors for large-scale projects was diminished by the financial crisis of 2007–2008, and the price competiveness of electricity generated by the EPR has been questioned. The Cour des Comptes (2012: 225) estimated a level between 70 and 90 €/MWh, roughly triple the prevailing consumer prices.

Longitudinal analysis has revealed that the underlying competitive position of French nuclear power is weaker than claimed. Grubler (2010) calculated the costs of building France's 58 nuclear reactors and found that unit costs per megawatt *increased* over time in real terms, which he called a "negative learning" effect. Rangel and Lévêque (2013) corroborated this finding, noting that between the first and the last NPP costs escalated by a factor of 1.5 (although not 3.5 as estimated by Grubler). The increase in unit costs means that scale economies were *not* achieved as reactors grew larger. Moreover, the same pattern of unit cost escalation per megawatt of build has been identified in relation to the Olkiluoto and Flamanville EPRs. A major and consistent cause of cost escalation has been the need to enhance safety and security.

## Safety and Security Norms

In the post-Chernobyl and post 9/11 contexts, the bar was raised for operational safety and site security. Areva made extensive claims about the capacity of the EPR to resist severe threats. "Passive safety" systems were incorporated to reduce the probability of operator error, whilst a "core-catcher" was installed under the reactor pressure vessel to retain radioactive elements in the event of a meltdown. In addition, the reinforcement of the EPR dome was deemed sufficient to withstand an airliner crashing into it, as well as to resist earthquakes (Nian and Chou 2014).

Fukushima, however, fundamentally changed the safety and security requirements for nuclear power. The initial reactions proved inadequate, not only in terms of the management of the accidents onsite, but also in regard to the response from the French nuclear sector. Anne Lauvergeon, then CEO of Areva, claimed that "if there had been EPRs in Fukushima, there would have been no radioactive leaks into the environment" (quoted in AFP 2011). This statement was neither helpful nor plausible. The Daiichi plant lost emergency power and external electricity for 11 days, and this blackout was the immediate contributory factor to the nuclear accidents. Yet the EPR (as originally designed) was also vulnerable to multiple shocks, as revealed by the Hirsch report (2011), which highlighted the short time period (24 hours) during which emergency electricity supply would be available to maintain systems control.

The drawing of lessons from Fukushima has been fraught and protracted. The Japanese parliamentary inquiry was extremely critical of the operator, TEPCO, because of its attempt to minimize its responsibilities by placing exaggerated emphasis on the impact of the tsunami and instead concluded that "the accident was clearly manmade" (National Diet of Japan 2012: 16). It catalogued a list of deliberate actions that led to the accidents, stressing that

> despite the fact that constant vigilance is needed to keep up with evolving international standards on earthquake safeguards, Japan's electric power operators have repeatedly and stubbornly refused to evaluate and update existing regulations, including backchecks and backfitting.
>
> (National Diet of Japan 2012: 43)

Thus, whereas the earthquake and tsunami that hit Fukushima were natural catastrophes beyond human control, the *nuclear accidents* (e.g. involving three core meltdowns) were *not* inevitable but the result of operational and regulatory failures. According to Srinivasan and Rethinaraj (2013: 733), "the Fukushima crisis has highlighted the problems of regulatory capture by the industry resulting in collusive ties between the regulators and the industry that can seriously compromise public health and safety." Fam et al. (2014: 200) identified this "nuclear village" as consisting of "pro-nuclear advocates from Japan's Diet, prefectural governors, bureaucracy such as the Ministry of Economy, Trade and Industry and other regulatory agencies, nuclear vendors, the financial sector and large corporations represented by Keidanren."[3] The "nuclear village" not only promoted nuclear power, but equipped itself to obstruct proposals for improved safety and security standards through a process of regulatory capture. The ultimate consequence of this process was the Daiichi nuclear disaster. The conclusion drawn by Wang et al. (2013: 143) is that "the Fukushima accident could have been averted with effective regulation."

A similar vision informed the post-Fukushima search to reinforce safety and security norms through remediation of both the technicalities of design failure and the pitfalls of institutional failure. In the EU, the European Council of March 24–25, 2011, initiated a risk assessment of all 143 NPPs in member states in relation to earthquakes, flooding, or other extreme events. These "stress tests" produced recommendations to improve robustness including the idea of a "hardened core," defined as "a limited set of material and organisational measures, designed to ensure basic safety functions in extreme situations" (ENSREG 2012: 47). A key aim is to maintain a permanent power supply (inter alia to systems control and cooling systems), whatever the circumstances. France participated fully in the EU stress tests, but identified no grounds for suspension or closure of any plants. In June 2012, the French nuclear regulator issued 19 recommendations to improve safety and security, including the establishment of a hardened core and the creation of a rapid intervention force at the national level (ASN 2012). As the improvements will be experimental and expensive, the authorities are feeling their way forward over the long term. Associated investments will be made in three phases: 2014–2015, 2015–2019, and beyond 2019 (Cour des Comptes 2014: 189).

Fukushima clearly provided a wake-up call to the French nuclear sector. As three of the world's great nuclear powers (the USA in 1979, the former USSR in 1986, and Japan in 2011) have each experienced a major nuclear accident, will France also suffer one? To explore this eventuality, in 2012 the French Institute of Radiological Protection and Nuclear Safety ran a series of extreme nuclear accident scenarios, and found that costs would run to a *median* figure of €430 billion. Because these findings came from a state institute (rather than an anti-nuclear NGO), a shock-wave went through the nuclear sector (IRSN 2012; Nucleonics Week 2013). These unwelcome discoveries have not transformed France's commitment to nuclear power, but have contributed to a change in the domestic energy debate (as explored below).

Fukushima also provoked a wider international response. The World Energy Council set up a taskforce to improve nuclear governance. Its global survey revealed support for harmonized nuclear safety norms, but acknowledged that "there also seems to be comparatively lower support for the international enforcement of safety standards" (WEC 2012: 22). No appetite exists for the establishment of an international regulator having the authority to draw up common standards and require national conformity with them. Thus, while aspects of the EU stress tests were mirrored in the US, Russia, and China, no common approach was agreed to, given the context of competing reactor designs and secrecy over their operations.

Neither has leadership emerged from any individual country, including France. In its renaissance discourse, the French nuclear lobby promulgated a narrow conception of leadership, limited to technological, economic, and passive safety claims. However, its engagement with the wider governance issues highlighted by the Fukushima nuclear accident has been limited. This stems from a governance structure, which historically was very similar to Japan's nuclear village. In an earlier configuration, it was composed of EdF, the Atomic Energy Commission, the reactor manufacturer Framatome, the fuel company COGEMA, and the Ministry of Industry, which coordinated the sector (Szarka 2009). The problem of regulatory capture did not arise because there was no regulator to capture. It was only in 2006 that an independent regulator—the *Autorité de sûreté nucléaire* (ASN)—was created. Hence the ASN has limited experience. The Flamanville EPR, started in 2007, is the first new build it has supervised—in relation to a new reactor design, and in the new post-Fukushima environment where it must invent a "hardened core" to implement EU policy. International leadership exercised by the ASN is implausible so long as it remains focused on such reactive "learning by doing." France is not able to promote an exemplary regulatory model when it is a follower, rather than a leader.

In summary, the international search post-Fukushima to strengthen safety and security norms has embraced both the technicalities of design failure and the consequences of regulatory failure. However, it is proving easier to correct technical shortcomings than solve institutional deficits. Hence we cannot be sure that the lessons have been fully drawn by either established users of nuclear power or newcomers. The most evident sign of this is that 46 reactors of Japan's nuclear fleet were still shut down at the time of writing.

## Post-Fukushima Prospects for a Nuclear Revival

In the aftermath of Fukushima, nuclear expansion encountered increased hostility. If a leading industrialized nation such as Japan could not manage nuclear power, the prospects were necessarily diminished elsewhere. Germany suspended its eight oldest reactors with immediate effect in 2011 (see Jahn and Stephan in this volume), with closure of the remaining nine planned to occur by 2022 (WNO 2012). Switzerland committed to not replacing its reactors (see Crettaz von Roten in this volume). In Italy, government proposals to end the nuclear moratorium were abandoned after rejection by 94% of voters in a referendum (BBC 2011). Neither did the revival persist in the US, albeit for different reasons. Although both the Bush and Obama administrations backed nuclear power and provided loan guarantees, new starts have been limited to the Vogtle plant in Georgia (Felder 2013). The combination of low gas prices (driven by the shale gas boom) and reduced electricity consumption (due to the recession) made NPP construction unattractive.

The UK, however, bucked the trend and ordered new reactors. EdF and Areva have promised to build four EPRs in the UK. Two are proposed at Hinkley Point in Somerset at a cost of £14 to £16 billion. Contrary to its neoliberal ideology, the Cameron government promised to underwrite new nuclear build with hefty subsidies paid by consumers. The 2013 agreement with EdF was based on "contract for difference" whereby the generator is effectively paid a guaranteed price. This "strike price" was set at £92.50/MWh (approximately €116/MWh), increasing annually in line with the Consumer Price Index for the 35-year duration of the contract and providing a post-tax rate of return of around 10%. The strike price was over double the prevailing wholesale price of electricity, and more costly than a number of renewable energy sources (Ambrose 2013). Econometric modeling, as used by analysts such as Linares and Conchado (2013), had already revealed that new NPPs are not cost-competitive in liberalized markets. The UK case provides a "real life" demonstration that they cannot be built without major subsidies. This reveals the limited prospects for a French-led nuclear renaissance.

Its fading may offer an opportunity to rivals. In 2013, of the 72 reactors under construction worldwide over half were in just two countries, China (29) and Russia (10) (IAEA 2014). The Russian nuclear industry is thus a candidate for leading a future revival. The other candidate is China, where the government strategy is to encourage the domestic nuclear industry to design its own reactors and to compete in the global market (Yi-chong 2014: 5; see also Fang in this volume). Chinese firms first bid independently for the UK Hinkley Point contract, and eventually participated in the Areva/EdF consortium, holding 30–40% of the equity and undertaking construction work. One question for the future will be the extent to which the French and Chinese nuclear industries will collaborate or compete for new orders.

Yet whilst Fukushima dramatically aggravated the problems faced by the French nuclear industry, it should not eliminate attention to their antecedents. France's nuclear revival ran into serious difficulties before the disaster in Japan, and the existence of these difficulties was acknowledged by major players *prior* to Fukushima. A number of strategic recommendations were made by the Roussely report (2010), including the need to offer a wider product range (with smaller reactors than the EPR), and improve project management during new build. The month before Fukushima, the French Council for Nuclear Policy acted on the report's recommendations (Présidence de la République 2011). It consecrated EdF as "consortium leader" in French bids for turnkey reactor build and required closer partnership between EdF and Areva on optimization of the EPR. Development of new, smaller nuclear reactors was encouraged, dethroning the EPR from its "one size fits all" ranking. A business model based on selling the same product in Europe, in the US, and to newcomer countries was acknowledged as flawed. Subsequently, the French nuclear industry deepened its cooperation with Asian rivals. In 2012, Areva, EdF, and China Guangdong Nuclear Power Company formed a partnership to develop a midsize reactor (Nucleonics Week 2012). In 2013, the first order for the Mitsubishi/Areva Atmea-1 was placed by Turkey (WNO 2014).

Within France, the terms of the nuclear debate have shifted. François Hollande, elected President in 2012, committed in his election manifesto to complete the Flamanville EPR, but also to reduce the share of nuclear power in electricity generation from 75% to 50% by 2025 (Hollande 2012). As the share of nuclear power is programmed to decline, and with no start date set for further EPRs, a key finding is that the domestic nuclear "renaissance" has been shelved.

## Conclusions

The landscape of the nuclear sector has been changed by Fukushima, but whether it has been transformed remains to be seen. Hence a question mark hangs over the nuclear revival: has it been terminated or merely postponed? Optimistic pronouncements about new generation reactors coming online circa 2010 were made in the early 2000s. They have proved false. But more sober scenarios assumed expansion in the 2020s. Given the existence of multiple indicators—and the uncertainties surrounding several—prudence is required in assessing the scope, timing and trajectory of a global nuclear revival.

Assessment of the prospects for a Russian or Chinese revival is beyond this chapter's remit, but the French-led revival has clearly stalled. The discourse of nuclear renaissance embodied French ambitions to convert domestic nuclear arrangements into an international template. However, the grand design to "normalize" the national exception by turning it into an international model has failed. Most of the factors that are hobbling the French nuclear export drive in the 2010s were already in place *prior* to the 2011 Japanese disaster. These included the limited scope for market expansion, organizational disharmony within the French "team," and a single product (the EPR) whose size, complexity, and cost produced a misfit in many export contexts. With grid-connection still a distant prospect, the EPR's operational performance remains unproven, but construction delays and cost overruns are well-publicized. Although four EPRs were under construction in 2014, and more proposed for the UK, it is now unlikely that the EPR can become the leading reactor internationally, manufactured on a standardized basis to replicate the French success formula of the 1980s.

The domestic evolution is even starker. Within France, numbers of reactors built have dropped dramatically across the four production series: 34 x 900 MW, 20 x 1300 MW, 4 x 1450 MW, and just 1 x 1650 MW EPR plants. As of 2014, there are no concrete plans to build another EPR in France. Instead, attention is focused on reconciling lifetime extensions of existing reactors with the safety and security hurdles raised by Fukushima. Meanwhile, wider options exist in relation to renewable energy options and gas, particularly when viewed over a 20-year horizon. Whilst the Hollande presidency has yet to set out its energy policy in detail, two directions are clear: technological lock-in means that France will persist with nuclear power, but nuclear will provide a decreasing share of electricity generation.

The key claims made in the 2000s regarding a nuclear renaissance proved to be hollow. New usages are always mooted, never implemented. Hardly any newcomers are building reactors. The claim that nuclear power is cost-competitive in liberalized markets has been proved false. In the US and the UK, new build is undertaken only with large government-backed subsidies. The economic fundamentals of nuclear power do not stack up. It is more expensive than other options (including many renewable energy options), as learnt from experience at Olkiluoto and Flamanville and by the "strike price" reached for Hinkley Point. Unit costs for nuclear have *increased* over time, demonstrating a lack of scale economies (Grubler 2010; Rangel and Lévêque 2013). Cost escalation has been driven by the quixotic search to make a dangerous technology safe, secure, and socially acceptable. The arguments in its favor now turn on its contribution to climate policy and energy security. However, no EPR or other generation three+ reactors have been completed. Their contribution is therefore nil, and will remain so for some years (at best). These developments are, however, independent of the "Fukushima effect."

Where the "Fukushima effect" can be identified is in relation to safety and security. The claimed superiority of the passive safety design has already been found wanting, notably in the context of a total power loss (as occurred at Daiichi). The consequence has been a major rethink, flagged by the quest for a hardened core to reinforce existing and new build. This has necessitated further increases in construction time and costs at Olkiluoto and Flamanville, and the decommissioning of at least some existing reactors—in Japan, Germany, and beyond. Further, the dismal regulatory failures in Japan, a country famed for its high technology industries and organizational expertise, beam a "stop" light to nuclear deployment in emerging economies with lower technological development and institutional capacity. Fukushima revealed that a genuine nuclear renaissance must be based on a strong institutional framework, with an independent and powerful regulator, able to withstand industry capture and uphold state-of-the-art standards. But where nuclear power is proposed for weak or failing states, a strong regulator is implausible while international regulation remains underdeveloped. Nation-states guard their sovereignty in technology choice, and resist third-party inspection and sanction (as demonstrated by Iran).

The safety and security domain continues to be characterized by inadequate leadership. We have seen that France tried and failed to play a leading role in updating the norms of global nuclear governance. This was because the marketization of nuclear power in the hands of Areva and EdF produced merely another commercial package. In regard to regulatory stringency, the evolution of the ASN (created in 2006 to correct the shortcomings of France's own "nuclear village") put France only marginally ahead of Japan, with no tangible claim to world leadership. Nuclear reactors are not simply another market product: they require a unique institutional framework based on a complex set of interacting norms.

This chapter has sought to explicate those norms, and analyze how the French nuclear lobby hoped to modify them for the purposes of a nuclear renaissance. It has demonstrated that the inadequacy of French attempts to renew the norms of global nuclear governance was exposed by the "Fukushima effect." An important finding, therefore, is that France's contribution to norm redefinition, being guided by instrumental rationality, proved too partial and self-serving to improve global nuclear governance. The post-Fukushima landscape continues to be marked by questionable regulatory capacity and limited ability to manage operational risk (including the disposal of nuclear waste). Fukushima proved that so long as the commercialization of nuclear power remains in conflict with safety and security requirements, talk of a nuclear renaissance will continue to mislead the public.

## Notes

1  See, for example, http://media.cns-snc.ca/media/toocheap/toocheap.html
2  For a detailed account, see Szarka (2013)
3  Editor's note: this usage of the term "nuclear village" is added to by Kainuma (2011), who described "nuclear villages" as "local areas hosting nuclear power stations" (cited in Juraku 2013: 42).

## References

AFP (Le Figaro Magazine). "Lauvergeon: on est dans l'urgence ['Lauvergeon: one is in a hurry']." *Le Figaro Magazine*, March 16, 2011, accessed May 13, 2015, http://goo.gl/3DljS1.

Ambrose, J. "UK, EDF Reach Deal on Hinkley Point C," *Nucleonics Week*, October 24, 2013.

Areva. "EPR Reactor Fact Sheet," n.d., accessed May 17, 2015, www.areva.com/EN/operations-5444/epr-reactor-fact-sheet.html.

ASN (*Autorité de sûreté nucléaire*). "Extraits du rapport de l'ASN sur l'état de la sûreté nucléaire et de la radioprotection en France en 2012 [Extract from the ASN Report on the state of nuclear safety in France in 2012]," 2012, accessed July 25, 2014, http://tinyurl.com/oeh6bty.

BBC. "Italy Nuclear: Berlusconi Accepts Referendum Blow," June 4, 2011, accessed July 7, 2011, www.bbc.co.uk/news/world-europe-13741105.

Cour des Comptes. *Les Coûts de la filière électronucléaire* [*The costs of nuclear electricity*], 2012, accessed February, 2, 2012, http://tinyurl.com/kxjuyxx.

Cour des Comptes. *Le Coût de production de l'électricité nucléaire* [*The production cost of nuclear electricity*]. 2014, accessed July 25, 2014, http://tinyurl.com/mhj3ngm.

EdF (Electricité de France). "EdF Will Start Selling the First KWh Produced by the EPR at Flamanville in 2016," July 20, 2011, accessed September 8, 2011, http://goo.gl/vfpvBU.

Elster, J. "Social Norms and Economic Theory," *Journal of Economic Perspectives* 3 (1989): 99–117.

ENSREG (European Nuclear Safety Regulators Group). *Final Report on the Peer Review of EU Stress Tests*, April 26, 2012, accessed July 29, 2014, www.ensreg.eu/node/407.

Fam, S.D., et al. "Post-Fukushima Japan: The Continuing Nuclear Controversy," *Energy Policy* 68 (2014): 199–205.

Felder, F.A. "Nuclear Power in the Second Obama Administration," *The Electricity Journal* 26 (2013): 25–31.

Findlay, T. *Nuclear Energy and Global Governance: Ensuring Safety, Security and Non-Proliferation*. London: Routledge, 2011.

Finon, D. "Force et inertie de la politique nucléaire française: une co-évolution de la technologie et des institutions ['Strength and inertia in French nuclear policy: The co-evolution of technology and institutions']." In *Etat et énergie XIXe-XXe siècle*, edited by A. Beltran, 183–215. Paris: Editions Comité pour l'histoire économique et financière de la France 2009.

Finon, D., and C. Staropoli. "Institutional and Technological Co-evolution in the French Electronuclear Industry," *Industry and Innovation* 8 (2001): 179–199.

Grubler, A. "The Costs of the French Nuclear Scale-up: A Case of Negative Learning by Doing," *Energy Policy* 38 (2010): 5174–5188.

Hayashi, M., and L. Hughes. "The Fukushima Nuclear Accident and Its Effect on Global Energy Security," *Energy Policy* 59 (2013): 102–111.

Hecht, G. *The Radiance of France: Nuclear Power and National Identity after World War II*. Cambridge, MA: MIT Press, 1998.

Hirsch, H. *Selected Aspects of the EPR Design in the Light of the Fukushima Accident, Report for Greenpeace International*, 2011, accessed April 4, 2015, www.greenpeace.org/france/pagefiles/266521/epr_report_greenpeace.fr.pdf.

Hollande, F. "Le changement, c'est maintenant—mes 60 engagements pour la France ['Change is now: My 60 pledges for France']," 2012, accessed May 13, 2015, http://goo.gl/ICi9PO.

IAEA (International Atomic Energy Agency). *IAEA Milestones Approach and Support Services*, 2011, accessed May 17, 2015, http://goo.gl/l1oggN.

IAEA. *Nuclear Power Reactors in the World*. Vienna: IAEA, 2014.

IRSN (*Institut de radioprotection et sureté nucléaire*). "Les rejets radiologiques massifs diffèrent profondément des rejets contrôlés [Massive radiological emissions differ profoundly from controlled emissions']," 2012, accessed July 6, 2014, http://tinyurl.com/o5c8xf6.

Juraku, K. "Social Structure and Nuclear Power Siting Problems Revealed." In *Nuclear Disaster at Fukushima Daiichi: Social, Political and Environmental Issues*, edited by R. Hindmarsh, 41–56. New York: Routledge Studies in Science, Technology and Society, 2013.

Kainuma, H. *"Fukushima" Ron: Genshiryokumura wa Naze Umareta no Ka* [*On Fukushima: Why the nuclear power village was born*]. Tokyo: Seitosha, 2011

Katzenstein, P.J. "Introduction: Alternative Perspectives on National Security." In *The Culture of National Security: Norms and Identity in World Politics*, edited by P.J. Katzenstein, 1–32. New York: Columbia University Press, 1996.

Lajoinie, A. *L'Energie: repères pour demain* [*Energy; Milestones for tomorrow*]. Paris: Assemblée nationale, 2001.

Lepage, C. *La Vérité sur le nucléaire: le choix interdit* [*The truth about nuclear: Choice prohibited*]. Paris: Albin Michel, 2011.

Linares, P., and A. Conchado. "The Economics of New Nuclear Power Plants in Liberalised Electricity Markets," *Energy Economics*, 40 (2013): S119–S125.

MacLachlan, A. "Sarkozy's Victory in France Seen as a Win for Nuclear Power," *Nucleonics Week*, May 10, 2007.

MacLachlan, A. "Sarkozy Launches Second French EPR, Citing Higher Fuel Prices," *Nucleonics Week*, July 10, 2008.

MacLachlan, A. "Areva's Olkiluoto-3 Manager Says Engineering Judgment Undermined," *Nucleonics Week*, March 26, 2009.

MacLachlan, A. "Lauvergeon: French lost UAE bid Because of Expensive EPR Safety Features," *Nucleonics Week*, January 14, 2010a.

MacLachlan, A. "Areva, Npcil Agree on Terms for Jaitapur EPR Project," *Nucleonics Week*, December 9, 2010b.

Mez, L. "Nuclear Energy—Any Solution for Sustainability and Climate Protection?" *Energy Policy* 48 (2012): 56–63.

Mühlenhöver, E. *L'Environnement en politique étrangère: raisons et illusions. Une analyse de l'argument environnemental dans les diplomaties électronucléaires française et américaine* [*The environment in foreign policy. An analysis of the environmental argument in French and US nuclear power diplomacy*]. Paris: L'Harmattan, 2002.

National Diet of Japan. *The Official Report of the Fukushima Nuclear Accident Independent Investigation Commission: Executive Summary*, 2012, accessed July, 24, 2014, http://tinyurl.com/k6vbk4a'.

Nian, V., and S.K. Chou. "The State of Nuclear Power Two Years after Fukushima—the ASEAN Perspective," *Applied Energy* 136 (2014): 838–848.

Nucleonics Week. "Areva Labour Council Requests Copy of French-Chinese MOU," *Nucleonics Week*, November 29, 2012.

Nucleonics Week. "IRSN Estimated Worst-Case Nuclear Accident at Eur5.8 Trillion," *Nucleonics Week*, March 14, 2013.

Ostrom, E., and X. Basurto. "Crafting Analytical Tools to Study Institutional Change," *Journal of Institutional Economics* 7 (2011): 317–343.

Platts Energy Economist. "Prospects for Nuclear Power in 2012," 2012, accessed April 22, 2015, www.commodities-now.com/reports/power-and-energy/9806-prospects-for-nuclear-power-in-2012.html.

Présidence de la République. "Compte rendu du Conseil de Politique Nucléaire du 21 février 2011 [Minutes of the Nuclear Policy Council, 21 February 2011]," 2011, accessed 10 August 2011, http://tinyurl.com/p9w2upb.

Rangel, L.E., and F. Lévêque. "Revisiting Nuclear Power Construction Costs Escalation," 2013, accessed April 22, 2015, http://goo.gl/h3BZnB.

Renard, J-D. "Ça bouge sur le nucléaire [Something's stirring in nuclear power]," *Sud Ouest*, 22 Mars, 2011.

Rosendahl, J., and G. De Clercq. "Finnish Nuclear Plant Delayed Again as Areva, TVO Bicker," February 28, 2014, accessed 6 August, 2014, http://tinyurl.com/nptf5sv.

Rougeau, J.-P. "The Importance of Nuclear Energy to Sustainable Development." In *Environment and Nuclear Energy*, edited by B.N. Kursunoglu, S.L. Mintz and A. Perlmutter, 129–134. New York: Plenum Press, 1998.

Roussely, F. *Avenir de la filière française du nucléaire civil [Future of the French civil nuclear sector]*, 2010, accessed May 13, 2015, http://goo.gl/ZzQRVB.

Sains, A. "Olkiluoto digital I&C System Still Needs More Work: Finnish Regulator," *Nucleonics Week*, April 17, 2014.

Schneider, M. "Nuclear Power in France: Trouble Lurking Behind the Glitter." In *International Perspectives on Energy Policy and the Role of Nuclear Power*, edited by L. Mez, M. Schneider and S. Thomas, 55–83. Brentwood, Essex: Multi-Science Publishing, 2009.

Srinivasan, T.N., and T.S.Q. Rethinaraj. "Fukushima and Thereafter: Reassessment of Risks of Nuclear Power," *Energy Policy* 52 (2013): 726–736.

Szarka, J. "Environmental Foreign Policy in France: National Interests, Nuclear Power and Climate Protection." In *Climate Change and Foreign Policy: Case Studies from East to West*, edited by P.G. Harris, 117–133, London: Routledge, 2009.

Szarka, J. "From Exception to Norm–and Back Again? France, the Nuclear Revival and the Post-Fukushima Landscape," *Environmental Politics* 22 (2013): 646–663.

Thomas, S. "Competitive Energy Markets and Nuclear Power: Can We Have Both, Do We Want Either?" *Energy Policy* 38 (2010): 4903–4908.

United Nations. "Treaty on the Non-Proliferation of Nuclear Weapons," 1970, accessed April 22, 2015, http://tinyurl.com/nqg74m4.

Wang, Q., Chen, X., and X. Yi-chong. "Accident like Fukushima Unlikely in a Country with Effective Nuclear Regulation: Literature Review and Proposed Guidelines," *Renewable and Sustainable Energy Reviews*, 17 (2013): 126–146.

WEC (World Energy Council). *Nuclear Energy One Year after Fukushima*, 2012, accessed April 22, 2015, http://tinyurl.com/knqnw64.

WNN (World Nuclear News). "Areva Lands World's Biggest Ever Nuclear Power Order," 2007, accessed May 13, 2015, www.world-nuclear-news.org/newsarticle.aspx?id=14454&LangType=2057.

WNO (World Nuclear Organisation). "Nuclear Power in Germany," 2012, accessed May 13, 2015, www.world-nuclear.org/info/Country-Profiles/Countries-G-N/Germany/.

WNO (World Nuclear Organisation). "Nuclear Power in Turkey," 2014, accessed April 22, 2015, http://tinyurl.com/orxby6y.

Yi-chong, X. "The Struggle for Safe Nuclear Expansion in China," *Energy Policy* 73 (2014): 21–29.

# 12 A Question of Confidence

## Nuclear Waste and Public Trust in the United States after Fukushima

*William J. Kinsella*

As disaster unfolded at the Fukushima Daiichi nuclear power plant in March 2011, much of the commentary at the time involved the threats posed by radioactivity present in the plant's reactor cores. However, some observers noted that used nuclear fuel stored in water-filled pools at the plant posed equally severe, or possibly greater, risks (Bradsher and Tabuchi 2011; Parenti 2011).[1] Although the worst possibilities involving the used fuel did not come to pass, the disaster brought new attention to this intensely radioactive product of nuclear power production.

In the United States, which stores the world's largest inventory of used nuclear fuel (Feiveson et al. 2011), the disaster further complicated controversies regarding its management and disposal. The controversy examined here has a long history prior to the events at Fukushima, but has been exacerbated by the "Fukushima effect" as manifested in new turbulence in the US nuclear regulatory system, industry concerns regarding possible regulatory changes, renewed public awareness regarding nuclear risks, and re-energized opposition to nuclear power. At the same time, the controversy provides a case in which the broader implications of Fukushima are being interpreted, negotiated and enacted.

In June 2012, a US federal court invalidated a long-established legal principle known as "waste confidence," which addresses the storage and disposal of irradiated nuclear fuel. Although the ruling responded to events that preceded the Fukushima disaster, it nevertheless led to an action that critics of nuclear power had called for following the disaster: a suspension by the US Nuclear Regulatory Commission (USNRC) of all pending licensing decisions for power reactors and used fuel storage facilities. This development complicated the nuclear industry's ongoing expansion efforts and reopened crucial questions. Without assurance of safe storage and disposal of used fuel, is nuclear power viable? What constitutes safe storage and disposal, and who should be involved in defining such safety? What lessons does the Fukushima disaster present for current and proposed used fuel storage methods? These questions intersect with controversies surrounding the continuing lack of a high-level nuclear waste repository in the US (USGAO 2011); and with recommendations of the Presidential Blue Ribbon Commission on America's Nuclear Future (BRC 2012) related to establishing new regional storage facilities for the wastes accumulating at plants throughout the nation.

This chapter examines the waste confidence controversy in relation to several interconnected themes: the role of the Fukushima disaster in US debates and policy responses regarding nuclear power; the effects of those responses on the US nuclear power infrastructure; implications for nuclear power regulation and governance; and critical-theoretical perspectives on related questions of public trust, confidence, and institutional legitimacy. Focal topics include power relations and decision-making involving regulatory, industry, technical, and public communities; associated argumentative strategies; public engagement as a communicative process; trust and confidence as products of communication; and implications for nuclear safety and democratic technology governance. The analysis is informed by perspectives drawn from communication studies and rhetoric, science and technology studies (STS), and the policy-in-action approach as conceptualized by Hindmarsh (2013a) in relation to the Fukushima disaster.

I begin with a prologue illustrating some of the connections among nuclear power production, nuclear waste, and risk in nuclear power contexts. I then provide an overview of the US nuclear regulatory system and a brief history of the waste confidence controversy, and outline a critical-theoretical approach addressing questions of trust, confidence, and institutional legitimacy. The analysis that follows draws upon government documents, discourses of the nuclear industry and critical public interest groups, and video records and transcripts of a series of USNRC public meetings related to the waste confidence issue to examine the policy formation process for nuclear regulation in the US.[2] Throughout, the waste confidence controversy provides an analytical window into larger questions of nuclear safety and risk, rhetorical strategies of the nuclear industry and its critics, inclusion and exclusion in democratic policy debate, and technology governance in a risk society, all in the context of the Fukushima disaster.

## Prologue: A Non-Encounter with Fukushima

In August 2010, less than seven months before the Tōhoku earthquake and tsunami and the onset of the catastrophic events at Fukushima Daiichi, I passed within 60 kilometers of the plant. Three colleagues and I were onboard the *Shinkansen*, Japan's high-speed train, on our way to visit the nuclear fuel reprocessing site at Rokkasho, at the northern end of Honshu Island. Despite our shared interest in nuclear matters, no one mentioned the power plant on the Pacific coast, just east of our route. Although it was one of the world's largest such facilities, at that time the plant was an ordinary and unremarkable artifact of the global nuclear system. It was the unique and more controversial facility at Rokkasho that claimed our attention.

When I heard about the earthquake on March 11, 2011, the state of affairs at Rokkasho was among my initial concerns. (As it turned out, Rokkasho suffered only minor damages.) I still had not learned of the existence of the Fukushima Daiichi plant and its neighboring plant, Fukushima Daiini, but that would soon change. Within hours, questions emerged regarding not only the six reactors at Daiichi, but also the storage pools containing used nuclear fuel: one in each reactor building and a seventh common pool located nearby. Years later, the damaged pools remain a focus of concern and ongoing remediation efforts are expected to continue beyond 2020.[3] Like other nuclear power nations, including the US, Japan has accumulated large quantities of used fuel at its power plants and has not resolved the problem of how to manage that material more safely.

This story of my brief, personal non-encounter with Fukushima illustrates a number of interrelated points. First, whereas the more exceptional components of the nuclear system receive regular attention, risks posed by more ordinary components tend to become normalized over time. In the hierarchy of typical risk perceptions, a nuclear fuel reprocessing plant trumps ongoing reactor operations, which, in turn, trump the perennially deferred question of safeguarding the reactors' used fuel.[4] But in fact, all of these operations are inherently risky, and any of those risks can suddenly manifest.

Second, the many components of the nuclear system are interconnected. The Rokkasho reprocessing plant was designed as a solution to the problem of hazardous waste generation that accompanies any choice to engage in nuclear power production. Ideally, after spending some time in local storage, irradiated fuel at facilities such as Fukushima Daiichi would be sent to Rokkasho, where uranium and plutonium would be extracted for re-use in reactors across Japan. The most intensely radioactive contents of the used fuel, products of the previous round of nuclear fission, would be separated and isolated for permanent disposal. In principle, this "closed cycle" arrangement would reduce the hazards associated with used fuel stored at reactor sites throughout the nation.

However, a third related point is that nuclear planning does not always proceed as envisioned. The reprocessing plant I visited in 2010 has been under construction since 1993, thus far costing more than the equivalent of US$23 billion. The plant's commissioning has been delayed repeatedly, and it remains a focus of controversies regarding cost, operational safety, seismic risks, and its potential as a terrorism target, a source of materials for nuclear weapons proliferation, and a provocative resource for domestic nuclear weapons production should that path be taken in the future. The plant has insufficient capacity to reprocess Japan's present inventory of irradiated fuel, not to mention additional quantities the nation may produce in the future. Further, no permanent repository site has been identified for the intensely radioactive materials that would be removed from the used fuel as part of its reprocessing (Feiveson et al. 2011). Critics of the Rokkasho project suggest that, rather than reducing the risks associated with nuclear power, it introduces further risks of its own (Takubo 2008).[5] For reasons such as these, other nations, including the US, have chosen not to pursue the reprocessing option.

Scholarship across multiple fields has long recognized that technological risks are complex, systemic, and intertwined, challenging official claims that such risks can be identified accurately, evaluated rigorously, and managed effectively (Luhmann 1993; Perrow 1984). Accordingly, STS scholars and communication scholars have argued that such risks are not only matters for expert analysis and official decision-making, but also demand broad public debate involving multiple, diverse voices (Fisher 1989; Wynne 1991). The present case study extends that more inclusive model of risk governance, examining one controversy that emerged in the US subsequent to the onset of the disaster at Fukushima.

## The Waste Confidence Principle and US Nuclear Regulation

With approximately 99 reactors operated by 23 private companies at 62 sites across 31 states, nuclear power provides approximately 19% of the US electricity supply (Schneider et al. 2014; USNRC 2014e).[6] Nineteen additional reactors are undergoing decommissioning, a process that can take as long as six decades.[7] More than 65,000 metric tons of used fuel from operating and closed commercial reactors is stored in water-filled pools at reactors and in dry casks at independent spent fuel storage installations (ISFSIs, which may be, but are not necessarily, located at reactor sites and are licensed separately). According to the US Government Accountability Office (2011: 8), the "volume of commercial spent nuclear fuel is expected to more than double by 2055—assuming currently operating reactors receive license extensions and no new reactors are built—and is currently accumulating at 75 sites in 33 states."

Dry cask storage is considered safer than storage in pools but adds financial costs for the operating companies. Neither of these approaches provides a safe option for permanent disposal, and no site has yet been authorized for that purpose. The proposed Yucca Mountain site in Nevada remains deeply controversial, as do recommendations for new "consolidated interim storage" sites that would manage used fuel for extended time periods (BRC 2012). Figure 12.1 provides a map of US storage sites as of 2011, and Figure 12.2 illustrates one of a variety of dry cask arrangements.

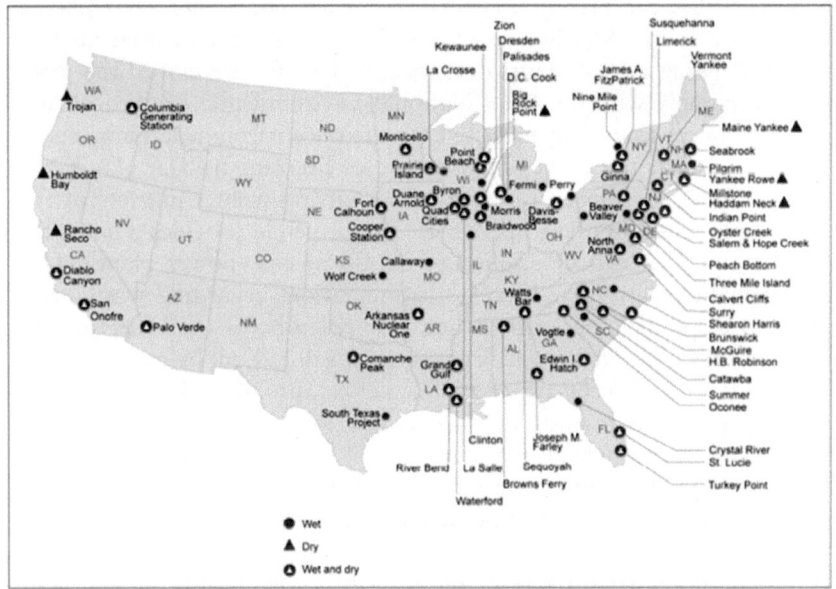

*Figure 12.1* US Used Nuclear Fuel Storage Sites and Proposed Yucca Mountain
Repository Site, as of 2011

Source: US Department of Energy (reprinted in USGAO 2011, www.gao.gov/products/
GAO-11–229)

*Figure 12.2* Used Fuel Dry Casks

Source: US Nuclear Regulatory Commission

Since 1984, the principle known as "waste confidence" has played a formal role in US policy regarding the management of used nuclear fuel. Following a 1979 court decision, the USNRC developed the principle to comply with the 1969 National Environmental Protection Act, which requires federal agencies to evaluate the environmental impacts of their actions. As the agency responsible for licensing nuclear plants, the USNRC is thus required to evaluate the environmental impacts of the wastes generated by their operation. The principle provides a so-called generic rule that can be applied to all US nuclear power plants. This ruling makes many case-by-case determinations regarding waste storage and disposal unnecessary when new reactors, or used fuel storage facilities, are licensed or their licenses extended. Indeed, by making those determinations unnecessary, the rule, in effect, prohibits them. The principle has provided an important foundation for the US nuclear industry's expansion efforts, which acquired new momentum during the decade preceding the Fukushima disaster. Without it, each new licensing action would potentially provide opportunities for challenges by opponents of particular plants and of nuclear power more generally. The USNRC reviewed and updated the principle in 1990, reviewed it again and reaffirmed it without change in 1999, and revised it in 2010, with those revisions becoming effective in January 2011.

The 2011 version of the waste confidence rule (USNRC 2010) comprised five formal "findings" paraphrased here, where quotations reflect the rule's official language:

1   Safe disposal of used nuclear fuel in an underground "geologic repository" is "technically feasible."
2   "Sufficient capacity" in at least one repository will be available "when necessary" (the 1984 version specified availability "by 2007–2009," extended to 2025 in the 1990 version and left unspecified in the 2011 version).
3   Used nuclear fuel and other high-level nuclear wastes can be "safely managed" until a repository becomes available.
4   In the interim, used nuclear fuel can be "stored safely and without significant environmental impacts" for at least 60 years beyond the licensed lifetimes of reactors (previous versions of the rule stated a 30-year timeframe for safe interim storage).
5   Interim storage facilities for used nuclear fuel can be made available at reactor sites or elsewhere "as needed."

Consistent with commonplace notions of "confidence," these findings were presented as products of technical and organizational inquiry conducted at levels of rigor and reliability adequate for "reasonable assurance" of the stated conditions.[8] However, not all interested parties share that sense of confidence. Early in 2011, four US states (Connecticut, New

Jersey, New York, and Vermont), the Prairie Island Indian community (located adjacent to a nuclear plant), and a number of nongovernmental organizations filed legal challenges to the revised rule based on the changes made to the second and fourth findings. The presiding court consolidated the multiple cases in an order dated March 10, 2011—the day preceding the earthquake and tsunami that precipitated the Fukushima disaster (US Court of Appeals 2011). The subsequent legal documents related to the court challenge make no reference to Fukushima, but I argue below that a "Fukushima effect" has played a significant role in an ensuing, more prolonged controversy, and, conversely, this domestic controversy has informed US understandings of the meanings and implications of the events at Fukushima.

The challenge to the waste confidence principle was argued in court on March 16, 2012 and decided on June 8, 2012. In its decision (US Court of Appeals 2012), the court determined that there was insufficient evidence warranting confidence that a permanent repository would be available "when necessary" (finding #2) or that used fuel can be stored safely for an interim period of 60 years (finding #4). The present chapter takes that court decision as a starting point and examines subsequent responses by the USNRC, the nuclear industry, and critical public interest and environmental groups. The following section provides a framework for that analysis.

## Confidence, Trust, Risk, and Responsibility

The ambiguous term *confidence* occupies a formal and central place in the case examined here. From an STS perspective, waste confidence constitutes an "obligatory passage point" (Callon 1986) or "trial of strength" (Latour 1987): a condition that must be satisfied if the US nuclear power enterprise is to advance. In terms of rhetorical theory (Kennedy 1994; Prelli 1989), waste confidence provides a "stasis point," a focal issue around which broader questions related to nuclear power are organized. As demonstrated below, common-sense notions of confidence intersect with, and arguably are displaced by, highly technical and legalistic arguments in the waste confidence controversy. Scholars of communication and STS argue that such rhetorical displacements relegate issues affecting broad public communities to the narrowly restricted domain of expert decision-making, excluding affected parties from meaningful involvement (Fisher 1989; Goodnight 1982; Wynne 1991).

Related, but not identical, to the notion of confidence is the notion of *trust*. Distinguishing between trust and confidence, Luhmann (1988) proposes that confidence involves assuming an envisioned future (in this case, one in which a permanent solution to the nuclear waste problem exists), thus ignoring or dismissing alternative possibilities (such as a continued failure to find a solution). Trust, on the other hand, involves recognizing that the envisioned state of affairs may or may not come to pass. Thus trust involves accepting *risk*, and should things go badly, acknowledging responsibility for one's choice:

> If you choose one action in preference to others in spite of the pos-sibility of being disappointed by the action of others, you define the situation as one of trust. In the case of confidence you will react to disappointment by external attribution. In the case of trust you will have to consider an internal attribution and eventually regret your trusting choice.
>
> (Luhmann 1988: 97–98)[9]

Luhmann's distinction helps clarify differences that emerged across parties in the waste confidence controversy. The USNRC's notion of "confidence" warrants continued nuclear power plant licensing by assuming, in advance, a future state in which used fuel disposal issues are resolved. Should that assumption prove wrong in the future, responsibil-ity for damages to public health, safety, and the environment could be rhetorically externalized: "unforeseen" events could be held responsible, rather than the USNRC or the nuclear industry. Kinsella (2012, 2013) and Paté-Cornell (2012) argue that just such a process unfolded in the case of Fukushima, as some members of the nuclear community sought to shift responsibility to "natural hazards" regarded as beyond the "design basis" for the plant. Indeed, the widespread use of the term "accident" to describe failures such as those at Fukushima, despite contributions made by errors in design, risk analysis, management, regulation, and policy, demonstrates the pervasiveness of such rhetorical displacements of responsibility.

Public understandings are often quite different: critics charge that by asserting confidence in the prospects for safe storage and disposal, the USNRC has dismissed legitimate concerns, abrogating responsibility for threats that can be envisioned realistically but are nevertheless excluded from consideration. Whereas the USNRC formally adopted the notion of confidence, critics more often speak of trust. The latter principle expands the USNRC's scope of responsibility and foregrounds questions regard-ing the agency's good intentions and competencies. At the same time, following Luhmann's argument, the principle also implies responsibility on the part of those who agree to trust the USNRC, making critique of the agency not only a right, but an obligation.

Earle and Siegrist have developed Luhmann's distinction further, iden-tifying social-psychological foundations for confidence and trust:

> Confidence is defined as the belief, based on experience or evidence, that certain future events will occur as expected . . . The basis for confidence is past performance or institutions designed to constrain

future performance . . . [T]he empirical antecedents of confidence are varied, including familiarity, evidence, regulations, rules/procedures, contracts, record keeping/accounting, social roles, ability, experience, control, competence, and standards.

(Earle and Siegrist 2006: 386)

In contrast:

We define trust as the willingness to make oneself vulnerable to another based on a judgment of similarity of intentions or values . . . The antecedents of trust . . . include social relations, ingroup membership, morality, benevolence, integrity, inferred traits and intentions, fairness, and caring.

(Earle and Siegrist 2006: 386)

Following this distinction, confidence involves tangible historical evidence and appraisals of prevailing structures and processes, whereas trust involves the perceived quality of a relationship between parties. Earle and Siegrist (2006: 386) characterize these warrants as "performance-relevant" and "morality-relevant," respectively. Thus, in the waste confidence controversy, trust aligns more closely with common-sense and experiential forms of knowledge, whereas confidence aligns more closely with claims of objective, expert knowledge. Nevertheless, as Miller (2003) has argued in the case of a notoriously contested USNRC risk analysis, such expert claims are often grounded less in rigorous analysis and more in tacit professional knowledge and institutional authority. As demonstrated below, the USNRC and its critics have understood the concepts of confidence and trust quite divergently throughout the waste confidence controversy.

## Analysis: Enacting and Contesting Confidence and Trust

"IT ALL BOILS DOWN TO—DO WE TRUST THE NRC?"
Title on flyer created by Nuclear Energy Information Service, distributed at USNRC public meetings and submitted as formal public comment

Literary critic Kenneth Burke (1969: 15) identified "the logically reductive notion of 'what it all boils down to' " and "the narrative notion of 'how it all ends up' " as two basic resources for human understanding. It is unclear whether the flyer quoted above, distributed by a non-profit organization opposed to nuclear power, deliberately alluded to the boiling down of cooling water that poses a dreaded scenario at used fuel storage pools and was a focus of concern at Fukushima.[10] If so, then the flyer presents both a narrative about dangerous possibilities and a logically reductive statement regarding how those possibilities might

develop. By trusting the USNRC, parties affected by nuclear power production delegate responsibility for their safety, the safety of others, and
environmental protection to the agency. However, some parties in the
waste confidence controversy argue that such delegation implies another
form of responsibility: to trust the USNRC is to trust one's own judgment
of the agency's abilities and intentions. I now examine how questions of
confidence, trust, risk, and responsibility were rhetorically enacted and
negotiated following the court's order to re-examine the prospects for
safe management and disposal of used nuclear fuel.

### Timeline and Process

Table 12.1 presents a timeline of related events beginning with the court's
June 8, 2012 decision.[11] Ten days later, a coalition of 24 public interest groups petitioned the USNRC, calling for the suspension of licensing actions for nuclear power plants and used fuel storage facilities
throughout the US. Two months after the court decision the Commission
suspended further licensing actions, while allowing activities supporting application reviews to continue. The five commissioners then voted
unanimously to approve a proposed approach for reexamining the waste
confidence principle. The agency created a Waste Confidence Directorate
to develop a Generic Environmental Impact Statement (GEIS) and a formal regulatory rule within two years, governing the disposition of used
nuclear fuel and addressing three deficiencies the court had found in the
rejected Waste Confidence rule.[12]

*Table 12.1* Milestones in the Waste Confidence Review Process

| | |
|---|---|
| June 8, 2012 | Decision by US Court of Appeals, District of Columbia |
| June 18, 2014 | Petition submitted to USNRC by 24 environmental and public interest groups |
| June 25, 2012 | USNRC staff provides answer to petition |
| July 9, 2012 | USNRC memo to staff re: approach for addressing Waste Confidence policy issues |
| August 7, 2012 | USNRC suspends licensing actions |
| August 10, 2012 | USNRC vote re: proposed approach |
| August 11, 2012 | USNRC creates Waste Confidence Directorate |
| September 6, 2012 | Staff requirements memo re: developing draft GEIS (2-year process) |
| October 25, 2012 | Generic Environmental Impact Statement (GEIS) scoping period begins |
| January 2, 2013 | GEIS scoping period ends |
| March 5, 2013 | Scoping Summary Report published |

| June 7, 2013 | Commissioners vote on proposed rule |
| September 13, 2013 | Draft GEIS (585 pages) and Proposed Waste Confidence Rule published |
| September 13, 2013 | Public comment period begins |
| December 20, 2013 | Public comment period ends |
| January 3, 2014 | Secretary of Energy proposes setting nuclear industry waste fees to zero |
| March 21, 2014 | Commissioner briefing by USNRC staff and invited parties |
| July 21, 2014 | Staff recommendation to Commissioners to publish Draft Final GEIS and proposed rule; name change to "Continued Storage" |
| July 24, 2014 | Draft Final GEIS and proposed rule made publically available in advance of commissioners' vote; not for formal public comment |
| August 26, 2014 | Commissioners vote 4–0 (with one minor partial disagreement), approving GEIS and final rule and lifting suspension of licensing decisions |
| September 19, 2014 | Continued Storage rule and notice of issuance of final GEIS published in *Federal Register*, to become effective 20 October 2014 |
| October 3, 2014 | Original projected publication date for final GEIS and Continued Storage rule |
| October, 24 2014 | Three states file petition for review of rule, GEIS, and licensing resumption |
| October 29, 2014 | Nine environmental groups and Natural Resources Defense Council petition for review of rule and GEIS |

The development of a GEIS is a multi-stage process, following guidelines provided by the National Environmental Policy Act (NEPA). The "NEPA process," as it is known across multiple US policy settings, often plays a key role in the development of environmental regulations, contributing in advance to defining the power and limits of regulatory rules.[13] Early in the waste confidence GEIS process, the range of issues to be considered was specified during a "scoping period." The agency presented an initial set of possibilities, solicited and reviewed public comments, and then defined the scope of issues in a "scoping summary report" (USNRC 2013a). In principle, such scoping decisions respond to concerns raised during the public comment period, but critics often question the degree of responsiveness achieved. Newcomers to the process often underestimate the importance of the scoping stage, as it provides a basis for including or excluding issues that may later be invoked by parties seeking to warrant or challenge administrative and legal decisions. In its scoping summary report, the USNRC identified 88 "major topics and issues of concern" raised by commenters, of which 20 were classified as "out of scope." Two topics, neither of them ruled out of scope, explicitly included the term "Fukushima": "concerns about accidents similar to the Fukushima Dai-ichi event" and "incorporation of lessons learned from Fukushima."

Following the scoping stage, USNRC staff developed a draft GEIS and a proposed regulatory rule, and another formal public comment period ensued. After reviewing those comments, the staff produced a revised "draft final GEIS" (USNRC 2014a) and a "revised rule" (USNRC 2014b). In July 2014, the final document and rule, not subject to further public comment, were presented to the five commissioners. To the dismay of critics, the commissioners voted promptly, approving the new rule on August 26 and lifting the suspension of licensing decisions. As detailed below, some critics found the language of the new rule profoundly troubling, regarding it as little more than a semantic adjustment allowing industry activities to resume without the burden of assertions of "confidence" in the long-term safety of nuclear wastes. On October 24, 2014, three states filed a petition for a review of the USNRC's determinations, and on October 29, 10 environmental and public interest organizations filed two similar petitions.

A comprehensive analysis of the waste confidence GEIS and rulemaking process would entail a book-length project. In the space available here, I examine some illustrative examples of how various parties enacted and challenged notions of trust, confidence, risk, and responsibility at a number of stages throughout the process, with the Fukushima disaster as a point of reference.

## The Court's Assessment of Waste Confidence

A suitable starting point is the 21-page document in which the court rejected the waste confidence principle (US Court of Appeals 2012). Characterizing the issue as "another in the growing line of cases involving the federal government's failure to establish a permanent repository for civilian nuclear waste" (p. 3), the court foregrounded the central argument offered by critics of the nuclear industry and of the USNRC: without assurance that used nuclear fuel can be permanently eliminated as an environmental and human health hazard, its continued production should be prohibited. An implication of that argument is that all nuclear power production should cease, but in the context of the waste confidence controversy most critics limited their demands to the cessation of licensing decisions.

A recurrent theme expressed by participants in the controversy is that the most important questions at stake are not technical, but "social and political." Responding to the petitions that prompted its review of the waste confidence principle, the court framed the positions of both the petitioners and the USNRC in those same terms:

> Though a number of commenters suggested that the social and political barriers to building a geologic repository are too great to conclude that a facility could be built in any reasonable timeframe, the [Nuclear Regulatory] Commission believes that the lessons learned from the Yucca Mountain program and the Blue Ribbon Commission on America's Nuclear Future will ensure that, through "open and transparent" decision making, a consensus would be reached.
>
> (US Court of Appeals: 10)

Linking the case to the intractable controversy regarding the proposed repository site in Nevada—and to the report of a presidential commission convened to explore alternatives following the elimination of funding for that project—the court highlighted a fundamental divergence between the understandings of the USNRC and its critics. The report of the presidential Blue Ribbon Commission (BRC) asserted confidence that a "consent-based" process for selecting a repository site could be developed and successfully implemented. Consent by a potential host community, however, requires trust in the intentions and commitment of the agency that would design, build, and operate a facility to contain high-level nuclear wastes for tens of thousands of years. No such agency has yet been identified, let alone been evaluated in terms of trustworthiness in relation to the task proposed. The BRC did not specify whether the US Department of Energy (the sponsor of the Yucca Mountain project), or another existing agency, or an agency yet to be created, would be the best choice for the task, although its report generally favored the creation of a new agency. If, as Earle and Siegrist (2006) suggest, confidence requires a tangible basis in historical or institutional facts, then critics can and do assert that no such facts exist. Aligning with that perspective, the court stated bluntly: "The Commission apparently has no long term plan other than hoping for a geologic repository" (US Court of Appeals: 13).

### One Commissioner's Concept of Confidence

In comments accompanying his vote to approve the USNRC plan for addressing the court's decision, one Commissioner nevertheless expressed disagreement with the court's reasoning. During his tenure from 2010 through 2014, Commissioner William Magwood was often a magnet for criticism by opponents of nuclear power, who viewed his stance as insufficiently independent from industry influence. Although his comments were particularly direct, they represent a view that critics attribute to most of the Commissioners:

> I reject the notion that it is necessary and appropriate for the NRC to analyze a scenario in which the United States Government utterly fails in its responsibility to safely dispose of high-level radioactive wastes. Whatever the short-term policy challenges this effort faces, the essential fact remains that deep geologic disposal is widely recognized as a technically appropriate and achievable strategy to deal with spent fuel and high-level wastes . . . It is not the role of this agency to address these policy issues, but merely to assess whether there is sufficient cause to maintain confidence that the US Government will properly dispose high level radioactive wastes at an appropriate time in the future.
>
> (USNRC 2012: 8)

Three themes are evident in this statement, all of which are challenged by critics. First, the statement expresses trust in the US government's commitment to resolve the problem of permanent nuclear waste disposition, and confidence in the government's ability to do so. Second, in characterizing the impasse over waste disposal as a "short-term policy challenge" the statement disregards the demonstrated intractability of the issue and its persistence over decades since its explicit recognition in the 1982 Nuclear Waste Policy Act. Third, the statement's emphasis on the issue's technical aspects further distances the Commissioner from complex and nuanced public concerns regarding equity, stakeholder voice, and social and environmental justice in the selection of a permanent repository site. In terms of the criteria suggested by Earle and Siegrist (2006), the Commissioner's statement lacks morality-relevant warrants for trust including identification, shared values, benevolent intentions, fairness, and caring, and performance-relevant warrants for confidence including historically demonstrated performance criteria and the existence of reliable institutional structures. Concerns regarding regulatory failures at Fukushima, widely attributed to insufficient independence of officials from industry influence (see Hindmarsh 2013a, 2013b), provide further grounds for skeptical responses to the Commissioner's expressed confidence.

## Conflicting and Conflicted Notions of Confidence

Although the principle of regulatory independence appears regularly in discourses of nuclear safety, the USNRC is in fact faced with the task of negotiating conflicting and conflicted interests, as evidenced in the agency's responses to the Fukushima disaster (Kinsella 2013; Kinsella et al. 2015). The five USNRC commissioners were selected through a process of presidential nomination and confirmation by the US Senate, with the constraint that no more than three may be affiliated with the same political party. Underlying premises regarding the necessity of nuclear power make it unlikely that a candidate expressing fundamental opposition to that energy source would be appointed as a commissioner. The agency's mandate has been discursively restricted to ensuring that the technology operates safely, without questioning the notion that such an outcome is, in fact, feasible. Despite efforts made by nuclear critics, it appears that the Fukushima disaster has not led to an institutional reconsideration of these basic understandings regarding US nuclear regulation.

In tension with the expectations of nuclear critics are expectations of the industry, which regards the agency's mission as one of supporting its continued operations and expansion in the context of a consumption-driven economy and ever-increasing energy demands (Kinsella 2015). The industry brings its own notions of confidence and trust to its engagement with the USNRC, and has developed an effective network of resources for that engagement. Industry concerns include reducing "regulatory uncertainty" to help ensure a predictable business environment, limiting the "regulatory burden" associated with compliance activities, and addressing "cumulative effects of regulation" as new expectations are added to existing ones. The USNRC's existing and possible future actions taken in response to the Fukushima disaster present a site of continuing negotiation between the industry and the agency (Kinsella 2013).

Thus, USNRC commissioners and staff often find themselves positioned between nuclear industry advocates and critics, each group asserting that the agency favors the opposing interests. Those two sets of interests were articulated differently during the waste confidence public engagement process. Seeking to provide venues for broad inclusion, USNRC staff conducted 13 open public meetings, three on weekday afternoons at its headquarters near Washington, DC and 10 on weekday evenings at locations near operating or closed nuclear power plants. The headquarters meetings were webcast, with the final meeting devoted entirely to receiving public comments.[14] The other meetings followed a standard template beginning with an hour-long "open house" followed by a USNRC presentation and oral comments provided by interested parties.

Additional teleconferences took place in parallel with those meetings, and interested parties were also invited to send written comments via post, email, or a website link. Industry comments were generally submitted in writing, and tended to be formal in tone and technically and administratively focused. Public comments ranged from individualized and personal in tone, to carefully crafted statements drafted by public interest groups and often submitted *en masse* as postcards or via organizations' websites. According to the USNRC (2014c):

> Over the course of the 98-day public comment period, the NRC collected over 1,600 pages of transcribed written testimony from

approximately 500 speakers, and received approximately 33,100 written comment submissions (approximately 32,000 of these were form letters). Since the close of the comment period on December 20, 2013, the NRC catalogued, reviewed, and responded to all comments.

In a publically-webcast meeting three months after the close of the public comment period, the agency's commissioners were briefed by staff members and heard comments from representatives of one Native American community (the Prairie Island Indian Community) and one state (New York) that had initiated the original challenge to the waste confidence rule, two industry advocacy groups, and one environmental advocacy group. I now examine selected themes that emerged throughout the comment period, including the role of the Fukushima disaster in shaping public concerns and the content of the GEIS and regulatory rule.

## Seeking the Fukushima Effect in the Waste Confidence Controversy

After receiving comments on its 585-page draft GEIS, the USNRC released a 1708-page document recording and categorizing statements received through the venues discussed above. The document provides the text of every comment judged by the agency's staff as "unique," and the texts of the form letters, mass emails, and web submissions coordinated by advocacy groups. A simple numerical comparison between those documents is suggestive, although its significance needs further clarification. A search for the term "Fukushima" indicated 974 occurrences in the comments received, approximately one per 1.8 pages of text and on the average, almost one per unique comment. In contrast, the term appears 19 times in the text of the draft GEIS and its eight appendixes (excluding five occurrences in reference lists), approximately once per 31 pages. Of course, this comparison requires further analysis in context; for example, the term is used in connection with a wide range of claims, and when it appears in a particular statement in either document it is often used repeatedly. Nevertheless, as a general indicator of salience the comparison suggests that the relevance of Fukushima was understood quite differently by USNRC staff members and concerned commenters.

A systematic analysis of the comments is not possible in the space of a single chapter; instead, I present a few excerpts from comments that explicitly invoke the topics of confidence, trust, and Fukushima. These excerpts are only a small sample of such statements, selected on the basis of conciseness, clarity, and relevance for this chapter's analytical purposes. In lieu of page numbers, parenthetical citations following the comments indicate their searchable locations in the USNRC summary document (USNRC 2014d), where their authors are identified.

In soliciting comments, the agency highlighted specific questions on which it sought guidance, based on previous interactions with interested parties. One question, regarding the possibility of changing the name of the regulatory rule, provided a rare topic on which industry critics and advocates agreed. One commenter remarked:

> The problem I think the public has with this Waste Confidence rule is the word Confidence . . . [T]he Japanese people were promised this as well in Fukushima and look what happened . . . So I think that that word is pitting the public against the NRC. And I think we need to work together here as NRC's supposed to represent the public and not the industry.
>
> (0112–9–3)

Another commenter's tone was less moderated:

> That you dare use the word confidence when talking about safeguarding rad waste for hundreds of times longer than the entire Christian era is preposterous. . . For a hazard that will last for thousands of years, Waste Confidence is an oxymoron. For a hazard that will last for hundreds of thousands of years, use of the word confidence is simply moronic.
>
> (0112–11–2 and 0112–11–8)

While suggesting that confidence is indeed warranted, an attorney for the Nuclear Energy Institute, an industry advocacy group, expressed concern about the practical effects of the term:

> Although the record amply supports a continued finding of reasonable confidence that safe disposal will become available, there is much confusion about that to which the term refers. To avoid the confusion and the mischaracterizations that the term "Waste Confidence" seems to engender, we strongly recommend that the Rule be retitled something along the lines "Storage of Spent Nuclear Fuel for the Period after License Term of Reactor Operation." Simple, straightforward, and descriptive. The GEIS should be similarly renamed.
>
> (0246–14–1)

Not all critics would consider the proposed new title simple, straightforward, or descriptive, and many would take issue with its erasure of the question of permanent disposal. Nevertheless, in releasing its draft final GEIS and revised rule, the USNRC announced a new title that achieves simplicity while avoiding issues of permanent disposal:

> It is clear from the comments that using the historical term "waste confidence" in the title has caused some confusion. The Commission

agrees that a title that more accurately reflects the Rule content is more appropriate. Therefore, the NRC has changed the title of the Rule to "Continued Storage of Spent Nuclear Fuel." The title of the GEIS was also changed accordingly.

(USNRC 2014b: D-12)

Ironically, the new title further highlights what is for many critics a clear lesson of the Fukushima disaster: that continued storage of used nuclear fuel is a highly risky activity. Such critics may find it remarkable that a number of statements emphasizing that point appeared among the 121 remarks, totaling 20 pages of text, that the USNRC placed in a section of the public comment document (USNRC 2014d) titled "Out-of-Scope Comments—Fukushima."[15] Despite the salience of Fukushima throughout the public engagement process, categorizing such comments as out of scope provides a procedural warrant for not addressing them in the final GEIS and regulatory rule. Such comments include:

The horrid mess at Fukushima is good evidence that nuclear power can become very dangerous when something goes wrong. I have NO CONFIDENCE in the ability of corporate nuclear power plant owners to safely store spent nuclear fuel.

(0013–1)

Fukushima should be a wake up call to the possibilities of a spent fuel pool fire. Due to prevailing winds, had Fukushima been on the West Coast of Japan the human and economic impacts would have been magnitudes greater.

(0826–25)

As Fukushima continues to demonstrate, fuel stored in spent fuel pools is not safe from environmental hazards such as earthquakes and floods . . . Clearly, storing spent fuel rods in pools does not insure long term safety, yet some storage pools in the US already hold as much as 9 times the amount of spent fuel they were designed to hold.

(0086–2)

In the view of some critics, the USNRC's draft final GEIS and proposed rule sought to resolve the question of confidence by semantic deletion. Nevertheless, questions of both confidence and trust remain, perhaps exacerbated by the name change and the contents of the documents produced following the extended public engagement process. Another statement ruled out of scope in the comment document encapsulates the connections among confidence, trust, and Fukushima that seem so clear to critics of the nuclear industry and its regulators:

> What we are pointing at is the Fukushima Effect. When the power went out there, we all know what happened there, within 8 hours. MUL-TIPLE Fukushima nuclear plants melted down, and blew up, along with their spent fuel pools, because they lost their power source FOR 8 HOURS. The pro nuclear apologists promised everyone that this would NEVER happen . . . Do you trust them, after all of the broken promises of free power, safe energy source, and an easy solution for the nuclear wastes? These same nuclear 'experts' knew 20 years ago that a tsunami was coming, but they ignored the warnings. These supposed nuclear experts had plenty of time to get ready for both a tsunami and an earthquake, but they depended on luck more than preparation.
>
> (0498–18)

Multiple comments linked the principle of trust with the theme of regulatory failure that has appeared in many accounts of the Fukushima disaster. Although the USNRC and the industry argue vigorously that the US system ensures regulatory independence, critics have pointed to the agency's decisions to approve new reactor licenses following the Fukushima disaster, and to its record of approving all requests to date for extensions of existing licenses beyond their initial, 40-year durations:

> Of 73 license extensions for 20 additional years of operations sought, all 73 have been approved by the Nuclear Rubberstamp Agency . . . If any applicant approaches NRC for approval for 80 years of operations, a rubberstamp is all but assured. NRC is itself a rogue, captured agency, captured by the industry it is supposed to regulate. This is very frightening, and dangerous. The Japanese Parliament concluded that the root cause of the Fukushima nuclear catastrophe was collusion between industry, regulator, and elected officials. We have that in spades here!
>
> (0919–3–9)

The anger, irony, and sense of betrayal expressed in many of the comments contrast starkly with the detached, analytical, and confident tone of the USNRC policy documents. The agency's dismissal of so many Fukushima-related comments as "out of scope" risks exacerbating those critical views of the agency and the nuclear industry.

## Fukushima, Nuclear Confidence, and Collective Risk

What effects has the Fukushima disaster had on questions of used nuclear fuel safety, regulation and governance, and the viability of nuclear power in the US? If the outcome of the waste confidence public process, to date, is taken as the sole indicator, then arguably, the effects of Fukushima have been minimal. The USNRC's characterization of so many public comments regarding Fukushima as "out of scope" may suggest, as critics allege, that the agency is more responsive to the industry it regulates than to the public community it is charged with protecting. Alternatively, or as well, an "it could never happen here" worldview may be evident in the Commission's approach to the GEIS and rule revision process. One reason given for discounting Fukushima-related comments in the waste confidence process was that they are being addressed in other USNRC venues, but similar patterns are visible in those venues as well (Kinsella 2012, 2013).

Other readings of the process are also possible. One outcome may be that the most recent legal challenges to the Commission's GEIS and revised rule, prompted by a lack of confidence in the waste confidence process, may lead to yet another reconsideration of the implications of Fukushima. In that case, a body of commentary is now on record that can become the foundation for a new round of engagement. The waste confidence controversy has highlighted a deep discrepancy between public concerns generated by the Fukushima disaster and the relative inertia of official responses in the US. In relation to questions of public confidence regarding nuclear safety and regulatory institutions, that lesson may be the most instructive of all.

Concerns remain regarding the adequacy of prevailing safety analyses, regulatory processes, and procedures for used fuel storage (USNRC 2015), as well as the prospects for establishing a permanent used fuel repository and maintaining the facilities and institutional structures necessary for safe interim storage (USGAO 2011). Beyond those relatively near-term challenges, used nuclear fuel poses hazards that extend for millennia, exceeding the duration of all human institutions to date. If the Fukushima disaster has prompted a more critical public conversation about those hazards and how to address them, then some benefits may emerge as part of its legacy.

## Notes

1 The adjectives "used," "irradiated," and "spent" are all used to characterize fuel removed following "burn-up" in reactors. Except when quoting other sources, this chapter avoids the term "spent fuel" as it inaccurately suggests diminished potency. In fact, used reactor fuel poses greater radiological and weapons proliferation hazards than does fresh fuel.

2 Kinsella et al. (2013) examine another aspect of US nuclear regulatory policy in a study of state-level public hearings.

3 See http://goo.gl/0v8NkQ, accessed November 24, 2014. Updates are available at:
http://goo.gl/u7Ab5c, accessed November 24, 2014.

4 Differences exist across national contexts regarding the salience of nuclear waste issues (see Greenberg 2013; Kinsella 2011; Kinsella et al. 2015).

5 See also http://goo.gl/JiPOAG, accessed November 26, 2014.

6 Numbers current as of February 2015; see www.nrc.gov/reactors/power. html.

7 See http://goo.gl/0RHyAG. Extended decommissioning timelines reduce costs for plant owners while posing continued safety risks.

8 The term "reasonable assurance" originated in legal interpretations of the 1954 Atomic Energy Act and its subsequent amendments (see Cavers 1962), cited in the 1979 court order directing the USNRC to evaluate the safety of used fuel.

9 For a more fully developed conceptualization of trust, see Luhmann (1979).

10 On the "uniquely dreaded" characteristics of nuclear power, see Butler et al. (2013) and Hindmarsh (2013a).

11 Associated documents are archived at http://goo.gl/hg0kuS, accessed November 26, 2014.

12 See http://goo.gl/lQp2Ft, accessed November 26, 2014. The deficiencies cited were insufficient analysis of the implications of failure to develop a permanent used fuel repository, potential radiological leaks at used fuel storage pools, and potential releases due to storage pool fires.

13 See http://goo.gl/ezq4zH, accessed November 26, 2014, and Council on Environmental Quality (2007).

14 Official meeting summaries are available at http://goo.gl/GPelgm; webcasts are archived at http://video.nrc.gov/#archivedwebcasts.

15 Other categories deemed out of scope (there are seven such categories) include "Reactor Accidents," "Yucca Mountain," and "Opposition to Nuclear Power." Comments regarding those topics were only deemed within scope when, in the judgment of USNRC staff, they directly addressed the primarily technical and administrative issues included in the draft GEIS.

# References

Ahlers, M.M. "Fukushima Shines Light on U.S. Problem: 63,000 Tons of Spent Fuel." *CNN*, March 31, 2011, accessed November 25, 2014, www.cnn. com/2011/US/03/30/spent.nuclear.fuel/.

Blue Ribbon Commission on America's Nuclear Future (BRC). *Report to the Secretary of Energy*. Washington, DC: US Department of Energy, 2012.

Bradsher, K., and H. Tabuchi. "Greater Danger Lies in Spent Fuel than in Reactors," *New York Times*, March 18, 2011: A12.

Bratspies, R.M. "Regulatory Trust," *Arizona Law Review* 51 (2009): 575–631.

Burke, K. *A Rhetoric of Motives*. Berkeley: University of California Press, 1969.

Butler, C., Parkhill, K.A., and N.F. Pidgeon. "Nuclear Power after 3/11: Looking Back and Thinking Ahead." In *Nuclear Disaster at Fukushima Daiichi: Social, Political, and Environmental Issues*, edited by R. Hindmarsh, 135–153. New York: Routledge, 2013.

Callon, M. "Elements of a Sociology of Translation: Domestication of the Scallops and the Fishermen of St. Brieuc Bay." In *Power, Action and Belief: A New Sociology of Knowledge?*, edited by J. Law, 196–233. London: Routledge, 1986.

Cavers, D.F. "Administrative Decisionmaking in Nuclear Facilities Licensing," *University of Pennsylvania Law Review* 110 (1962): 330–370.

Council on Environmental Quality. *A Citizen's Guide to the NEPA: Having Your Voice Heard*. Washington, DC: Council on Environmental Quality, 2007.

Earle, T. C., and M. Siegrist. "Morality Information, Performance Information, and the Distinction between Trust and Confidence," *Journal of Applied Social Psychology* 36 (2006): 383–416.

Feiveson, H. et al. (eds.). *Managing Spent Fuel from Nuclear Power Reactors*. Princeton, NJ: International Panel on Fissile Materials, 2011.

Fisher, W.R. *Human Communication as Narration: Toward a Philosophy of Reason, Value, and Action*. Columbia: University of South Carolina Press, 1989.

Goodnight, G.T. "The Personal, Technical, and Public Spheres of Argument," *Journal of the American Forensic Association* 18 (1982): 214–227.

Greenberg, M.R. *Nuclear Waste Management, Nuclear Power, and Energy Choices: Public Preferences, Perceptions, and Trust*. London: Springer, 2013.

Hindmarsh, R. "Nuclear Disaster at Fukushima Daiichi: Introducing the Terrain." In *Nuclear Disaster at Fukushima Daiichi: Social, Political, and Environmental Issues*, edited by R. Hindmarsh, 1–21. New York: Routledge, 2013a.

Hindmarsh, R. (ed.). *Nuclear Disaster at Fukushima Daiichi: Social, Political, and Environmental Issues*. New York: Routledge, 2013b.

Kennedy, G. *A New History of Classical Rhetoric*. Princeton: Princeton University Press, 1994.

Kinsella, W.J. "Research on Nuclear Energy in an International Context," *Technikfolgenabschätzung: Theorie und Praxis* 20, no. 2 (2011): 84–89.

Kinsella, W.J. "Environments, Risks, and the Limits of Representation: Examples from Nuclear Energy," *Environmental Communication* 6, no. 2 (2012): 251–259.

Kinsella, W.J. "Negotiating Nuclear Safety: Responses to the Fukushima Disaster by the US Nuclear Community." STS Forum on the 2011 Fukushima/East Japan Disaster, University of California-Berkeley, 2013, http://go.ncsu.edu/kinsella-fukushimaforum2013.

Kinsella, W.J. "Rearticulating Nuclear Power: Energy Activism and Contested Common Sense," *Environmental Communication* 9, no. 3 (2015): DOI:10.10 80/17524032.2014.978348.

Kinsella, W.J., Andreas, D.C., and D. Endres. "Communicating Nuclear Power: A Programmatic Review." In *Communication Yearbook 39*, edited by E. Cohen, 277–309. London: Routledge, 2015.

Kinsella, W.J., Kelly, A.R., and M. Kittle Autry. "Risk, Regulation, and Rhetorical Boundaries: Claims and Challenges Surrounding a Purported Nuclear Renaissance," *Communication Monographs* 80, no. 3 (2013): 278–301.

Latour, B. *Science in Action*. Milton Keynes: Open University Press, 1987.

Luhmann, N. *Trust and Power: Two Works*, edited by T. Burns and G. Poggi, translated by H. Davis, J. Raffan, and K. Rooney. New York: Wiley, 1979.

Luhmann, N. "Familiarity, Confidence, Trust: Problems and Alternatives." In *Trust: Making and Breaking Cooperative Relations*, edited by D. Gambetta, 94–107. Oxford: Basil Blackwell, 1988.

Luhmann, N. *Risk: A Sociological Theory*, translated by R. Barrett. New York: de Gruyter, 1993.

Miller, C.R. "The Presumptions of Expertise: The Role of Ethos in Risk Analysis," *Configurations* 11, no. 2 (2003): 163–202.

Parenti, C. "Fukushima's Spent Fuel Rods Pose Grave Danger," *The Nation*, March 15, 2011, accessed November 22, 2014, www.thenation.com/article/159234/fukushimas-spent-fuel-rods-pose-grave-danger.

Paté-Cornell, E. "On 'Black Swans' and 'Perfect Storms': Risk Analysis and Management When Statistics Are Not Enough," *Risk Analysis* 32, no. 11 (2012), 1823–1833.

Perrow, C. *Normal Accidents: Living with High-Risk Technologies*. Princeton, NJ: Princeton University Press, 1984.

Prelli, L. *A Rhetoric of Science: Inventing Scientific Discourse*. Columbia: University of South Carolina Press, 1989.

Schneider, M., et al. *World Nuclear Industry Status Report 2014*. Paris: Mycle Schneider Consulting, 2014.

Takubo, M. "Wake Up, Stop Dreaming: Reassessing Japan's Reprocessing Program," *Nonproliferation Review* 15, no. 1 (2008), 71–94.

US Court of Appeals, District of Columbia Circuit. *Opening Brief for Petitioners, Petition for Review of Final Administrative Action of the United States Nuclear Regulatory Commission* (Case number 11–1045). Washington, DC: US Court of Appeals, September 15, 2011.

US Court of Appeals, District of Columbia Circuit. *State of New York, et al. versus Nuclear Regulatory Commission and United States of America* (Case number 11–1045). Washington, DC: US Court of Appeals, June 8, 2012.

US Government Accountability Office (USGAO). *Commercial Nuclear Waste: Effects of a Termination of the Yucca Mountain Repository Program and Lessons Learned*. Document GAO-11–229. Washington, DC: USGAO, 2011.

US Nuclear Regulatory Commission (USNRC). "10 CFR Part 51: Consideration of Environmental Impacts of Temporary Storage of Spent Fuel after Cessation of Reactor Operation," *Federal Register* 75, no. 246 (23 December 2010): 81032–81037.

US Nuclear Regulatory Commission (USNRC). *Commissioner Comments on COMSECY-12–0016: Approach for Addressing Policy Issues Resulting from Court Decision to Vacate Waste Confidence Decision and Rule*. Document ML12250A036. Washington, DC: USNRC, 2012.

US Nuclear Regulatory Commission (USNRC). *Waste Confidence Generic Environmental Impact Statement Scoping Process Summary Report*. Washington, DC: USNRC, 2013a.

US Nuclear Regulatory Commission (USNRC). "10 CFR Part 51: Waste Confidence—Continued Storage of Spent Nuclear Fuel; Proposed Rule." *Federal Register* 78, no. 178 (13 September 2013b): 56776–56805.

US Nuclear Regulatory Commission (USNRC). *Generic Environmental Impact Statement for Continued Storage of Spent Nuclear Fuel*. NUREG–2157, ML14188B749. Washington, DC: USNRC, 2014a.

US Nuclear Regulatory Commission (USNRC). *Continued Storage of Spent Nuclear Fuel*. 7590–01-P, 10 CFR Part 51 [NRC-2012–0246], RIN 3150-AJ20, SECY-14–0072. Washington, DC: USNRC, 2014b.

US Nuclear Regulatory Commission (USNRC). "Frequently Asked Questions about the Waste Confidence Generic Environmental Impact Statement (GEIS) and Rulemaking," 2014c, www.nrc.gov/waste/spent-fuel-storage/wcd/faq.html.

US Nuclear Regulatory Commission (USNRC). *Comments on the Waste Confidence Draft Generic Environmental Impact Statement and Proposed Rule*. ML14154A175. Rockville, MD: USNRC, 2014d.

US Nuclear Regulatory Commission (USNRC). *Information Digest, 2014–2015.* Rockville, MD: USNRC, 2014e.

US Nuclear Regulatory Commission (USNRC), Office of the Inspector General. *Audit of NRC's Oversight of Spent Fuel Pools.* OIG-15-A-06. Rockville, MD: USNRC, 2015.

Wynne, B. "Knowledges in Context," *Science, Technology, & Human Values* 16, no. 1 (1991): 111–121.

# 13 The Fukushima Effect in New Zealand

## A Historical Perspective from a "Nuclear-Free" Country

*Rebecca Priestley*

On Saturday, March 12, 2011—less than a year after I completed my PhD thesis on New Zealand's nuclear history—a newspaper journalist called me, wanting information for a story he was writing on the unfolding Fukushima nuclear disaster. In response to his question as to whether New Zealand was in any immediate danger, I suggested he speak to a nuclear physicist and an atmospheric scientist. I recommended New Zealand's Institute of Geological and Nuclear Sciences (GNS Science), the National Radiation Laboratory (NRL), and the Meteorological Service (MetService) as places he should call to find these scientists. In an email response, sent later that day, I wrote about the radioactive fallout received in New Zealand through the 1960s, when bombs were being tested in nearby Pacific waters, and pointed out that atmospheric circulation patterns meant that radiation released in the Northern Hemisphere would be unlikely to threaten New Zealand.

To my surprise, the following Monday a page two newspaper article, under the heading "New Zealand 'Should Be Safe' from Nuclear Fallout," quoted me as the source of that reassuring information. Along the top of the page I was quoted as saying: "I would be a whole lot more concerned if I was living downwind of Japan right now . . . Air masses tend to travel around the globe along east-west lines, rather than along north-south lines" (Easton 2011). Later that day, the headline of an online version of the story quoted me and referred to me as an "expert." At my request, the headline was changed, although the news story still called me an "expert."[1]

As a recent PhD graduate in the history and philosophy of science, and weekly science columnist for current affairs magazine *The Listener*, I was considered enough of an expert to speak on any danger the Fukushima disaster posed to New Zealand. After the newspaper had identified me as an expert, requests followed for radio and television interviews. I refused further media requests and contacted the Science Media Centre to ask them to find suitably qualified people to talk to the media. In my own magazine article about the Fukushima disaster, written that week, I quoted a nuclear physicist from the University of Auckland, a health physicist from the NRL, a meteorologist from the MetService, and the head of the IAEA (Priestley 2011).

"Expert" opinion on nuclear science and technology can come from local and international scientists; from government agencies; from politicians; from activists; and occasionally, it seems, from a science historian/journalist. In this chapter I look at New Zealand's response to the Fukushima disaster in the context of New Zealand's nuclear history and responses to past nuclear disasters: Windscale (1957), Three Mile Island (1979), and Chernobyl (1986). How was news of these disasters reported and interpreted in New Zealand, which was, variously, looking forward to a nuclear future (1957), in the position of having recently rejected nuclear power as an option for the immediate future (1979), a nation in the process of asserting its "nuclear-free" policy (1986), and, most recently, a nation with a firm "nuclear-free" identity (2011). With a focus on response to Fukushima, who were the "experts" quoted by the media? Which other experts, or publics, added their voices to the discussion or debate, and what did they say? Did the Fukushima disaster further cement and validate New Zealand's nuclear-free identify and rejection of nuclear power?

In this chapter I first outline my investigative approach, then give an overview history of New Zealand's 20th century plans for and attitudes towards nuclear power, then outline response to 20th century nuclear disasters before moving on to analyze public attitudes towards nuclear power in the 21st century both before and after the Fukushima disaster. I end with a discussion of my findings and a conclusion.

## Investigative Approach

Given that New Zealand is a "nuclear-free" country, with a policy against nuclear power, I focus attention on pro-nuclear voices—people whose attitudes threaten a change to the status quo—and how they have or have not changed in response to Fukushima 3/11. Looking at any Fukushima effect in New Zealand provides an interesting contrast to the other chapters in this volume, which all focus on nuclear nations.

In my overview of attitudes to nuclear power through the 1950s to 1970s, my main sources are media accounts and published government reports. In looking closely at attitudes to nuclear power in the first decade of the 21st century, key sources are media accounts, including opinion pieces, and online discussion fora. Post-Fukushima, key sources are media reports, online discussion fora, my own experience, and interviews with five high-profile New Zealanders who made pre-Fukushima statements in support of nuclear power.

In looking at the response to the Fukushima disaster in the context of New Zealand's nuclear history, I follow a narrative history of science approach and look for any Fukushima effect decades after the debates over the possible introduction of nuclear power. This chapter aligns with the Science, Technology and Society Studies (STS) subfields of science communication and public understanding of science and the STS theme of experts and expertise.

## Background of Nuclear Power in New Zealand

New Zealand has an international reputation for being "nuclear-free." In 2013, the main electricity sources were hydro (54.5%), gas (19.4%) and geothermal (14.5%), with coal, wind, and bioenergy making minor contributions (Ministry of Business, Innovation & Employment 2014). The country's nuclear-free legislation, introduced in 1987, prohibits nuclear weapons and waste from New Zealand's air, land, and water (New Zealand Nuclear Free Zone, Disarmament, and Arms Control Act 1987). The legislation does not mention nuclear power, but all major political parties have policies against nuclear power.

As covered in my book *Mad on Radium* (Priestley 2012), New Zealand was once as enthusiastic about nuclear power as any nation, with many public scientists speaking out in support of nuclear power. *The Dominion* of January 13, 1955 reported Professor of Physics Charles Watson-Munro as saying atomic power was "a coming force that would be comparable to the first onslaught of electricity on civilization." The *Te Puke Times* of June 17, 1955 reported Ernest Marsden, director of New Zealand's Department of Scientific and Industrial Research (DSIR), as predicting that by the 21st century, "electricity from nuclear sources will be supplied to houses and small industries under much the same conditions as the present water supply." Demand for electricity was starting to exceed supply and new sources of electricity were needed.

While nuclear power was yet to be offered commercially in either the UK or US, there was a lot of rhetoric about the nuclear future that lay ahead. Scientists and managers at the DSIR, the State Hydroelectricity Department, and the Ministry of Works—the organizations that would be involved in commissioning and running a nuclear power station—kept an open mind about nuclear power, but were not prepared to commit to it unless it was the most economic option. Francis Farley, a physicist at Auckland University College, was reported in the *Auckland Star* of April 26, 1956 as saying that "the real obstacle to nuclear power in New Zealand is the over-cautious play-safe attitude that is adopted in Wellington," predicting that if the "present increase in power demand continues, we might expect to have 10 nuclear power stations by 1975 to 1980" (see Figure 13.1).

Many newspapers sided with the scientists speaking out in favor of nuclear power, and criticized government officials for being over-cautious. "These scientists must not be ignored," said the *Auckland Star* editorial of April 27, 1956. The editorial of May 2, 1956 added that the "people . . . see atomic energy as a probable, and perhaps the only effective, alternative to a future darkened by power shortages." Even the left-wing newspaper *The People's Voice* chipped in, with a front-page article on May 9,

TIME FOR A TRADE-IN

*Figure 13.1* In this Auckland Star cartoon, Prime Minister Sidney Holland, Who Was Criticized for Going Slowly on Nuclear Power, Is Seen Driving the 1911 Hydroelectric Model Vehicle. Cartoon by Neil Lonsdale, Auckland Star, April 27, 1956

1956. Under the headline "Atomic Power for Industry," the story reiterated the pro-nuclear statements of Farley and others, and concluded by saying, "It is in the interests of workers, farmers, housewives and industrialists of New Zealand to have available a source of reliable power, and it is the responsibility of the government to provide it."

In 1964, nuclear power appeared, for the first time, in the New Zealand Electricity Department (NZED) report on electric power development in New Zealand. Serious consideration, it said, should be given to locating a nuclear power station north of Auckland to meet that city's growing demand for electricity. The plan, outlined in the 1975 report, was for two 600 MW nuclear reactors at a site on Kaipara Harbour. With a planned commissioning date of 1988, a decision would be needed by 1977 on whether or not to proceed.

## Royal Commission of Inquiry into Nuclear Power Generation in New Zealand

In 1976, the government set up a Royal Commission of Inquiry into Nuclear Power Generation in New Zealand. In the two decades that New Zealand had been considering nuclear power, the public mood had changed. Nuclear power plants were now operating in nearly 20 countries, but because indigenous power sources had always been cheaper than nuclear, New Zealand had deferred its nuclear decision until arguments for and against nuclear power were more sophisticated.

In response to public protest and rising costs, many Northern Hemisphere countries were cancelling or postponing their own plans for new nuclear power stations. In 1975, *The Listener's* Boyce Richardson described nuclear reactors as now symbolizing "an impersonal future world of computers, robots, explosive violence and uncontrollable technology, rather than the cornucopia overflowing with goods and pleasures that they once promised" (Richardson 1975).

In 1976, a coalition of environmental and anti-nuclear groups formed the Campaign for Non-Nuclear Futures, and set a goal of collecting half a million signatures in a petition against the introduction of nuclear power to New Zealand. In November that year they presented a petition to Parliament with 333,088 signatures.

Most of the 141 submissions to the Royal Commission were against the introduction of nuclear power, which, the report noted, respondents found "an expensive, dangerous, imported technology." A survey of New Zealanders around the time of the inquiry found that only 24–25% of New Zealanders favored nuclear power (Royal Commission of Inquiry into Nuclear Power Generation in New Zealand 1978: 36).

Some scientists had changed their attitude to nuclear power. Percy Burbidge, an Auckland university physicist who spoke out in favor of nuclear power in the 1950s, wrote in his submission to the Royal Commission: "The danger to our population and the potential damage to our industrial production are too great to justify the introduction of reactors for electrical power." Nuclear power was "a partially developed technology that is not acceptably safe," wrote ex-mining engineer Gordon Williams (Submissions 1978).

But during the course of the inquiry, nuclear power stations were dropped from the NZED's 15-year plan. Electricity demand forecasts had been revised downwards. The discovery of new gas fields and a reassessment of coal reserves meant New Zealand now had sufficient indigenous resources to continue to develop new power stations without resorting to nuclear power.

When the Royal Commission reported in 1978, they described the history of nuclear power in New Zealand as one of "official enthusiasm, early public acceptance or apathy, and then of rising opposition." They concluded that "nuclear power is not justified for New Zealand until about the turn of the century or perhaps even later" (Royal Commission of Inquiry into Nuclear Power Generation in New Zealand 1978: 35, 45).

### Response to Nuclear Accidents: Windscale, Three Mile Island, and Chernobyl

How did this non-nuclear country, with its public opinion moving from "official enthusiasm," to "public acceptance or apathy," and then to "rising opposition," respond to nuclear disasters through the twentieth century?

The first three major nuclear accidents that got media attention in New Zealand took place on the other side of the world, in Windscale in the UK in 1957, at Three Mile Island on the east coast of the US in 1979, and at Chernobyl in the Soviet Union in 1986. There was little coverage of the Windscale disaster in New Zealand papers, although the following year, in weekly current affairs magazine *The Listener*, editor Monte Holcroft wrote about the "accident in the plutonium factory at Windscale," saying that "the whole of Cumberland . . . was subjected to radiation as if from a small atomic bomb." He concluded by saying that, in spite of all the expert opinions and assurances, a suspicion was growing that "the scientists are using forces they do not fully understand" (Holcroft 1958).

The Three Mile Island nuclear disaster of 1979 occurred less than a year after the Royal Commission on Nuclear Power Generation in New Zealand reported. While nuclear nations were in turmoil, with protesters demanding safety assessments of their nuclear plants and governments cancelling plans for new nuclear power stations, New Zealand was secure in its position of having no nuclear power stations and no plans for them. Media coverage focused on the unfolding events at the Pennsylvania power station, with little reference to any implications for New Zealand, although *The Dominion* of April 2, 1979 quoted Robert Mann, director of the Environmental Defence Society and one of the lead campaigners in the 1970s anti-nuclear movement, saying this was a warning to New Zealand if it ever introduced nuclear power. "The few people who want nuclear power here have claimed the chances of this happening are one in a million . . . The public can now judge for itself."

In April 1986, in response to the Chernobyl nuclear disaster, most New Zealand news reports—as well as recounting details of the disaster—focused on news of New Zealand tourists in the area and possible impact on trade. Murray Matthews, head of the environmental monitoring section of the National Radiation Laboratory (NRL), said the laboratory would be monitoring fallout from Chernobyl in New Zealand. But, the news report reassured, "if any radiation was detected it would be of scientific interest only and well below levels which would be dangerous to health" (Clark 1986). An editorial in the *New Zealand Times* on May 4, 1986 described nuclear power as "a gamble that is not worth the risk." Although, as the Soviet embassy's head of information pointed out from Wellington, "unlike New Zealand, the Soviet Union has no option but to develop nuclear-powered energy. Half of the country is frozen solid for half of the year" (NZPA 1986).

## Pre-Fukushima Attitudes to Nuclear Power

Although New Zealand's 1987 nuclear-free legislation did not cover nuclear power—its focus is nuclear weapons and nuclear propulsion—by

2011 all major political parties had adopted policies against nuclear power. In the first decade of the 21st century, though, public attitudes seemed to be changing. Human-induced climate change and the need for new sources of energy—demand for electricity was increasing, some dry years had led to shortfalls in supply of hydroelectricity, and a proposed major wind farm was met with NIMBYism (not-in-my-backyard) from Otago residents—led to calls for nuclear power to be revisited as an option for New Zealand. At the same time, politicians from both the left and the right suggested that New Zealand reverse the decision—enshrined in the 1987 nuclear-free legislation—to ban nuclear-powered ships from ports.

In the first decade of the 21st century, several high profile New Zealanders made statements in support of nuclear power as an option for New Zealand. In 2003, for example, economist and philanthropist Gareth Morgan wrote, in a newspaper opinion piece, in support of nuclear power for reasons of $CO_2$ emissions, cost, and environmental impact, saying, "The contribution a nuclear plant or two could make in New Zealand is enormous" (Morgan 2003). In a July 2004 opinion piece, energy-sector engineer John De Bueger used climate change predictions to promote nuclear power in New Zealand: "New Zealanders have got to get past emotive French Pacific atmospheric testing and the nuclear ship ban lunacy. We have nothing to lose by following Canada's lead and going mainline hydro/nuclear. We don't need hideous wind turbines on every second hill" (De Bueger 2004).

In his 2004 book *Nuclear New Zealand* (McEwan 2004), and in his own newspaper opinion piece (McEwan 2005), retired health physicist Andrew McEwan provoked further discussion around nuclear power with statements in support of nuclear power and reassurances that the problem of radioactive waste was "political rather than technical" (2004: 49). McEwan said that the advantages of nuclear were "no greenhouse gas emissions; cost competitiveness, particularly if carbon taxes are imposed; and reduction in transmission costs if appropriately sited" (2005: 14). In the concluding paragraph to this book he wrote that:

> Warm fuzzy "nuclear-free-ism" may be a fine philosophy to embrace on some remote Pacific atoll, but sensible management of the world's resources requires a more global view; a recognition, that worldwide, nuclear power plays a very important part in both electricity supply and limitation of green house gases. . . .
>
> (p. 265)

These were not lone voices. In August 2004, the Canterbury regional council, ECan, voted to include nuclear power in a debate about future energy sources. A 2005 *New Zealand Herald* poll said that 64% of CEOs surveyed said nuclear power "should be investigated as a means to help secure New Zealand's long-term energy supply" (Taylor 2005).

Readers responding in a "Your say" spot on the Stuff.co.nz news website in 2007 were split on their response to nuclear power. Although most of the 12 respondents were against nuclear power, three were very supportive of nuclear as an option for New Zealand, with statements like "NZ needs to get its head out of the sand and join the rest of the world" and "We've all been brainwashed by decades of anti-nuclear activists."[2]

In a 25-minute radio program in 2005, presenter Eric Frykberg canvassed pro-nuclear attitudes and interviewed nuclear advocates in New Zealand. Philip Ross, a consulting engineer and chairman of an organization called the New Zealand Atomic Energy Advocacy Council, said New Zealand "should have some nuclear to round out and balance its electricity generation grid." Nuclear, he added, "is an obvious one for generating baseloads." As for concerns about safety, Ross said the waste is "relatively easily dealt with." Since Chernobyl, he added, "there has been a gradual understanding that nuclear does have a place in this society. It allows a form of electricity generation which doesn't generate a lot of carbon dioxide, very important in these days of global warming."

Other interviewees spoke in support of his views. Alan Poletti, retired professor of nuclear physics and a Fellow of the Royal Society of New Zealand, said that nuclear power had an "extremely good safety record" and the nuclear power industry was now "perhaps the safest generator of electricity we have." The only "reliable baseload generator which does not generate carbon dioxide is nuclear power," he added. Roy Hemmingway, chairman of the Electricity Commission, said he had "always been a fan of nuclear." In terms of environmental effects, he said, nuclear was "more manageable and more acceptable" than many alternative energy sources. But, he added, "I don't believe nuclear is really a realistic option for New Zealand," in part because a 1000 MW plant was too big for New Zealand and because of the lack of infrastructure, "you don't just build a plant, you build a whole nuclear industry."

But Ross had the practical details worked out. He suggested New Zealand could choose a "Westinghouse 600 megawatt light water reactor or, for example, the now shelved Candu 3 which was 450 megawatts." Services such as fuel enrichment, waste reprocessing and even regulatory oversight, he said, could be outsourced.[3] In 2008, Grahame Sydney, an acclaimed landscape painter and Otago resident active in the campaign against wind farms in Otago, argued, in a newspaper opinion piece, that a single nuclear power station was preferable to a nationwide assault of wind turbines: "Nuclear power is carbon emissions-free, comparatively cheap and base-load dependable, and it would mean that the extraordinary national treasure of our outstanding landscapes are preserved from the devastating assaults of further wind farms and hydro inundations," he said (Sydney 2008). In another article, which also quoted Sydney, McEwan added that the key issue with nuclear "is economics, not safety or blind adherence to the nuclear-free catchcry" (Macdonald 2008).

In a 2008 Internet Survey by the New Zealand Business Council for Sustainable Development, 19% of 3546 people said nuclear power was the best electricity option for New Zealand in the next 10 years (wind power got the most positive responses). The result surprised the Council's chief executive, who said he believed it was because people were searching for solutions to climate change. "If you had asked that question 10 years ago, I'd have been surprised if you had more than 3 or 4 per cent," he said (Fawkes 2008). A survey later that year, by Research New Zealand, asked 501 people aged 15 and over if they thought, "given concerns about the environment and the prospect of ongoing power crises and price rises," New Zealand should consider nuclear energy as a viable energy source. Most said no, but a significant 36% of respondents answered yes. David Parker, Labour Minister of Energy, however, rightly criticized the survey for having a "loaded question" (Gorman 2008).

As well as reporting on survey results, journalists responded to the issue with articles with headlines such as "Are we ready to join the nuclear family?" (Macdonald 2008), "Why not nuclear power?" (Bone 2004), and "Interest in nuclear power as costs fall" (Anderson 2004) and the issue became a topic for political cartoons (see, for example, Figure 13.2). The general theme of these pieces was that nuclear power should no longer be a taboo topic for discussion, and should instead be part of the

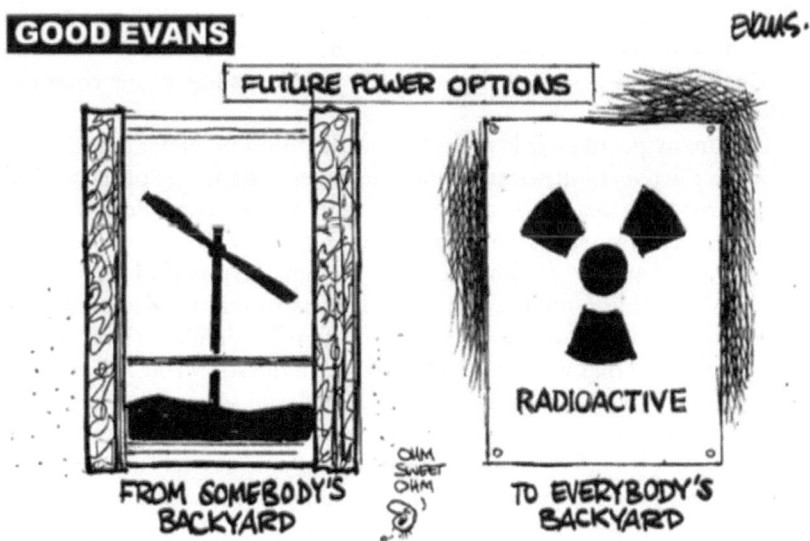

*Figure 13.2* In 2008, Plans for an Extensive Wind Farm in Otago Led to Calls for Nuclear Power as an Alternative, and Less Visually Intrusive, Form of Energy

Source: Cartoon by Malcolm Paul Evans, May 29, 2008

consideration for New Zealand's energy future. This introductory paragraph to a 2009 article in *e.nz magazine* is indicative of the content of many of these articles:

> For the past two decades, New Zealand's nuclear-free legislation has been a source of pride and a key part of our nuclear identity. It has generally been considered unthinkable that nuclear power could be a genuine option to meet the country's energy needs. But faced with a steadily growing demand for electricity, and increasing concern about greenhouse gas emissions, could it be time to revisit our anti-nuclear stance?
>
> (Shelton 2009)

## The Fukushima Effect in New Zealand

> The latest nuclear disaster in Japan makes me so glad that our country is nuclear-free, and I pay tribute to the many thousands of New Zealanders whose active campaigning from 1960 to 1985 made us so. I pay particular tribute to the Campaign Half Million petitioners in the 1970s, who knocked on the head plans to have a nuclear power plant in New Zealand.
>
> (Locke 2011)

Green Party Politician Keith Locke probably expressed the immediate sentiment of many New Zealanders with his statement, in the wake of the Fukushima nuclear disaster, that he was "so glad our country is nuclear-free."

Like many people worldwide, I spent the morning of March 12, 2011 glued to the television set, watching the incredible footage of the tsunami wave breaching sea walls alternating with scenes from the Fukushima Daiichi nuclear power plant. When the *Dominion Post* journalist called me, I was happy to talk, but only, as I thought, to point him in the right direction regarding finding suitable scientists to talk to. My embarrassment about being quoted as an "expert," though, didn't curb my interest in the story and that week, at the magazine's request, I wrote a 1500-word article for *The Listener* magazine, for which I am science columnist.

In a reflection of the New Zealand news media—the industry is small and cuts, amalgamations, and international takeovers over the past decade have reduced the number of working journalists—many of the newspaper articles on Fukushima over the following week came from international wire services, such as Reuters, AP, and AFP, or were previously published in other media such as the *Guardian*, *Washington Post*, and *The Times*.

Whereas other countries were beginning to question the safety of their own nuclear power plants, stories written by New Zealand journalists tended to focus on New Zealanders living in Japan or the potential impact on New Zealand in terms of trade or radiation release (Japan's nuclear power plants are, along with those in Taiwan, the closest nuclear power plants to New Zealand).

New Zealand has close links with Japan. The two island nations, while in different hemispheres, are physically similar in terms of their size, climate, geology, and latitude. At the time of the disaster, Japan was New Zealand's fourth largest trading partner, and there were more than 6500 New Zealanders living in Japan.

But there was another reason New Zealanders were particularly bonded to this disaster. The Tōhoku earthquake and tsunami, which led to the Fukushima disaster, hit Japan just 17 days after the February 22 Christchurch earthquake, which killed 185 people and decimated the city of Christchurch. "New Zealanders have watched in horror and disbelief the scenes of destruction in Japan following the earthquake and tsunami," said Prime Minister John Key in Parliament on March 15. "We have recently experienced our own deadly earthquake, so it was with very raw emotion that we heard of the devastation and loss in Japan. Unfortunately, we know all too well the pain, stress, and heartbreak that can be caused by a natural disaster such as this" (New Zealand 2011: 17167). As a first response to the quake, New Zealand sent a 48-strong Urban Search and Rescue team to Japan: reciprocating what Japan had done for New Zealand in response to the Christchurch earthquake (The Dominion Post 2011).

A problem for journalists wanting to write about the nuclear disaster—as illustrated in the introduction to this chapter—was finding appropriate local experts. Even the Science Media Centre relied primarily on expert commentary compiled by the UK and Japan Science Media Centers.[4] As the director of the Science Media Centre put it, "New Zealand's thin on the ground in terms of people who can walk you through what's happening in a nuclear reactor."[5] One scientist who tried to address this need was theoretical physicist Professor Shaun Hendy, from Victoria University of Wellington. Hendy's March 16, 2011, blog post on "How to shut down a nuclear reactor" (Hendy 2011) got more than 5000 hits in its first week online and was picked up and run by the *New Zealand Herald* the following day. Scientists from the NRL—an obvious source of informed scientists who could speak on any radiation risk to New Zealand from Fukushima—did not, or were unable to (their premises were badly damaged in the Christchurch earthquake), front up for media interviews. Rather, the NRL provided advice direct to government, with public information provided only through carefully worded statements on their website.

The initial response from GNS Science—host to the National Isotope Centre, "New Zealand's premier provider of isotope science expertise"[6]—was reluctance to talk to the media, both on the part of individual scientists and at an organizational level.[7] As well as the fact that the incident—an accident at a nuclear power station—was well outside their area of expertise, the organization had worked to distance itself from the word "nuclear," in part because it led to public misconceptions, and unfounded concerns, about what they did.[8] As the incident unfolded, and it became clear that there was growing public concern about airborne and seaborne radioactivity, they felt they could contribute. So, four days into the disaster, GNS Science issued a press release, which was picked up by some news media.[9]

One of the most experienced local scientists to talk to the media was Peter Roberts, retired from GNS Science, who contacted the Science Media Centre and was able to provide reassuring information about radiation risk for New Zealanders in Japan. "No one should be exposed to any unnecessary dose of radiation, and a precautionary approach near the Fukushima plant is essential," he wrote. "However, if general radiation levels a little above background are received for a short period, then the increased risk is within the natural variation in cancer risk. At doses approaching those where people are acutely sick but survive, the future cancer risk is a few percent."[10] He also appeared on the current affairs television program *Campbell Live*, saying that, in terms of impact, he expected the disaster would turn out to be somewhere between Three Mile Island and Chernobyl.[11] One scientist who spoke regularly to the media (and who I interviewed for my own magazine article) was Professor David Krofcheck, an American physicist working at the University of Auckland.

Much to my relief, scientists providing information to the media confirmed the information I had given on day two of the disaster: New Zealand was at no immediate risk of radiation. On March 16, *The Dominion Post* (Chug 2011), drawing on the GNS Science press release, quoted senior scientist Bernard Barry saying that "even in the 'worst possible event' of radioactive fallout at the stricken Fukushima plant, the southern hemisphere would not be affected . . . because the atmosphere's circulation around the equator formed a natural barrier." The *New Zealand Herald* published their own reassuring story on March 23, 2011, referring to a statement issued by the MetService and the NRL that said radiation releases would be confined to the Northern Hemisphere and quoting Peter Kreft, chief forecaster, who said "people are concerned for no particularly good reason" (Donnell 2011). A March 30, 2011 story contained comments from both Barry and Peter Abernethy, an NRL "spokesman" (media relations manager at NRL's parent organization the Ministry of Health, rather than an NRL scientist). Barry said "even if the partial breakdown of the reactors worsened, particles would not travel deep into the Southern Hemisphere" and Abernethy added if any radiation did reach New Zealand, "it would take months, maybe years, and only be in scientists' interest" (Davison 2011).

Meanwhile, members of the public made their views known online. On the Public Address news and blogsite (publicaddress.net) in an extensive and expansive conversation about the Christchurch earthquake, news started creeping in about the disaster in Japan. "Anyone for Nuclear power in Aotearoa???" wrote one contributor at the end of a post. "We would have ended up the long white shroud after an event like that."[12] (The Maori word commonly used for New Zealand, Aotearoa, roughly translates as "the land of the long white cloud.")

In the more than 50 responses to the issue of nuclear power as an option for New Zealand, many people presented ostensibly rational arguments why it was not something that should be feared. Other power sources were statistically deadlier than nuclear, it was pointed out, along with references to the dangers of coal mining (in 2010 an accident at New Zealand's Pike River coal mine killed 29 men) and to the vulnerability of hydro dams to earthquakes. "I think being downhill from a hydro dam could be more devastating," wrote one commentator. Other commentators wrote that the Fukushima nuclear accident was not that bad— "What's happened in Japan has been a real testament to the modern design of nuclear plants . . . There has been no dangerous leak, despite the fact that it was hit by a massive earthquake AND a massive tsunami." Another argument was that if New Zealand ever did embrace nuclear power it would, of course, use newer and better technology than Fukushima. Despite these comments in defense of nuclear power, the overall mood of the blogsite was that even if nuclear wasn't that bad, it was not economically feasible for New Zealand, and, even if it was, there were social and engineering barriers to its adoption.

The left wing political blogsite The Standard (http://thestandard.org. nz) presented another view. "This event ends whatever remote possibility that ever existed of NZ heading down the nuclear power path," wrote ROB a few days after the triple disaster. "Coming as it does on the heels of the Christchurch earthquake, it is all too easy to imagine the same scenario playing out here. What price our agricultural exports, tourism industry, and 'clean green' image then? NZ would be finished," he wrote.[13]

Most comments in response to the article were in agreement, pointing out, for example, that New Zealand's geology meant there was nowhere to "store long life waste," that nuclear power here would be "absurd and uneconomical." Some comments, though, continued the line that if New Zealand ever got nuclear power, it would have newer and better plant than at Chernobyl and Fukushima, so would not be at risk. For example: "Nuclear will adapt, will become safer, and will be coming to NZ. And we will never be building the type of reactor that has just gone into meltdown even if we were nuclear slap happy tomorrow," wrote ZeeBop.[14] Lanthanide wrote "modern nuclear reactors are much smaller, safer and cheaper than these old ones (built in the 60s) that Japan is using. Any reactors commissioned in NZ would obviously be of the new design, and therefore much less at risk of being dangerous."[15] Bazar continued the sentiment, writing that "Nuclear power is by far the greenest, cheapest power we have to mankind. NZ isn't a terrorist nation, so as long as don't import weapons grade uranium we'll be fine. Australia is both a great source of raw uranium and storage space, if they were willing to commit, we'd have most of the logistics cleared. NZ will have nuclear power, it's just a matter of time."[16]

While punters gave their opinions, and nuclear scientists communicated mostly through carefully worded press releases, earth scientists spoke to the media about the disaster. Although lacking in nuclear experts, New Zealand does have a great many scientists expert in earthquakes and tsunamis, and local scientific commentary focused on the earthquake and tsunami in Japan—which was of particular interest to New Zealanders because of our seismic similarity and the recent Christchurch earthquake.

### Public Opinion in 2013: In Search of a "Fukushima Effect"

In 2013, the stuff.co.nz website gave another opportunity for readers to give their opinions on nuclear power. This time, of the eight blog-style responses, only two were against nuclear power: five were actively pro-nuclear for New Zealand, and one was cautiously pro-nuclear, saying it should only be used as "a last-resort" (Duggan 2013). Pro-nuclear contributions included statements such as:

> The history of nuclear power worldwide doesn't seem appreciably more or less safe than other forms of power. Modern reactor designs do not have the same safety problems of Fukushima or Chernobyl.
>
> (Campbell 2013)

And

> For New Zealand, operating a couple of reactors instead of coal plants would create a truly green, stable and reliable energy system coupled with solar and wind.
>
> (Rens 2013)

Philip Ross, this time not identifying his association with the New Zealand Atomic Energy Advocacy Council (following the death of a key member, the organization is no longer active, he now says), also posted a response, in which he wrote that compared to a concrete hydro dam, a nuclear power plant was "relatively safe," that the waste problem was "relatively easily solved" and that coal-fired plants emit more radioactive material into the environment than nuclear power stations.

> Nuclear power could and should be part of New Zealand's mix of generation resources. Nuclear stations are good at providing stable

base load generation, ideal for load centres such as the Tiwai Point aluminium smelter or a major city such as Auckland.

(Ross 2013)

In response to concerns about New Zealand's seismicity, he pointed out that "the problems at Fukushima followed a tsunami that flooded the emergency core cooling system, shutting it down, rather than the earthquake."

These responses should not be considered representative of the population—New Zealand has a strong policy against nuclear power, so perhaps there is a greater incentive for those who want to change the status quo to speak out—but the level of support for nuclear power is still surprising.

In July 2013, an article in monthly current affairs and lifestyle magazine, *North & South*, argued that New Zealand's anti-nuclear stance was "an 80s throwback that hinders the sensible pursuit of our own interests. We need to . . . put nuclear power back on our national agenda" (Rapson 2013).

### Political Response: Nuclear Power "Never Going to be an Option"

Despite these statements in support of nuclear power, politicians have been steady in their opposition to nuclear power. In 2004, in response to De Bueger's pro-nuclear opinion piece in the *Otago Daily Times*, Green Party co-leader Jeanette Fitzsimons said that climate change was a serious threat, "and we do have to quickly cut coal and oil use. But nuclear power is not the way to do it, especially for New Zealand" (Fitzsimons 2004).

In 2007, Prime Minister and Labour leader Helen Clark pointed out that New Zealand was one of the richest countries on earth for renewable energy, and said she didn't see nuclear energy being acceptable to New Zealand.[17]

Even New Zealand's current Prime Minister, National leader John Key, has rejected arguments in favor of nuclear power. In a June 2014 interview, in response to US Secretary of State John Kerry's lauding of nuclear energy in his speech at the New Zealand Embassy in Washington, Key stated that, "We have anti nuclear legislation and New Zealanders wear it as a badge of honour. There ain't any time in the future of [New Zealand] that we're ever going to nuclear power, nuclear weapons . . . or nuclear anything; it's just not happening." He added that, given New Zealand's high level of seismicity, nuclear power was "never going to be an option."[18]

### Continued High-Profile Support for Nuclear Power

What of the high-profile New Zealanders who spoke out in favor of nuclear power in the decade before Fukushima? Did the Fukushima nuclear disaster affect their views? I approached Gareth Morgan, Andrew McEwan, Alan Poletti, Grahame Sydney, and Philip Ross—each of whom spoke out in favor of nuclear power pre-Fukushima—for comment. All replied that Fukushima has not changed their views on nuclear power.

"Confronted with global warming, the destruction of those four reactors at Fukushima with the consequent very low exposure to ionizing radiation of the general population in no way vitiates the case for the continued and increasing exploitation of nuclear power," said Poletti.

"No one died from radiation exposure as a result of the accident, but about 20,000 died from the earthquake and tsunami," added McEwan.

"I still support nuclear power. Fukushima was relatively minor alongside the earthquake and the tsunami. It need not have happened at all, as the core cooling systems could easily have been situated in a location where a tsunami would not be able to reach them and prevent them operating," said Ross.

They don't all endorse nuclear power for immediate consideration for New Zealand, but generally agree it should be considered as an option. "The possibility of nuclear power in New Zealand should not be forgotten and we in this country should seek to keep abreast of international developments," said Poletti.

Although he said that the Fukushima disaster had not changed his opinion towards nuclear power, in terms of New Zealand, "I don't expect it to be a necessary option for consideration in the foreseeable future," said Morgan, a comment in line with McEwan's statement that the introduction of nuclear power to New Zealand is "unlikely to happen in the foreseeable future."

Grahame Sydney added his opinion that "the Fukushima disaster was not a failure of the nuclear plant. It was a failure of that plant's original siting," and "I hold no fear at all for the safety of NZers if a nuclear plant were to be constructed within the next decade or two . . . but I doubt it will be necessary."[19]

## Discussion

In March 2011, in response to a media enquiry, I—to use Helga Nowotny's (2003) terminology—unwittingly transgressed beyond my specific area of competence, or expertise, which is New Zealand's nuclear and radiation *history*, and strayed into the field of radiation safety. My media-assigned status as an "expert" could be seen as an example of what Nowotny describes as the "inherent 'transgressiveness' of expertise," although I was uncomfortable with this inappropriate (and thankfully short-lived) status. Perhaps this sort of unwitting transgression is more frequent in areas of publicly controversial science. Atmospheric chemist Rhian Salmon (2013) reflects on how the very broad and generic title "climate scientist," which is frequently assigned to her by non-specialists, "inaccurately conveys my training and invites questions on topics on which I do not have expertise."

In the case of nuclear in New Zealand, one of the reasons the media is looking for "experts" who can speak broadly about nuclear and radiation issues is the lack of public scientists with appropriate expertise who are available to talk to the media.

Through New Zealand's history, many scientists have engaged with the media and the public on issues related to nuclear technology and radiation safety. But many of the scientists with expertise in radiation biology and nuclear physics, who monitored environmental radioactivity from nuclear testing and advised on health impacts of proposed nuclear power stations, have since retired. Today's nuclear science workforce focuses on other things, such as radiocarbon dating and medical and occupational radiation safety.

Furthermore, even these nuclear scientists are generally not permitted to talk directly to the media. New Zealand is distinctive in that half of its publicly funded scientists work in government-owned research institutes outside of the university sector (Gluckman 2015). These institutes, including GNS Science and ESR—the new home of the former National Radiation Laboratory, now known as the National Centre for Radiation Science—rely on contestable government funding and commercial contracts, and scientists are instructed not to talk to the media without permission. In response to the Fukushima disaster, rather than provide scientists to talk freely to the media about the radiation effects and, crucially, to answer questions, both GNS Science and the NRL provided written online media statements from which journalists could quote.[20] My success in interviewing an NRL scientist for my 2011 article on Chernobyl involved me sending a series of questions to their media relations manager. The answers provided by the scientist were then cleared before being emailed back to me. This process took nearly 48 hours, which, although acceptable for a magazine article with a long lead time, is not useful for newspaper journalists under pressure to file a story that day.

The role of public scientists in engaging with the public has been a recent topic of discussion and debate in New Zealand, with notable comment from Professor Shaun Hendy (Hendy 2014), and chemist and New Zealand Association of Scientists (NZAS) president Nicola Gaston (Gaston 2014, 2015), prompted in part by the Prime Minister's chief science advisor's call for a code of practice for New Zealand scientists on public engagement (Gluckman 2014). In response to this topical issue, the NZAS annual meeting in April 2015 had the theme *Going Public: Scientists Speaking Out on Difficult Issues*. According to a summary of the conference, "discussion ranged from empowering scientists to speak out, to ensuring scientists are trusted communicators, to the responsibilities scientists hold when they engage with the public."[21]

As for the public response to the Fukushima disaster, New Zealand is in line with other developed nations in its exceptionalism when it comes to the prospect of nuclear safety. As one of the Stuff.co.nz readers opined in his response to Stuff's "Should we stay nuclear free?" question, the Chernobyl and Fukushima nuclear disasters happened because of "poor planning and management." If New Zealand were to adopt nuclear power, he wrote, optimistically, "we would have to ensure that the systems used were world class, infallible, with every possible worst case scenario being taken into account" (Bree 2013). The pro-nuclear argument in New Zealand has always held that if and when we did embrace nuclear power, because we were so late to the game, we would have the advantage of being able to use the latest and best available technology.

## Conclusion

As of April 2015, the World Nuclear Association says that "Nuclear power remains an option for New Zealand, using relatively small units of 250–300 MW . . . Nuclear is a sustainable option, able to enhance the country's desired image. With minimal aesthetic impact, it would provide the power for Auckland's continued growth, including energy-intensive industry."[22]

Majority public opinion, political opinion, and economic realities are all currently against the introduction of nuclear power to New Zealand anytime in the foreseeable future. But the Fukushima disaster did little, if anything, to change pre-existing narratives about nuclear power in New Zealand: there is still a significant minority who think it would be a good idea.

When, at some time in the future, in response to a massively growing population or changing economics of nuclear power, a New Zealand government proposes to introduce nuclear power to New Zealand—perhaps a small, new generation reactor to provide baseload supply to Auckland, the country's largest and fastest growing city—we will need scientists who can speak out on the issue. Some might offer evidence why nuclear power is a safe, economical, and rational option for New Zealand. Others might provide evidence to show that nuclear power is not suitable for New Zealand—a tectonically active country with abundant sources of renewable energy and extensive scope for improving efficiency of energy transmission and use.

Given the size and nature of New Zealand's science communication ecosystem (Salmon and Priestley 2015), scientists—when they do speak out on an issue—can have a significant influence. New Zealand has a science-literate public which is generally interested in, and has positive attitudes towards, science and technology (Nielsen 2014). To enable public participation in any future decision-making about nuclear power—to enable our democracy to function—we will need public scientists willing to answer questions from the media and the public and take part in an informed dialogue on the topic. Nuclear power attracts controversy everywhere, but is likely to do so more in "nuclear-free" New Zealand than elsewhere. Any scientists participating in a future national conversation about nuclear power would need not only media training, but also a basic awareness of some key STS literature around controversial science. There are many models for the roles a scientist can take on public issues, such as, for example, Roger Pielke's *honest broker, issue advocate, pure scientist,* or *science arbiter* (Pielke 2007) and individual scientists and the organizations they work for would also need to give consideration to what role they would take in any public and political debate on nuclear power.

If and when New Zealand does have a debate on nuclear power, the "expert" opinions cannot be left to politicians, environmentalists, and business people, or even science historians. If the public is to make an informed choice, scientists must be allowed to engage with the public and contribute to the discussion and debate.

In the words of an anonymous respondent to the 2014 NZAS survey of New Zealand scientists:

> There is already too much fear and silencing of voices that seek to challenge government policy. I think that scientists and academics have an obligation to speak out on issues that inform public debate and challenge uninformed and ideological policy and legislation.[23]

## Notes

1 "Japanese Officials Say Nuclear Disasters Can Be Prevented," March 14, 2011, accessed April 24, 2015, http://tinyurl.com/4gwzluj.
2 "Your Say: Time for NZ to Consider Nuclear Energy?" September 6, 2007, accessed April 24, 2015, http://tinyurl.com/nq4ky6u.
3 Eric Frykberg (presenter), on RNZ "Insight," Sunday April 24, 2005. Transcription by Newztel News Agency Limited.
4 See, for example, "Experts on Japan's Nuclear Crisis and Analysis of the Earthquake," March 15, 2011, accessed April 24, 2015, http://tinyurl.com/q6ps54g.
5 Peter Griffin, personal communication, April 24, 2015.
6 National Isotope Centre, accessed May 4, 2015, http://tinyurl.com/oymz7fn.
7 John Callan, GNS Science communications manager, personal communication, May 4, 2015.
8 When I worked there in the early 1990s some taxi drivers would refer to it as "the bomb factory" and there were rumors there was a nuclear reactor on site.
9 "Fukushima Nuclear Reactor Incident No Danger to New Zealand," March 15, 2011, accessed April 25, 2015, http://tinyurl.com/pzw5lmj.
10 "Outside the Evacuation Zone: Health Effects of Low Dose Radiation," March 18, 2011, accessed April 25, 2015, http://tinyurl.com/oqs83vm.
11 "Fukushima Radiation: How Bad Will It Get?" March 16, 2011, accessed April 24, 2015, http://tinyurl.com/oc9xtme.

12  Ross Mason, accessed April 25, 2015, http://tinyurl.com/ps63lj8. Responses to this comment can be found on this and the subsequent web pages.
13  ROB, "Nuclear Free," March 14, 2011, accessed April 25, 2015, http://tiny url.com/4qc7dew.
14  ZeeBop, March 14, 2011, accessed April 25, 2015, http://tinyurl.com/ lq5mz6c.
15  Lanthanide, March 14, 2011, accessed April 25, 2015, http://tinyurl.com/ l9qt94c.
16  Bazar, March 14, 2011, accessed April 25, 2015, http://tinyurl.com/km8nkd4.
17  "PM Affirms Nuclear Power Stance," September 7, 2007, accessed April 25, 2015, http://tinyurl.com/k7qcynl.
18  Tracy Watkins, "Key Emphasizes NZ's Anti-nuclear Stance," June 20, 2014, accessed April 24, 2015, http://tinyurl.com/m7f2ern.
19  Interviews conducted by email in April and May 2015 under Victoria University of Wellington Human Ethics Number 21949.
20  "Advice on Exposure to Radiation Arising from the Nuclear Accident in Japan Provided by the National Radiation Laboratory," n.d., accessed May 15, 2015, http://tinyurl.com/pffvlr8.
21  "Going Public at the New Zealand Association of Scientists Annual Conference," April 16, 2015, accessed May 14, 2015, http://tinyurl.com/kqx5kfo.
22  "Nuclear Energy Prospects in New Zealand," updated June 2014, accessed May 9, 2015, http://tinyurl.com/pnb7stj.
23  "Survey on the Proposed Code of Public Engagement," November 1, 2014, accessed May 8, 2015, http://tinyurl.com/ppz62fj.

## References

Anderson, J. "Interest in Nuclear Power Rises as Costs Fall." *The National Business Review*, August 13, 2004: 8.
Bone, A. "Why Not Nuclear Power?" *North & South*, October 2004: 70–81.
Bree, D. "Learning from Nuke Mistakes," *Stuff Nation*, December 12, 2013, accessed April 24, 2015, http://tinyurl.com/kl8x6k3.
Campbell, M. "It's Time to Consider Nuclear Power," *Stuff Nation*, 2013, accessed April 24, 2015, http://tinyurl.com/ojqd7ky.
Chug, K. "NZ Safe from Japanese Radiation," *The Dominion*, March 16, 2011: A3.
Clark, L. "Scientists Ready for NZ Fallout," *The Dominion*, May 20, 1986: 1.
Davison, I. "NZ Safe from Radiation, Say Experts," *New Zealand Herald*, March 30, 2011: A5.
De Bueger, J. "Nuclear Power Opposition Misconceived," *Otago Daily Times*, July 23, 2004: 15.
Dominion Post. "Key Sends Usar Team to Search For Survivors," *The Dominion Post*, March 14, 2011: 3.
Donnell, H. "Radioactive Material Won't Reach NZ—Scientists," *New Zealand Herald*, March 23, 2011, accessed May 20, 2015, http://tinyurl.com/mt9tqut.
Duggan, A. "Nuclear Power a 'Last Resort'," *Stuff Nation*, 2013, accessed April 24, 2015, http://tinyurl.com/o5rdf6k.
Easton, P. "NZ 'Should Be Safe' from Nuclear Fallout," *The Dominion Post*, March 14, 2011: 2.
Fawkes, B. "More Kiwis Coming Round to Nuclear Power," *The Dominion Post*, April 7, 2008: 1.

Fitzsimons, J. "Many Reasons Nuclear Power Not for NZ," *Otago Daily Times*, July 30, 2004: 15.

Gaston, N. "Science and Democracy," *Speaker*, September 26, 2014, accessed May 14, 2015, http://publicaddress.net/speaker/science-and-democracy/.

Gaston, N. "Empowering Informed Voices," *Medium*, April 12, 2015, accessed May 18, 2015, http://tinyurl.com/ovql7yx.

Gluckman, P. "The Global Science System is Evolving Rapidly: Scientists Will Need to Adapt," *Stuff Nation*, March 3, 2014, accessed May 14, 2015, http://tinyurl.com/mg5q2at.

Gluckman, P. "Trusting the Scientist," April 10, 2015, accessed May 13, 2015, www.pmcsa.org.nz/blog/trusting-the-scientist/.

Gorman, P. "36% Indicate Support for Nuclear Power." *The Press*, May 27, 2008: A2.

Hendy, S. "How to Shut Down a Nuclear Reactor," *A Measure of Science*, March 16, 2011, accessed May 23, 2015, http://tinyurl.com/6f8jx3u.

Hendy, S. "Scientists Need to Hold Policy-Makers to Account," *A Measure of Science*, September 30, 2014, accessed May 14, 2015, http://tinyurl.com/lz4bgrd.

Holcroft, M. "The Cloud That Grew," *New Zealand Listener*, April 11, 1958: 10.

Locke, K. General debate March 16, 2011. New Zealand Parliamentary Debates, *670*, 2011.

Macdonald, N. "Are we Ready to Join the Nuclear Family?" *The Dominion Post*, March 29, 2008: E1.

McEwan, A. *Nuclear New Zealand: Sorting Fact from Fiction*. Christchurch: Hazard Press, 2004.

McEwan, A. "Nuclear New Zealand?" *New Zealand Geographic* August 2005: 12–14.

Ministry of Business, Innovation & Employment. *Energy in New Zealand*. Wellington, New Zealand, 2014.

Morgan, G. "Nuclear the Clean Answer." *The Press*, April 12, 2003: C3.

New Zealand Nuclear Free Zone, Disarmament, and Arms Control Act (1987).

Nielsen. *Report on Public Attitudes Towards Science and Technology*. New Zealand: Ministry of Business, Innovation and Employment, 2014.

Nowotny, H. "Democratising Expertise and Socially Robust Knowledge," *Science and Public Policy* 30 (200): 151–156.

NZPA. "Damaged Reactor Unstable," *The Dominion*, April 30, 1986: 7.

Pielke, R. *The Honest Broker: Making Sense of Science in Policy and Politics*. Cambridge: Cambridge University Press, 2007.

Priestley, R. "Crisis," *New Zealand Listener*, March 26, 2011: 14–17.

Priestley, R. *Mad on Radium: New Zealand in the Atomic Age*. Auckland: Auckland University Press, 2012.

Rapson, B. "Thorium on our Breath," *North & South*, July, 2013: 66–67.

Rens, J. "Nuclear Tech 'Worth Considering'," *Stuff Nation*, November 21, 2013, accessed May 20, 2015, http://tinyurl.com/kejbook.

Richardson, B. "The Nuclear Decision: A Debate about New Zealand's Future: Part 1," *New Zealand Listener*, November 22, 1975: 14–16.

Ross, P. "Time to Abandon Nuclear-free," *Stuff Nation*, November 25, 2013, accessed May 20, 2015, http://tinyurl.com/q9b2ce4.

Royal Commission of Inquiry into Nuclear Power Generation in New Zealand. *Nuclear Power Generation in New Zealand: Report of the Royal Commission of Inquiry*. Wellington: Royal Commission of Inquiry into Nuclear Power Generation in New Zealand, 1978.

Salmon, R.A. "Is Climate Science Gendered? A Reflection by a Female 'Climate Scientist'," *Women's Studies Journal* 27(1), 2013: 49–55.

Salmon, R., and R. Priestley. "A Future for Public Engagement with Science in New Zealand," *Journal of the Royal Society of New Zealand* 45, no. 2 (2015): 1–7.

Shelton, A. "Debating the Atom," *E.nz Magazine*, August 2009: 22–23.

Submissions. *Royal Commission on Nuclear Power Generation in New Zealand: Submissions, Vol.1–4*. Wellington: Royal Commission on Nuclear Power Generation in New Zealand, 1978.

Sydney, G. "NZ on Wrong Track with Energy Strategy," *Otago Daily Times*, June 5, 2008: 15.

Taylor, K. "Business Backs Look at N-power," *New Zealand Herald*, September 6, 2005: A4.

# 14 The Effect of the Fukushima Effect

## From Strong to Weak

*Richard Hindmarsh and Rebecca Priestley*

> The peoples of the earth have . . . entered in varying degrees into a universal community, and it has developed to the point where a violation of rights in *one* part of the world is felt *everywhere*.
>
> (Kant cited in Reiss 1991: 106–107)

In *Nuclear Disaster at Fukushima Daiichi: Social, Political and Environmental Issues* (Hindmarsh 2013), Richard Hindmarsh identified Fukushima as a "new type of major nuclear disaster marking the close interactivity of the social, technological and natural disaster" (Hindmarsh 2013: 217). Contributors to this follow-up volume, *The Fukushima Effect: A New Geopolitical Terrain*, have investigated the international effect of the Fukushima disaster three to four years out, when the longer-term effects are becoming more apparent. In doing so, they have identified varied downstream effects, some significant, on national histories, debates. and policy responses in the Asia Pacific region, Europe, and the US. A new geopolitical post-Fukushima terrain is apparent even in countries without a strong Fukushima effect on nuclear power development. According to the thrust of this book, contributors investigated the Fukushima effect in regard to the safety of nuclear energy, radiation risk, nuclear waste management, anti-nuclear protest movements, and civic impacts on nuclear energy, future energy mixes, nuclear power representations by a range of stakeholders, the role of media frames, and the implications of the revealed effects for nuclear power development.

As observed by Susan Molyneux-Hodgson and Marika Hietala in Chapter 8 of this volume, past nuclear disasters have also had variable impacts in different countries. Now it is the turn of Fukushima, which many argue has had an even more sweeping effect than Chernobyl. We also reflect on the view of Joskow and Parsons (2012: 13) that "the accident at Fukushima [would] not 'kill' the much discussed renaissance of nuclear power, but it adds one more negative pressure on the rate of growth globally." So what does this volume add to these views? While it is difficult to predict the future, and perhaps even to assess the current global state of play with regard to a Fukushima effect, our in-depth investigation has yielded insights and better understandings of the *why* and *how* of the Fukushima effect and its variability in different nuclear nations.

This volume also contributes authoritatively to a progressive longitudinal benchmark of the effects, issues, and implications of the Fukushima disaster. In this regard, *The Fukushima Effect* is positioned, like its predecessor, at the forefront of what will undoubtedly be a long interrogation of Fukushima, just as Three Mile Island and Chernobyl have long been interrogated and will continue to be so, despite the attempt of the global nuclear industry to overcome the "Chernobyl syndrome" (Schneider et al. 2011: 7–8). Now, there is the more immediate "Fukushima syndrome."

Of note is the significant impact upon the group of top nuclear nations, with Germany at the so-called forefront of change in relation to this new geopolitical terrain pressuring nuclear energy, as Detlef Jahn and Sebastian Stephan outline in Chapter 9 of this volume. Germany's notable decision to phase out nuclear power has been mirrored in the nearby but less-developed nuclear nations of Switzerland and Belgium. Perhaps most remarkable, but also quite understandable, is the freeze on Japan's nuclear industry and major setbacks to that country's planned future of nuclear energy development. Civic resistance is steadfast and it now seems that only a few reactors might survive a strengthened regulatory appraisal of their fitness to operate.

At the same time, Japan's neighbors have had variable reactions. Taiwan—also located in the seismically volatile landscape of the Pacific Ring of Fire (Connor 2011)—decided to shut down the construction of its Fourth Nuclear Power Plant. Perhaps this is not surprising, given that Japanese and Taiwanese reactors are located in areas of high seismic hazard (Cochran and McKinzie 2012; Tamman et al. 2011). In contrast, nearby Korea and China, and India, continue to expand their nuclear fleets but, post-3/11, with more caution and apprehension about nuclear safety and civic unrest, which is now more actively accompanying such expansion.

Returning to Europe, the Fukushima effect is also patchy, with the UK shrugging off the German reaction to continue its rebuild and France planning to reduce its current nuclear commitments by 25%, and decommissioning some of its older reactors, as part of a notable safety effect. Finland is also notable in its exceptionalist perspectives and politics to the nuclear safety issues raised by Fukushima, as are the countries of the former Soviet Union, which are also maneuvering around safety issues but facing broadening civic contestations.

Meanwhile, in the US, although the Fukushima effect also appears minimal, lurking behind the regulatory scene of both nuclear plant and waste operations are doubts over the claims that "it could never happen here," as noted by William Kinsella (Chapter 12), which is leading to a more critical conversation about the hazards of nuclear energy and how to address them. Such variability of national response appears to align somewhat with the notion of two broad policy pathways adopted following Fukushima: "(1) amplification of risk and withdrawal of policy support, and (2) safety review, then attenuation (reduction) of risk, followed by continued support" (Butler et al. 2013: 137).

Last, but not least, is the case of the non-nuclear nation New Zealand, investigated in Chapter 13, where a critical conversation about nuclear energy also emerged post 3/11. While it was somewhat typical of the global conversation—with many other non-nuclear nations, such as Austria and Italy, cementing their non-nuclear status post-Fukushima (Felt 2013)—it was evident that a significant minority still thought nuclear power might be a good idea for New Zealand. Interestingly, this state of affairs is similar to neighboring Australia (e.g. Bird et al. 2014; Falk et al. 2006), which, while it also has no nuclear power, does have a significant uranium mining industry, thereby fueling the global nuclear industry.

That said, Australia also posted a Fukushima effect, with the results of a nationwide survey before and after Fukushima revealing that "fewer Australians in 2012 were willing to accept the building of nuclear power stations, even if this would help tackle climate change. Although they believed nuclear power offered a cleaner, more efficient option than coal, an increased proportion believed that the risks of nuclear power outweighed the benefits" (Bird et al. 2014: 651). These authors concluded that "despite the shortcomings of the surveys . . . more Australians oppose nuclear power (2010: 31.7%; 2012: 41.4%) than support it (2010: 29.0%; 2012: 24.4%) (Bird et al. 2014: 652).

Notably, before Fukushima, the study advanced that Australian attitudes towards nuclear power had softened "in the absence of recent visible accidents . . ." and that some respondents had "viewed climate change as a greater risk than nuclear power" (Bird et al. 2014: 651). Support for nuclear power has clearly been challenged by Fukushima.

Also interesting was that a 2010 comparative Australian-UK survey found that "Australians were less likely to accept nuclear power as an option for mitigating the effects of climate change (34.4% in Australia versus 56.4% in the UK)." In response, Bird et al. (2014: 651) commented that the difference between Australia and the UK may have been "due to the UK already having an established nuclear industry and an influential pro-nuclear lobby who reframed the argument to obtain more social acceptance of nuclear power . . ." (also Blowers 2013: 183–184).

Informed by this interesting exposure of UK nuclear politics and policy in action, we now turn to what the contributors in this volume reveal about the international nuclear politics and policy in action that contributed to, or socially shaped, the varying international Fukushima effect worldwide. Across the Asia Pacific region, Europe, and the US, the Fukushima effects identified by contributors fit into three meta-themes: (i) media responses, public opinion, and social mobilization; (ii) nuclear power governance, policy, and development; and (iii) questioning nuclear renaissance. In the rest of this concluding chapter we identify the effects of Fukushima under these three themes that have emerged across multiple chapters. We follow this with a discussion and identification of implications for nuclear power development and future research.

## *Media Responses, Public Opinion, and Social Mobilization*

Although "the 1986 Chernobyl disaster marked the nadir of public support for nuclear power,"[1] Fukushima also had significant impact, with public acceptance of nuclear power dropping worldwide from 57% to 49% in the disaster's aftermath (Hayashi and Hughes 2013: 106). In the days and weeks after the disaster, international news media reported on the Fukushima Daiichi core meltdowns, the explosions, the citizen evacuations, and fears of exposed spent fuel rods reaching criticality and widespread radiation contamination. In contrast to the Chernobyl disaster, which saw limited media coverage, news of the Fukushima disaster was not only relayed by traditional media of TV, radio, and newspapers, but was also disseminated and discussed globally over the Internet. In some countries, such as China and India, social media was a powerful source of commentary, as noted by Xiang Fang and Anupam Jha in Chapters 5 and 6, respectively. As Friedman (2012: 56) commented,

> these sources yielded a deluge of Fukushima information, and they have changed the definition of mass media in many ways . . . Hundreds of Twitter conversations appeared under a variety of hashtags—such as #fukushima, #nuclear, and #meltdown—with people keeping up to date on events and where to find articles to read or videos to watch. Anyone who wanted a timeline for Fukushima events could also turn to Wikipedia, which compiled a day-by-day account, including radiation readings.

Indeed, as Friedman also related, some indication of the post-disaster interest of the global community was demonstrated by 73,700,000 Google hits in little more than four months for the search term "Fukushima" and 22,400,000 hits for the search terms "Fukushima and radiation" (Friedman 2012: 55). In Japan itself, in the absence of official radiation data in the immediate aftermath of the disaster, new and social media catalyzed citizen radiation monitoring by way of DIY Geiger counters and civic radiation maps (Kera et al. 2013; Morita et al. 2013), with Twitter also playing a large role in social communication about disaster information and how to personally manage the disaster (Kaigo 2012). Internationally, concerned citizens played an active role in pressuring governments to respond, particularly through protests and mass street demonstrations, bolstered by social media.

In Taiwan, the Fukushima disaster led to public fears about radiation pollution from Japan, nuclear safety in Taiwan's three nuclear power plants, and the possibility of a large earthquake causing a disaster in any of the three NPPs in close proximity to very high population centers, including the capital, Taipei. An empowered anti-nuclear movement organized mass demonstrations and other forms of social mobilization like the activities of the Sunflower Movement, which led to the cancellation of plans for a Fourth Nuclear Power Plant, with prospects of a likely future decline in nuclear power in Taiwan.

Korea also saw street protests. Many citizens were concerned about the impact of a possible local nuclear disaster, especially given that four million people lived in the equivalent of a Fukushima exclusion zone—a 30km radius—around the Kori-1 nuclear power plant. By July 2012, a national survey found that 67% of respondents favored shutting down Kori-1, and civic groups protested against the renewal of a license for another power station, Wolsong-1, as Hyomin Kim points out in Chapter 4. However, the Korean government is not committed to open dialogue on these topics, given its priority focus on the progress of nuclear power and its portrayed role in the "economic development and the well-being of the nation."

In China, reports Xiang Fang in Chapter 5, news about the disaster and the possibility of radioactive dust reaching China spread quickly through newspapers, TV news, and on Internet media platforms and social media, which helped inform people's risk awareness of nuclear power projects. As a result, citizens became more active in participating in anti-nuclear events. Particularly through social media and NGOs, they pressured for enhanced safety of nuclear power plants, and reduction of risks and hazards through more careful plant siting that takes note of local citizen concerns.

In India, in response to news reports and social media commentary about the Fukushima disaster, "a huge debate emerged about the pros and cons of nuclear power," reports Anupam Jha in Chapter 6. Protesters demanded the denuclearization of energy, and immediate safety audits of nuclear power plants, especially in respect to the threat of tsunamis and earthquakes occurring in the Indian Ocean. As time passed, public protest and other forms of social mobilization focused on the safety of nuclear power plants in relation to their surrounding natural habitat, operators' land acquisition policies, safety evaluation processes associated with siting nuclear power plants, and the willingness of the government and nuclear power operators to pay adequate compensation in case of nuclear disaster. In Germany, the Fukushima disaster catalyzed a strong anti-nuclear sentiment, galvanized Germany's strong risk-averse culture, and, subsequently, fast-tracked the country's move to phase out nuclear power, a decision supported by nearly 80% of the population. In Switzerland, a similar phase-out decision was supported by 87% of the population.

Overall, citizen awareness of, interest in, and concern about nuclear issues, have notably increased post-Fukushima, along with social mobilization and anti-nuclear movements. Even in countries without significant change, the effect on public opinion is quite evident, as numerous opinion polls conducted in many countries attest to. In the US, for example, as William Kinsella reports in Chapter 12, the Fukushima disaster has led to "renewed public awareness regarding nuclear risks, and re-energized opposition to nuclear power." In response to the draining of water protecting spent fuel rods at Fukushima, public interest groups have lobbied for better management of the US nuclear waste, but so far this has made little impact on current practice.

In contrast, in France—one of the largest nuclear nations, with 58 reactors, surveys taken in June 2011 "found 62% of French respondents supporting a gradual phasing out of nuclear power—over 20 to 30 years—and 15% calling for a rapid halt. Only 22% of people favored continued development of France's nuclear sector (Crumley 2012). Subsequently, part of the political platform of François Hollande's successful presidential campaign in 2012 was his endorsement to reduce the share of nuclear power in electricity generation from 75% to 50% in favor of renewable energy sources.[2]

In other countries, though, Fukushima seems to have had limited effect. For example, in their investigation of nuclear waste disposal issues in the UK and Finland (Chapter 8), Susan Molyneux-Hodgson and Marika Hietala found "that Fukushima did little to change the trajectories or scope of disposal debates." The impact of the disaster on disposal debates, they said, was practically invisible in the local news media around the anticipated repository sites. Significantly, "a disaster on the other side of the world proved insufficient to override existing narratives that had dominated the siting process for some years, for example, local narratives of job creation (in support of the site) and of landscape protection (against the site)."

Similarly, in New Zealand, which has no nuclear power, Rebecca Priestley's investigation of online and media commentary about nuclear power (Chapter 13) shows that the Fukushima disaster did little, if anything, to change pre-existing narratives about nuclear power: there remains a significant majority opposed to nuclear power, and a robust minority in favor.

## Nuclear Power Governance, Policy, and Development

In the areas of nuclear governance, policy and development, in our selected countries of analysis, we have identified what we refer to as a "spectrum of Fukushima effects from strong to weak" (as discussed below).

Nevertheless, over time, the strength of effects may change as the Fukushima effect continues to unfold, such as in regard to high investment costs in building (with associated insurance and liability) and decommissioning nuclear power plants and the long time spans involved, which may increasingly effect political and social acceptability. Of note is that the cost for the cleanup and decommissioning work at Chernobyl, which continues today, is now estimated at €2.15 billion (NucNet 2015a).

Accompanying the continuation and expansion of nuclear energy are also, of course, core "dread" issues which post-Fukushima have become revitalized: radioactive pollution, and by association, the long-lasting issues of unresolved nuclear waste management and lack of secure repositories (as pointed out in Chapters 8 and 12, respectively), as well as concerns about nuclear weapon proliferation. Subsequently, alongside high adverse public opinion about nuclear power, particularly on safety, there is favorable public opinion about renewable energy and energy conservation, opinions that are likely to strengthen as these energy industries grow.

We turn now to the construction of our Fukushima effect spectrum on nuclear power governance, policy, and development (see Figure 14.1), which we have devised as a conceptual device to illustrate the Fukushima effect in the countries we have investigated. It is also informed by other relevant literature, and includes Sweden and Belgium, which are included in the analysis of Germany (Chapter 9), but excludes New Zealand, which is a non-nuclear nation. The order of the countries may be debated, but seems to us the best fit at this time. In addition, the placements of the countries from strong to weak effect is not yet qualified by a long-lasting effect as the politics and policy in action of nuclear energy is characterized—as evidenced in the contributors' analyses—as a discursive (or dynamic changing) policy in action terrain. The contested nuclear power terrain features a cast of nuclear processes and power plants, institutions, sociotechnological networks, and the main policy actors of the administration (or governmentality, see below) of nuclear power development, and of those who resist or critically question nuclear power development.

Foucault (1990: 95) positions "governmentality" as involving complex processes of deploying political technologies or techniques to affect the

*Figure 14.1* The Fukushima Effect Spectrum on Nuclear Power Governance, Policy, and Development from Weak to Strong Effect (of Featured Countries), 2015

subservience of citizens to the state, which is achieved, or attempted, through state institutions in collaboration with allied sections of society, here, to normalize the nuclear energy endeavor for electricity production. One apparent example of this is the case of Sweden, as suggested in the account of Detlef Jahn and Stephen Sebastian in Chapter 9. These contributors outline that the difference between Germany and Sweden lay in the historical insufficiency of the anti-nuclear movement in Sweden to construct the sort of political opportunity window that Germany's anti-nuclear movement was able to build, and which finally led to Germany's transformed energy mix post-Fukushima.

From its beginning, these authors argue that the Swedish movement was weakened, as it "was subsumed by the political system to some extent, furthered by the anti-nuclear movement's failure to recruit the strong labor movement as an ally." Such incorporation of environmental concerns at the early stage of the nuclear power debate in Sweden can also be seen as deliberately crafted "absorption of protest." The major aim of this strategy of power is to maintain the stability of ruling elite power, not to eliminate or alter protest movements but rather to destabilize them and dissipate their threat (Etzioni-Halevy 1990: 215). A key maneuver is cooptation—the absorption of protest movements and/or their elites into the existing formal power structures, or policies entailing acceptance of some of the movements' demands (ibid: 215–216). This diffuses conflict (see also Selznick 1949), to reduce uncertainty as well as to moderate potentially disruptive radical elements.

In the nuclear power regime of governmentality revealed in this volume are state agencies, markets, research institutes, nuclear energy developers, and financial institutions. Other important actors are the media, non-governmental or civic organizations, and publics, both pro-nuclear and contesting ones. In this context, publics are more traditionally recognized as a pressuring influence driven by a desire for reformative change or by resistance to nuclear power development for any number of reasons.

It is clear from the contributors' analyses across a range of nuclear nations that the struggle over nuclear power development involves a range of discursive practices, as the example of Sweden above indicates, but also includes "practices of talk, text, writing, cognition, argumentation, and representation generally" (Clegg 1989: 151). Alternatively, Andrée (2002) refers to discourse as "a system of interwoven truth claims embedded in social relations and material practices." Here, "truth claims" can also be seen as "narratives," such as policy narratives or stories (and events), that provide policy meaning and orientation (see also Hindmarsh 2005).

Often, discourses—for example, those of "nuclear exceptionalism" in the case of the UK and Finland (Chapter 8) and post-Soviet regimes (Chapter 7), or reframed nuclear safety discourses involving revised "engineering logics" and the "core-catcher" (Chapter 7)—when considered powerful enough to persuade, are presented by coalitions of policy actors who have developed and/or coalesced around them (Barnes 1986: 58). In the post-Soviet case, the policy actors are nuclear policy-makers and engineers. Typically, such policy elites and experts act as brokers and relays of these discourses, translating and mobilizing their meanings (Hajer 1995: 58–59), sometimes opportunistically, to form the grounds for policy-making. In doing so in contested new geopolitical terrains, nuclear power interests are typically empowered through their location within long established technocratic systems and sites of government (for example, Fischer 1990), as Andrei Stsiapanau well relates in Chapter 7.

In addition, Hyomin Kim relates in Chapter 4 the strategy of the Korean government to "separate" in political and social space the local and general publics to best develop nuclear power. The aim is to enroll general publics, who live far from the nuclear power plants, to socially accept nuclear energy for the greater good of society. This was and still is, of course, at the expense of citizens living close to nuclear power plants, which (for this strategy to work) are conveniently located at distant, local coastal places largely invisible to the general public residing in centralized and large population centers such as Seoul. Typically, resisters in the local publics, who can become quite active in their opposition to either nuclear power plant or waste dump siting, as Kim narrates, are implicitly referred to, and negatively stereotyped as, NIMBYists. NIMBYists, as Kim infers, are most often depicted by developers and pro-development agencies of controversial facility siting as "illegitimate and anti-public interest 'NIMBYists' in standing for their own interests rather than the public interest" (e.g. Hindmarsh 2010: 555). Furthermore, they are often framed as being opposed to what the government portrays as "essential development" (McClymont and O'Hare 2008; Wolsink and Devilee 2009), regardless of well-qualified land use contestation issues.

Given these technopolitical contexts revealed in *The Fukushima Effect*, we can at this stage position Japan first at the strong end of our Fukushima effect spectrum on nuclear power governance, policy and development, as shown in Figure 14.1. This is, first, because Japan experienced a complete shutdown of nuclear energy, and will quite possibly in the near future have its nuclear energy fleet decimated by new safety regulations—if, of course, they are upheld rigorously by the incumbent government—such that a baseload for energy production, that is, sufficient power to satisfy minimum electricity demand, may not be achieved.

A key factor driving this effect is the steadfast public opposition by a majority of Japanese citizens to nuclear power, itself a keen effect of the disaster reflected by the sharp decline of public trust in government following the event of the disaster itself, and the debacle in the evacuation and emergency responses immediately following the disaster (as referred to by Akira Nakamura and Wataru Nishimura in Chapter 2). These are major Fukushima effects in Japan, as is the plight of evacuees, the cost of their compensation, and the long-lasting, tragic reminder of them and the disaster in the form of the wandering town of Futaba Machi. In this case, local government and residents sought safe refuge in regions away from the epicenter of the disaster, finally attempting to establish an independent political entity within the boundaries of a neighboring city government. Nakamura and Nishimura remark that "if the attempt succeeds, the outcome would represent another important Fukushima effect, one that Japanese officials and academics should consider as a possible option for a new role and function for local units of government in the aftermath of future natural and/or human made disasters."

This is a good point to highlight, with the Sendai-1 reactor (in Japan's southwest Kagoshima Prefecture) August 20, 2015, restart. This first reactor restart of the 48 reactors shut down following Fukushima had to survive a legal bid to prevent its restart (NucNet 2015b). Notably, the Japanese government was pressured during 2014 by the Keidanren or Japan Business Federation—which represents more than 1300 corporations, 112 industrial associations, and 47 regional economic organizations—"to accelerate rehabilitation efforts around the Fukushima-Daiichi nuclear station and to secure a stable energy supply by restarting nuclear reactors" (NucNet 2014). On October 15, Kyushu Electric Power Company restarted its Sendai-2 reactor.

Next, we position Germany as the standout nation in quickly deciding on a phase-out of nuclear energy by 2022, followed then by Switzerland and Belgium, with their own phase-outs. Taiwan is the final country in this strong Fukushima effect grouping, though it is in a considerably weaker position than the three countries above it (to the right of the spectrum in Figure 14.1). Although construction of the Fourth Nuclear Power Plant was abandoned by the Taiwanese Government, three NPPs remain in a highly seismic landscape in close proximity to major cities and population centers. However, Fukushima has strengthened the determination of Taiwan's anti-nuclear movement to continue contesting the terrain in its bid to phase out nuclear energy completely.

Positioned after Taiwan is France, where, in October, 2014, a moderate to strong Fukushima effect was also evidenced as the lower house of France's parliament voted in favor of reducing nuclear energy from 75% to 50% by 2025 and the closure of France's oldest nuclear power plant, Fessenheim, by the end of 2016 (World Nuclear News 2014). The reduction followed Hollande's promise in the 2012 presidential election, concerns raised and direct actions carried out by environmental groups about the safety of the country's nuclear plants (e.g. McKeating 2014), and 62% of the public favoring a gradual phase-out of nuclear energy in the aftermath of Fukushima (Crumley 2012), as mentioned above. The Deputies also voted to reduce 2012 energy consumption levels by half by 2050.[3]

Next, in perhaps reflecting a more moderate Fukushima effect on nuclear governance and policy, is India. With increased public pressure in the form of lobbying, protests, and other forms of social mobilization, policy advances were made on questions of liability in the event of nuclear disaster, and overall safety issues of nuclear power plants. At the same time, there is now increased institutional acceptance of social concerns related to land acquisition for NPPs, and the independence of regulatory bodies on nuclear safety. However, little advances have been made in a number of other areas, including public participation in environmental impact assessment and emergency preparedness in relation to NPPs, and transparency issues around government policy and decision-making, especially in regard to nuclear safety. Legislation, it appears, is being deployed as a political technology to "shield relevant information from the public," as Anupam Jha articulates in Chapter 8. However, the tempo of social concerns being mobilized about these issues has substantially risen since Fukushima, with continuing pressure for further reforms.

Moving more to the weaker end of the spectrum, we find countries of the former Soviet Union affected by Fukushima in relation to the evolution or revision of the engineering logic about the safety of nuclear plants, as Andrei Stsiapanau comments in Chapter 7, "to make the consequences of the nuclear disaster technically containable" through the core-catcher, but also "politically convertible to mitigate post-accident nuclear technopolitics," such as public pressure over safety and other social uncertainties, which, however, appears a common feature across nuclear nations, especially those experiencing quite a weak Fukushima effect (to the left of Taiwan in Figure 14.1).

China also evidenced a weak Fukushima effect, one that appears to bring in the social infrastructure of development to presumably better plan the siting of nuclear power and reduce social and environmental risks and hazards. This shift to apparent risk governance has been in response to rising public awareness of and pressure over safety and siting issues of NPPs since Fukushima, alongside China's aspirations to develop more a market-style approach to civil nuclear power development. However, stronger shifts to participatory risk governance for nuclear energy appear tenuous at this point.

Notably, while China is building more nuclear reactors than any other country, it is also fast developing renewable energy. According to the Earth Policy Institute, it "continued to get more electricity from the wind than from nuclear power plants in 2014. This came despite below-average wind speeds for the year. The electricity generated by China's wind farms in 2014–16 percent more than the year before—could power more than 110 million Chinese homes" (Roney 2015). In this energy mix, China thus seems to be aligning with the UK.

Likewise. Korea appears to be embracing citizen concerns about nuclear safety, but only superficially, amidst, as Hyomin Kim observes in Chapter 4, "increasing social demands for transparency and civic engagement in science and technology decision-making processes."

Korea seems more concerned with containing than addressing these concerns. Along with Korea, we have positioned Finland, Sweden, the UK, and the US as evidencing a quite minimal Fukushima effect on policy change, although with public interest concerns remaining and being bolstered (post-Fukushima) and expressed on many aspects of nuclear power development; as William Kinsella, for example, outlines in Chapter 12, from safety to regulation "and procedures for used fuel storage, as well as the prospects for establishing permanent used fuel repositories and maintaining the facilities and institutional structures necessary for safe interim storage." Such positions appear well-informed by notions of exceptionalism, which also inform post-Soviet countries' policy positions, of "it could never happen here" because of the differences in geophysical features and safety technologies supposedly superior to those of Japan. Notably, Japan resorted to the same exceptionalism narratives following Three Mile Island and Chernobyl (Hara 2013: 27).

## Questioning Nuclear Renaissance

The World Nuclear Association (WNA), an organization devoted to promoting nuclear power, claims that while Fukushima "set back public perception of nuclear safety"[4] the nuclear renaissance continues, with "nuclear power under serious consideration in over 45 countries which do not currently have it."[5] Most of the increased capacity in nuclear power, though, will come from existing nuclear nations, as predicted by Joskow and Parsons (2012), with China embarking upon a huge increase in nuclear capacity to 58 GWe by 2020 and India's attempting to add 20 to 30 new reactors by 2030. This is all driven, says the WNA, by increasing energy demand, concerns over climate change, and dependence on overseas supplies of fossil fuels.[6]

But the contributors to this book tell a different, more nuanced story. While some countries have continued their nuclear programs, seemingly unaffected by Fukushima, other countries have shut down, or intend to phase out either completely or partially, their nuclear power stations. And, as shown in the sections above, even in those countries that have chosen to continue with nuclear power, Fukushima has impacted on nuclear governance and public opinion to the extent that there are quite likely to be further impacts on nuclear power continuation and development.

In Japan, the epicenter of the disaster, the accident was followed by the shutdown or suspension of operation of all of Japan's remaining 48 nuclear reactors, leading to a new dependence on expensive imported fossil fuels, with associated impact on Japan's economy and carbon emissions. In Germany, Fukushima cemented national anti-nuclear sentiment and brought forward the decision to phase out nuclear power. Less than four months after Fukushima, eight nuclear reactors lost their operating licenses, and the remaining nine reactors are scheduled to be shut down by 2022. This decision was seen as part of the *Energiewende*—an energy transition from nuclear and fossil-fuel to renewables that has dominated Germany's nuclear and energy politics for decades.

In Switzerland, in a move strongly supported by public opinion, only six weeks after the disaster, the Swiss Federal Council—the executive government authority—announced its decision to phase out nuclear power by 2035. Existing power plants would operate until the end of their life expectancy and no new power plants would be constructed. With 41% of Switzerland's energy produced by nuclear power in 2011, the decision to phase out this energy source was accompanied by a campaign to develop hydropower and new renewable energy, to improve energy efficiency of buildings, appliances and transport, and to reduce energy consumption. A little while later, on October 30, 2011, Belgium announced its decision to phase out nuclear reactors after 40 years of operation, also a consolidation of pre-Fukushima plans.

In Taiwan, which has three nuclear power plants, a fourth nuclear power plant development has closed, and there is now continuing public pressure to shut down Taiwan's remaining three power plants. Other countries, such as China, India, France, and the post-Soviet countries, have continued their nuclear programs, but there is more to the story. In China, despite the WNA projection of a huge increase in nuclear capacity, Xiang Fang reports in Chapter 5 that "growing environmental awareness and negative perceptions of nuclear risk appear to be steadily growing and putting pressure on China's rapid expansion of nuclear power." In India, says Anupam Jha in Chapter 6, Fukushima's lasting effect is to continue informing and broadening "civic contestations around controversial nuclear power plant siting at the very time when government [is] pushing forward with a number of new nuclear power plant projects."

Reinforcing the findings of Jha is another recent study on India which found that "spontaneous resistance to nuclear energy has woken up the elitist establishment to new realities and ensured that the nuclear affairs can no longer be run behind closed doors. . . ." (Kumar 2014: 57). On the notion of a nuclear renaissance for India, Kumar argues that nuclear expansions plans face the prospect of being derailed in the wake of Fukushima and subsequent public anxieties about any expansion, such that among nuclear interests a "sense of pessimism" prevails (Kumar 2014: 57).

In Chapter 10, Joseph Szarka acknowledges that the landscape of the French nuclear sector has been changed by Fukushima, but whether it has been "transformed" remains to be seen. Szarka maintains that "a question mark hangs over the nuclear revival: has it been terminated or merely postponed?" He adds: "The French-led revival, at least, has clearly stalled . . . for reasons of the limited scope for market expansion, organizational disharmony within the French 'team,' and a single product (the first third generation nuclear reactor worldwide: the European

Pressurized Reactor) whose size, complexity, and cost produced a misfit in many export contexts." And, of course, there is France's decision to reduce nuclear generation of electricity by 25% to 50% due to Hollande's 2012 presidential election promise, and pressure from post-Fukushima public opinion and environmental groups. All together, these factors now appear quite convincingly to challenge former ideas of a nuclear renaissance.

Overall, on the notion of a nuclear renaissance (or rebuild) pre-Fukushima, that is, of continuing notable nuclear expansion especially informed by contexts of climate change and energy security, we find that this notion has arguably been more than just dampened (e.g. Joskow and Parsons 2012; Varrall 2012); instead, it appears shaken, perhaps immeasurably. Fukushima will certainly not be forgotten, and arguably the seeds of doubt the disaster sowed will continue to grow and increase opposition to nuclear expansion, aided by increasingly viable trends toward renewable energy and energy efficiency and conservation.

## Implications for Future Policy and Research and Concluding Remarks

A key aim of this book is to facilitate wide dialogue to better understand and respond to the Fukushima Daiichi disaster and its effects. We hope it will now be used for useful and enabling knowledge production, good governance, and social and policy learning, in Japan and globally, and to promote and develop long-term social and environmental well-being and sustainability transformations in relation to energy choices, safety, and futures.

At the time of the publication of this book, we are close to the fifth anniversary of 3/11. While many Fukushima effects are apparent, they promise to change more as the new geopolitical terrain Fukushima generated unfolds further political, temporal, and spatial effects. Questions about this unfolding effect may include: will Japan attempt an ambitious rebuild or turn more to alternative energy sources? In Germany, will the *Energiewende* fulfill its ambitions as a model for an industrial society that has taken a lead in shifting to renewable energy? Will "increasing energy demand, concerns over climate change and dependence on overseas supplies of fossil fuels" lead China to the predicted increase in nuclear capacity of 58 GWe by 2020, or will renewable energy prove more attractive in the longer-term? Likewise, will India add 20 to 30 new reactors by 2030 or find its planned nuclear renaissance too difficult to proceed with? Will the World Nuclear Association's lofty projection of 1100 to 3500 GWe of nuclear energy generation by 2060 (compared with 373 GWe today) be realized, or is this just false hope?[7]

Despite the status of current plans, frames, and projections, the nuclear industry, and its regulators, will continue to be pressured and influenced by media responses, public opinion, and social mobilization, and by nuclear power governance, politics, and policy, which in turn have been, and will continue to be, influenced by the Fukushima disaster, which, like the Chernobyl disaster, has its own legacy for the long-term future.

In the medium term, contributors to this volume have noted areas of particular concern or implication for future policy and research. In Switzerland, says Fabienne Crettaz von Roten in Chapter 10, more research is needed on energy transitions with a focus on "science in society," such as information on the efficiency of transition measures or scrutiny of public perceptions and behaviors related to energy. On January 15, 2015, the Swiss National Science Foundation (2015) announced it was launching two National Research Programs: "Energy Turnaround" and "Managing Energy Consumption." These projects, says Crettaz von Roten, will also continue to document the "significant and ground-breaking" effects of the Fukushima disaster in Switzerland.

In talking about New Zealand, Priestley raised concerns about the lack of availability of public scientists to talk to the media and the public about Fukushima and says "if and when New Zealand does have a debate on nuclear power, the 'expert' opinions can't be left to politicians, environmentalists, and business people, or even science historians. If the public is to make an informed choice, scientists must be allowed to engage with the public and contribute to the discussion and debate."

In Japan, say Nakamura and Nishimura, the growth of an independent political entity within the boundaries of a neighboring city government is a new and an interesting experiment. If the attempt to relocate the town of Futaba Machi to a new prefecture succeeds, this would be another important Fukushima effect, "one that Japanese officials and academics should consider as a possible option for a new role and function for local units of government in the aftermath of future natural and/or human made disasters."

In India and China, the case is made for increased participatory decision-making models that reflect open and knowledge-partnering between developers, experts and communities in relation to safety and siting aspects of nuclear power plants and rebuilding trust in relation to controversial facility siting. As Anupam Jha argues:

> Transparency, safety, and participation in relation to nuclear power development and management . . . need to be addressed in a more meaningful and engaging manner to address urgent livelihood concerns of the public. Only then will civil society develop any measure of faith in technoscientific procedures related to nuclear power.

Such arguments endorse those of the previous volume "that in this high-stakes, and highly sensitive and risky technological area is the need for a highly responsible and transparent civic regulatory style to better inform issues of complexity, risk, trust, civic inclusion and legitimacy, and, in turn, policy effectiveness" (Hindmarsh 2013: 216). However, participatory governance can go beyond such a specific focus on remedying nuclear ills to meaningfully inform national choices for energy mixes, which may or may not include nuclear energy, as illustrated in the Swiss case (Chapter 12).

Significantly, in their chapter on the UK and Finland, in reflecting on the very weak Fukushima effect on these countries, Molyneux-Hodgson and Hietala conclude by asking: "Is there any form of nuclear event that would prompt greater national reflections on the management of waste disposal programs?" The same might be said about nuclear power development in general. Although Fukushima has had an intense and immediate effect on some countries' nuclear power development, governance, and waste management, in many countries, business appears to continue as usual, despite the emergence of a powerful new geopolitical terrain encompassing both nuclear and energy development, marked by the disaster and its sweeping global impacts both obvious and not so obvious.

This situation begs a final question in relation to the horrendous Fukushima disaster: what size and scope of a future adverse nuclear event would prompt greater international reflection on the ongoing plans for expansion of nuclear power? We earnestly hope another big nuclear disaster does not have to be suffered to find out, and to better appreciate Kant's dictum at the front of this chapter that "a violation of rights in *one* part of the world is felt *everywhere*." Instead, we look forward to a safe, secure, and sustainable energy future with minimized risks of dangerous forms of pollution, particularly radioactive pollution that can impact the entire global community, an issue that Fukushima Daiichi has rivetingly drawn our attention to.

## Notes

1  See http://goo.gl/FBwxIm, accessed April 23, 2015.
2  See http://goo.gl/fwMQl5, accessed May 30, 2015.
3  See http://goo.gl/uM10CI, accessed May 30, 2015.
4  See http://goo.gl/FBwxIm, accessed April 23, 2015.
5  See http://goo.gl/tDFI4R, accessed May 27, 2015.
6  See http://goo.gl/Y1WjPB, accessed May 27, 2015.
7  See http://goo.gl/Y1WjPB, accessed May 27, 2015.

## References

Andrée, P. "The Biopolitics of Genetically Modified Organisms in Canada," *Journal of Canadian Studies* 37, no. 3 (2002): 162–191.

Barnes, B. "On Authority and Its Relationship to Power." In *Power, Action and Belief: A New Sociology of Knowledge?*, edited by J. Law, 180–195. London: Routledge & Kegan Paul, 1986.

Bird, D., et al. "Nuclear Power in Australia: A Comparative Analysis of Public Opinion Regarding Climate Change and the Fukushima Disaster," *Energy Policy* 65 (2014): 644–653.

Blowers, A. "The Future Is Not Nuclear: Ethical Choices for Energy after Fukushima." In *Nuclear Disaster at Fukushima Daiichi: Social, Political and Environmental Issues*, edited by R. Hindmarsh, 175–193. New York: Routledge Studies in Science, Technology and Society, 2013.

Butler, C., Parkhill, K.A., and N.F. Pidgeon. "Nuclear Power after 3/11: Looking Back and Thinking Ahead." In *Nuclear Disaster at Fukushima Daiichi: Social, Political and Environmental Issues*, edited by R. Hindmarsh, 135–153. New York: Routledge Studies in Science, Technology and Society, 2013.

Clegg, S. *Frameworks of Power*. London: Sage, 1989.

Cochran, T., and M. McKinzie. *Global Implications of the Fukushima Disaster for Nuclear Power*. US: Natural Resources Defense Council, 2012.

Connor, C. "A Quantitative Literacy View of Natural Disasters and Nuclear Facilities," *Numeracy* 4, no. 2 (2011): Article 2. DOI: 10.5038/1936-4660.4.2.2.

Crumley, B. "The Fukushima Effect: France Starts to Turn Against Its Much Vaunted Nuclear Industry." *Time*, January 2, 2012, accessed May 28, 2015, http://goo.gl/7gPuFt.

Etzioni-Halevy, E. "The Relative Autonomy of Élites: The Absorption of Protest and Social Progress in Western Democracies." In *Rethinking Progress: Movements, Forces, and Ideas at the End of the 20th Century*, edited by J.C. Alexander, and P. Sztompka, 202–226. Boston: Unwin Hyman, 1990.

Falk, J., Green, J., and G. Mudd. "Australia, Uranium and Nuclear Power," 63, no. 6 (2006): 845–857.

Felt, U. *Keeping Technologies Out: Sociotechnical Imaginaries and the Formation of a National Technopolitical Identity*, 2013, accessed August 15, 2014, http://tinyurl.com/mrlz3nk.

Fischer, F. *Technology and the Politics of Expertise*. London: Sage, 1990.

Foucault, M. *The Will to Knowledge: The History of Sexuality: Volume One*. London: Penguin, 1990 [French version 1976].

Friedman, S.M. "Three Mile Island, Chernobyl, and Fukushima: An Analysis of Traditional and New Media Coverage of Nuclear Accidents and Radiation," *Bulletin of the Atomic Scientists* 67, no. 5 (2012): 55–65.

Hajer, M. *The Politics of Environmental Discourse: Ecological Modernization and the Policy Process*. Oxford: Clarendon Press, 1995.

Hara, T. "Social Shaping of Nuclear Safety: Before and After the Disaster." In *Nuclear Disaster at Fukushima Daiichi: Social, Political and Environmental Issues*, edited by R. Hindmarsh, 22–40. New York: Routledge Studies in Science, Technology and Society, 2013.

Hayashi, M., and L. Hughes. "The Fukushima Nuclear Accident and its Effect on Global Energy Security," *Energy Policy* 59 (2013): 102–111.

Hindmarsh, R. "Genetic Engineering Regulation in Australia: An 'Archaeology' of Expertise and Power," *Science as Culture*, 14, no. 4 (2005): 373–392.

Hindmarsh, R. "Wind Farms and Community Engagement in Australia: A Critical Analysis for Policy Learning," *East Asian Science, Technology and Society: An International Journal* 4 (2010): 541–563.

Hindmarsh, R. (ed.). *Nuclear Disaster at Fukushima Daiichi: Social, Political and Environmental Issues*. New York: Routledge Studies in Science, Technology and Society, 2013.

Joskow, P.L., and J.E. Parsons. *The Future of Nuclear Power after Fukushima*. USA: MIT Center for Energy and Environmental Policy Research, 2012.

Kaigo, M. "Social Media Usage during Disasters and Social Capital: Twitter and the Great East Japan Earthquake," *Keio Communication Review* 34 (2012): 19–34.

Kera, D., Rod, J., and R. Peterova. "Post-apocalyptic Citizenship and Humanitarian Hardware." In *Nuclear Disaster at Fukushima Daiichi: Social, Political and Environmental Issues*, edited by R. Hindmarsh, 97–115. New York: Routledge, 2013.

Kumar, V.A. (2014) "India's Nuclear Energy Renaissance: Stuck in the Middle?" *Journal of Risk Research* 17, no. 1 (2014), 43–60.

McClymont, K., and P. O'Hare. "We're Not NIMBY's! Contrasting Local Protest Groups with Idealised Conceptions of Sustainable Communities," *Local Environment* 13, no. 4 (2008): 321–335.

McKeating, J. "Greenpeace Activists Occupy France's Fessenheim Nuclear Power Plant to Say 'Stop Risking Europe.'" Greenpeace blog, March 18, 2014, accessed May 30, 2015, http://goo.gl/dnoGc2.

Morita, A., Blok, A., and S. Kimura. "Environmental Infrastructures of Emergency: The Formation of a Civic Radiation Monitoring Map during the Fukushima Disaster." In *Nuclear Disaster at Fukushima Daiichi: Social, Political and Environmental Issues*, edited by R. Hindmarsh, 78–96. New York: Routledge, 2013.

NucNet. "Japan Business Federation Calls for Reactor Restarts," *NucNet* September 29, 2014.

NucNet. "'Substantial Funding Gap' Remains For Chernobyl Work, Says EBRD," *NucNet* March 31, 2015a.

NucNet. "Japan's Sendai-1 Scheduled for July 2015 Restart," *NucNet* April 30, 2015b.

Reiss, H.S. (ed.). *Kant: Political Writings*. New York: Cambridge University Press, 1991.

Roney, M.J. "Wind Power Beats Nuclear Energy Again in China," Earth Policy Institute, March 5, 2015, accessed June 1, 2015, www.earth-policy.org/data_highlights/2015/highlights50.

Schneider, M., Froggatt, A., and S. Thomas. *The World Nuclear Industry Status Report 2010–2011, Nuclear Power in a Post-Fukushima World: 25 Years after the Chernobyl Accident*. Washington, DC: World Watch Institute, April 2011.

Selznick, P. *TVA and the Grass Roots: A Study in the Sociology of Formal Organization*. New York: Harper & Ro, 1949.

Swiss National Science Foundation. "Managing Energy Consumption," Swiss National Science Foundation: National Research Programme NRP 71, January 15, 2015, accessed June 1, 2015, http://goo.gl/6YoJRa.

Tamman, M., Casselman, B., and P. Mozur. "Scores of Reactors in Quake Zones," *Wall Street Journal*, March 19, 2011, accessed May 27, 2015, http://goo.gl/CfVXAx.

Varrall, S. "The Next Nuclear Wave: Renaissance or Proliferation Risk?" *Global Change, Peace & Security: formerly Pacifica Review: Peace, Security & Global Change* 24, no. 1 (2012): 127–140.

Wolsink, M., and Devilee, J. "The Motives for Accepting or Rejecting Waste Infrastructure Facilities: Shifting the Focus from the Planners' Perspective to Fairness and Community Commitment," *Journal of Environmental Planning and Management* 52, no. 2 (2009): 217–236.

World Nuclear News. "French Parliament Approves Energy Transition," *World Nuclear News*, October 13, 2014, accessed May 30, 2015, www.world-nuclear-news.org/NP-French-parliament-approves-energy-transition-1310144.html.

# Contributors

**Dung-sheng Chen** is Professor of Sociology at National Taiwan University. He was president of the Taiwanese Sociological Association from 2010–2012, and served as director-general of the Department of Humanity and Social Sciences, National Science Council, Taiwan, for three years. His research interests include public engagement in science and technology, public perceptions of technology risk, and formation of local civil society. His publications include a case study of consensus conferences in Taiwan, and an article on the Fukushima disaster, both in *East Asian Science, Technology, and Society: An International Journal*. His current research project compares different trajectories of formation of local civil society in four places of Taiwan.

**Fabienne Crettaz von Roten** is Senior Lecturer and Head of the "Science-society relationship" research unit at the Faculty of Social and Political Sciences, University of Lausanne, Switzerland. Her research offers a symmetrical view of the relations between science and society: public perceptions of science and technology (in particular, of nuclear energy) and scientists' engagement with society (particularly with a gender perspective). She has authored and co-authored several books and book chapters in these fields, as well as numerous articles in journals including *Public Understanding of Science, Science, Technology and Society* and *Science Communication*.

**Xiang Fang** is Associate Professor at the Department of Sociology and Social Work, Sun Yat-Sen University, China. Her field is environmental sociology and STS. Recent publications include "Local People's Understanding of Risk from Civil Nuclear Power in the Chinese Context," in *Public Understanding of Science* 2014, and *Risk and Social Construction of Nuclear Power Development in China: Local People's Participation in Civil Nuclear Issues in China at the Start of the 21st Century* (Social Sciences Academic Press: China, 2014) in Chinese. Fang's current research is on risk governance of China's civil nuclear power and public engagement in environmental issues.

**Marika Hietala** is Doctoral Researcher in the Department of Sociological Studies, University of Sheffield. She is pursuing her PhD study on the differing cultures of nuclear waste management across European countries. Her project is funded by the Economic and Social Research Council as part of the Nuclear Societies training network. She originates from Helsinki, Finland, and has a BA in History from the University of Newcastle, UK, and a Masters in European Politics and Society from the University of Aberdeen, UK.

**Richard Hindmarsh** is (Snr) Associate Professor at Griffith School of Environment and Centre for Governance and Public Policy, Griffith University, Australia. His field is Environmental Politics and Policy, and STS. He is cofounder of the Asia-Pacific Science, Technology and Society Network, and has published in journals such as *Environmental Politics, Local Environment, Nature, Social Studies of Science, Science as Culture, and East Asian Science, Technology and Society*. His seven books include *Edging towards BioUtopia* (University of Western Australia Press 2008); *Genetic Suspects* (Cambridge University Press 2010, co-edited with Barbara Prainsack); and *Nuclear Disaster at Fukushima Daiichi: Social, Political and Environmental Issues* (Routledge 2013). His current research topics include new and social media, GM crops, energy (renewable and nuclear) politics and policy, and sustainability transformations.

**Detlef Jahn** is Professor of Comparative Politics at the Department of Political Science, University of Greifswald, Germany. His main research interests include political institutions, welfare, and environmental policies, as well as conceptual and methodological issues of processes of diffusion and public policy. His publications include "Globalization as Galton's Problem: The Missing Link in the Analysis of Diffusion Patterns in Welfare State Development," in *International Organization* (2006); "Conceptualizing Left and Right in Comparative Politics: Towards a Deductive Approach," in *Party Politics* (2011); and "The Three Worlds of Environmental Politics," in Duit, A. (ed). *Mapping the Politics of Ecology* (MIT Press 2014). Currently he is finishing a book on Environmental Performance and Politics (Cambridge University Press).

**Anupam Jha** is Reader at Law Centre-II, Faculty of Law, University of Delhi, India. His field is international governance, public international law, and the Indian legal system. He was a Commonwealth Academic Fellow in England during 2010–2012, and a Raman Fellow in the US (2014–2015). He has written extensively on matters related to international governance, including environmental governance, in national and international journals. He has also contributed chapters to three books published in India and presented many papers at national and international conferences. A recent publication is "Nuclear Energy Law and Decision-Making in India," in the *Journal of Risk Research* (2014).

**Hyomin Kim** is Assistant Professor at the Division of General Studies, Ulsan National Institute of Science and Technology, Korea. After receiving her PhD in Sociology (Science and Technology Studies) at University of Illinois at Urbana-Champaign, she has written on non-communicable lifestyle-related diseases, genetically modified foods, spent nuclear fuel management, and gender mainstreaming in science and engineering. Her research focuses on historical and cultural studies of technoscience linked with public engagement and policy making. A recent publication is: "Functional Foods and the Biomedicalisation of Everyday Life: A Case of Germinated Brown Rice," *Sociology of Health and Illness* (2013).

**William J. Kinsella** is Professor of Communication at North Carolina State University, where he directed the interdisciplinary program in Science, Technology and Society from 2009–2014. His research addresses communication in environmental, energy, organizational, and science and technology contexts, and interactions across technical, regulatory, and public communities. He is co-editor of *Nuclear Legacies: Communication, Controversy, and the US Nuclear Weapons Complex* (Lexington Books, 2007), and has published in a range of journals, books, and reference volumes. From 2000–2004 he served with the citizen advisory board for the US Department of Energy's Hanford nuclear site, and in 2010 he was US Fulbright Scholar at the Institute for Nuclear Energy and Energy Systems, University of Stuttgart, Germany.

**Susan Molyneux-Hodgson** is Senior Lecturer in Sociology at the University of Sheffield, UK, working in STS and specializing in the ethnographic study of technoscientific cultures and practices. She established the cross-disciplinary Nuclear Societies network in 2013. A core group of doctoral students, social scientists, and research engineers are developing new thinking on a range of nuclear topics. She has written on the cultures of research communities; engineering practices; the emergence of disciplines; and the performance of barriers to innovation in academia and industry and is published in sociology, social policy, and STS outlets.

**Akira Nakamura** has served as both Vice President and Dean of the Graduate School of Meiji University. In 2011, he was nominated as Professor Emeritus of the School of Political Science and Economics and the Graduate School of Governance Studies at Meiji University, Tokyo, Japan. He has had several academic appointments in the US and Japan, and is the incumbent President of the Japan Emergency Management Association. His publications include "A Quest for Improving Global Governance," "Preparing for the Inevitable: Japan's Ongoing Search for Best Crisis Management Practices," and "A Scenario for Local Government Autonomy" (in Japanese). In 2007, Nakamura was awarded a medal of honour (JMN) from the Government of Malaysia for contribution to the training of public officials, and in 2008 was accorded an honour of recognition from President Sadako Ogata of the Japan International Cooperation Agency.

**Wataru Nishimura** is Assistant Professor in Public Administration and Public Policy, Department of Political Science, School of Political Science and Economics, Meiji University, Japan. He completed his PhD in Political Science at Meiji University in 2007. From 2007–2012 he was a researcher at the Research Center for Crisis and Contingency Management, Meiji University. His current research interests include inter-governmental relationships in emergency, disaster contingency planning, and public administration, including public sector reform. He has authored or co-authored eight books on public administration and management, and public policy. His recent co-authored books are *Nippon no Kokyo-keiei: Atarasi Gyosei* (*Public Management in Japan: New Public Administration*) (Hokuju-syuppan 2014) and *Kikikanr-gaku: Syakai-unnei to Governance no Korekara* (*Crisis Management: Governance from Now on*) (Daichi-hoki 2014).

**Rebecca Priestley** is Senior Lecturer in the Science in Society Group, in the Faculty of Science at Victoria University of Wellington, where she teaches history of science, science communication, and creative science writing. Her most recent book, *Mad on Radium* (Auckland University Press 2012), told the radiation and nuclear history of New Zealand. Her academic articles have been published in journals such as the *Journal of the Royal Society of New Zealand*, *New Zealand Geographer*, and the *Journal of Environmental Studies and Sciences*. Rebecca is also an award winning science writer whose work has been published in a range of online and print media, including the *New Zealand Listener*, *Scientific American Online*, and *Griffith Review*.

**Christopher Rootes** is Professor of Environmental Politics and Political Sociology, and Director of the Centre for the Study of Social and Political Movements, at the University of Kent, Canterbury, England. He has published extensively on environmental movements, Green parties and protest, and edited/co-edited *The Green Challenge: The Development of Green Parties in Europe* (Routledge 1995), *Environmental*

*Movements* (Cass/Routledge 1999), *Environmental Protest in Western Europe* (Oxford University Press 2003), *Acting Locally: Local Environmental Mobilizations and Campaigns* (Routledge 2008), and *Environmental Movements and Waste Infrastructure* (Routledge 2010). He is editor-in-chief of the journal *Environmental Politics*.

**Sebastian Stephan** is a PhD Candidate and Researcher at the Department of Political Science at the University of Greifswald, Germany. His research interests include spatial econometrics, energy policy, and political economy. His publications include "German Exceptionalism: The End of Nuclear Energy in Germany!" in *Environmental Politics* (2012, with Detlef Jahn); "The Problem of Interdependence," in Braun, D. and Maggetti, M. (eds). *Comparative Politics: Theoretical and Methodological Challenges* (Edward Elgar 2015, with Detlef Jahn).

**Andrei Stsiapanau** is Associate Professor in Social and Political Sciences at the European Humanities University (Vilnius). In 2008–2011 he was a Fellow of the international project, "Politics and Society after Chernobyl, Belarus, Ukraine, Russia, Lithuania, and Germany in Comparative and Entangled Historical Perspective (1986–2006)." In 2013 he was a postdoctoral fellow in the Fernard Braudel Program (MSH, Paris) and a visiting scholar at the Center for the Sociology of Innovation (CSI, Mines ParisTech). His main research interests are comparative studies of different technopolitical regimes in Russia, Belarus, and Lithuania on nuclear power plant constructions. His publications include "The Chernobyl Politics in Belarus: Interplay of Discourse-Coalitions," in *Filosofija Sociologija* 2010; and *Scientific Discourse of Chernobyl: Laboratories of Political Decisions* (in Crossroads Digest N5/2010 European Humanities University).

**Joseph Szarka** was Reader in Policy Studies in the Department of Politics, Languages and International Studies at the University of Bath, UK, until 2014. He has published widely on politics and policy-making in France and the European Union. His books include *The Shaping of Environmental Policy in France* (2002) and *Wind Power in Europe: Politics, Business and Society* (2007). He is lead editor of *Learning from Wind Power* (Palgrave 2012).

# Index